Process Safety
Key Concepts and Practical Approaches

Process Safety
Key Concepts and Practical Approaches

by
James A. Klein and Bruce K. Vaughen

CRC Press
Taylor & Francis Group
Boca Raton London New York

CRC Press is an imprint of the
Taylor & Francis Group, an **informa** business

CRC Press
Taylor & Francis Group
6000 Broken Sound Parkway NW, Suite 300
Boca Raton, FL 33487-2742

First issued in paperback 2020

© 2017 by Taylor & Francis Group, LLC
CRC Press is an imprint of Taylor & Francis Group, an Informa business

No claim to original U.S. Government works

ISBN-13: 978-1-4665-6542-5 (hbk)
ISBN-13: 978-0-367-73617-0 (pbk)

Library of Congress Cataloging-in-Publication Data

Names: Klein, James A., 1958- author. | Vaughen, Bruce K., author.
Title: Process safety : key concepts and practical approaches / James A. Klein, Bruce K. Vaughen.
Description: Boca Raton : Taylor & Francis, a CRC title, part of the Taylor & Francis imprint, a member of the Taylor & Francis Group, the academic division of T&F Informa, plc, [2017]
Identifiers: LCCN 2016043429| ISBN 9781466565425 (hardback : acid-free paper) | ISBN 9781482282580 (ebook)
Subjects: LCSH: Chemical processes--Safety measures. | Industrial safety--Management. | Chemical plants--Safety measures.
Classification: LCC TP150.S24 K54 2017 | DDC 660.068/4--dc23
LC record available at https://lccn.loc.gov/2016043429

Visit the Taylor & Francis Web site at
http://www.taylorandfrancis.com

and the CRC Press Web site at
http://www.crcpress.com

Contents

Acknowledgments xiii
Disclaimer xv
About the Authors xvii
List of Abbreviations xix
Glossary xxi
Preface: A Different Approach xxiii

Section I   Key Concepts of Process Safety

1. Introduction to Process Safety 3
   1.1   Introduction 3
       1.1.1   Purpose and Scope.. 4
       1.1.2   Application 6
   1.2   Key Concepts 9
       1.2.1   What Is Process Safety?... 9
       1.2.2   Process Safety Is Not "Personal Safety" 14
   1.3   Why Is Process Safety Important? 16
       1.3.1   Business Impact... 18
       1.3.2   Code of Ethics... 19
       1.3.3   Conservation of Life 20
   1.4   Case Studies 20
       1.4.1   Large Companies 20
       1.4.2   Small Companies 21
       1.4.3   Universities 22
   1.5   Key Questions.. 23
   References 24

2. Process Safety Culture, Leadership, and Performance 27
   2.1   Introduction 28
   2.2   Key Concepts 31
       2.2.1   What Is Safety Culture? 31
       2.2.2   Characteristics of Safety Cultures...................... 34
       2.2.3   Employee Participation.. 37
   2.3   Evaluating and Improving Safety Cultures 37
   2.4   Process Safety Leadership.. 43
   2.5   Process Safety Performance 46
   2.6   Case Study.. 49
   References 51

3. Process Safety Systems 55
   3.1   Introduction 56
   3.2   Key Concepts 57
          3.2.1   Process Risk 57
          3.2.2   Managing Process Risk 59
          3.2.3   Protection Layer Model..64
   3.3   Process Safety Systems 66
   3.4   Barrier Protection Layer Models 71
          3.4.1   Swiss Cheese Model 72
          3.4.2   Bow Tie Diagrams...74
   3.5   Case Study..76
   References 79

4. Operational Discipline 81
   4.1   Introduction 82
   4.2   Key Concepts 83
          4.2.1   Benefits of Strong OD..83
          4.2.2   OD Program Characteristics...86
                 4.2.2.1   Organizational OD................................87
                 4.2.2.2   Personal OD 89
   4.3   Improving OD 92
   4.4   Evaluating OD..94
          4.4.1   General Approaches 94
          4.4.2   Evaluating Personal OD..96
   4.5   Leadership Processes...98
   4.6   Case Study..100
   References 102

Section II   Practical Approaches for Designing Safe Processes

5. Design Safe Processes...107
   5.1   Introduction 107
   5.2   Key Concepts 109
          5.2.1   Using Codes and Standards 110
          5.2.2   Documenting the Technology 111
          5.2.3   Sustaining the Information..111
   5.3   The Process and Equipment Design Basis................................112
          5.3.1   Applying Inherently Safer Process Design 112
          5.3.2   Understanding the Hazards..112
          5.3.3   Understanding the Process Risks 113
          5.3.4   Selecting the Equipment 113
          5.3.5   Selecting Safe Operating Limits for Process Equipment 118

5.3.6 Establishing Safe Equipment Layout 121
5.3.7 Establishing Equipment Integrity 121
5.3.8 Documenting the Process and Equipment Design
Specifications 122
5.4 Case Studies 124
5.5 Measures of Success 128
References 129

6. Identify and Assess Process Hazards 131
6.1 Introduction 131
6.2 Key Concepts 134
6.2.1 Physical Property Data 134
6.2.2 Completing the Intrinsic Hazards Assessment 134
6.3 Toxicity Hazards 137
6.3.1 Assessing Toxicity Hazards 138
6.3.2 Case Study 142
6.4 Flammability Hazards 144
6.4.1 Assessing Flammability Hazards 145
6.4.2 Assessing Explosion Hazards 148
6.4.3 Case Study 150
6.5 Combustible Dust Hazards 151
6.5.1 Assessing Combustible Dust Hazards 153
6.5.2 Case Study 155
6.6 Chemical Reactivity Hazards 155
6.6.1 Assessing Chemical Reactivity Hazards 158
6.6.2 Developing a Chemical Compatibility Matrix 160
6.6.3 Case Study 164
6.7 Other Process Hazards 165
6.8 Measures of Success 167
References 167

7. Evaluate and Manage Process Risks 171
7.1 Introduction 172
7.2 Key Concepts 173
7.2.1 What Is Process Hazards and Risk Analysis (PHRA)? 173
7.2.2 When Is a PHRA Done? 175
7.2.3 Who Does a PHRA? 177
7.2.4 How Is a PHRA Done? 178
7.3 Process Hazards Identification 182
7.4 Consequence Analysis 182
7.5 Hazardous Event Evaluation 189
7.5.1 HAZOP 190
7.5.2 What-If/Checklist 195
7.5.3 Other Methodologies 197
7.6 Human Factors 198

7.7   Facility Siting..200
7.8   Inherently Safer Processes...203
7.9   Protection Layers...204
7.10   Risk Analysis..205
       7.10.1 Qualitative Risk Analysis...205
       7.10.2 Layer of Protection Analysis..208
       7.10.3 Quantitative Risk Analysis 210
7.11   PHRA Revalidation..211
7.12   Measures of Success..213
References 214

Section III   Practical Approaches for Implementing Process Safety

8. Operate Safe Processes 219
8.1   Introduction...220
8.2   Key Concepts 220
       8.2.1   Using Procedures 220
       8.2.2   Providing Competency Training.....................................221
       8.2.3   Sustaining Safe Processes...221
8.3   Developing Procedures: Process Safety Systems.....................222
8.4   Developing Procedures: Safe Operations 223
8.5   Developing Capability: Process Safety Systems 230
8.6   Developing Capability: Safe Operations...................................231
8.7   Measures of Success..236
References 237

9. Maintain Process Integrity and Reliability 239
9.1   Introduction...239
9.2   Key Concepts...240
       9.2.1   Inspection, Testing, and Preventative Maintenance
               Programs 240
       9.2.2   The Difference between Integrity and Reliability 241
       9.2.3   Ensuring Equipment Fitness for Service 242
       9.2.4   Maintenance Capability and Competency...................242
9.3   Identifying Critical Equipment...243
9.4   Identifying Causes of Equipment Failure 245
9.5   Developing an Effective Maintenance System 247
       9.5.1   Questions Posed When Developing the
               Maintenance System..247
       9.5.2   Equipment Risk Based Inspection Programs 250
       9.5.3   Equipment Reliability Programs 251
       9.5.4   Equipment Quality Assurance Programs......................254

	9.5.5	Equipment Integrity Programs	255
	9.5.6	Combining Programs	256
9.6	Ensuring Equipment Fitness for Service		258
9.7	Addressing Equipment Deficiencies		258
9.8	Monitoring the Equipment Maintenance System		259
9.9	Case Study		260
9.10	Measures of Success		266
References			267

10. Change Processes Safely 271
10.1	Introduction		272
10.2	Key Concepts		273
	10.2.1	Defining Changes	273
	10.2.2	Managing Changes	273
	10.2.3	Communicating Changes	274
10.3	Case Study		275
10.4	Understanding Organizational Changes		276
10.5	Understanding Technology Changes		279
10.6	How Change is Managed		281
10.7	Measures of Success		287
References			288

11. Manage Incident Response and Investigation 289
11.1	Introduction		290
11.2	Key Concepts		291
	11.2.1	The Six Phases for Managing Incidents	292
	11.2.2	Effective Investigations	292
11.3	Classifying the Types of Responses and Incidents		293
	11.3.1	Defining Emergencies	293
	11.3.2	Defining Incidents	295
11.4	Planning and Responding to Emergencies		296
	11.4.1	Planning for Incidents	297
	11.4.2	Responding during the Incident	300
11.5	Recovering From and Investigating Incidents		301
	11.5.1	Recovering from the Incident	302
	11.5.2	Investigating after the Incident	304
	11.5.3	Combining Resources When Recovering and Investigating	309
11.6	Changing and Sustaining after the Incident		310
	11.6.1	Implementing the Changes	310
	11.6.2	Sustaining the Changes	310
11.7	Incident Investigation Methods		311
	11.7.1	Why-Trees	312

	11.7.2	Fishbone Diagrams	312
	11.7.3	Root Cause Analyses	314
	11.7.4	Fault Tree Analyses	315
11.8	Applying Bow Tie Diagrams to Investigations		315
	11.8.1	Searching for the Event-Related Causes	316
	11.8.2	Visualizing Barrier Weaknesses on the Bow Tie Diagram	316
	11.8.3	Searching for the Systemic Causes	317
	11.8.4	Visualizing Systemic Weaknesses on the Bow Tie Diagram	319
	11.8.5	Searching for the Foundational Causes	320
11.9	Case Study		321
11.10	Measures of Success		333
	References		334

12. Monitor Process Safety Program Effectiveness 337
12.1	Introduction		338
12.2	Key Concepts		339
	12.2.1	Sensitivity to Operations	339
	12.2.2	Learning Organizations	340
12.3	Process Safety Metrics (Key Performance Indicators)		344
	12.3.1	Development of Metrics	347
	12.3.2	Use of Metrics	349
12.4	Process Safety Audits		353
	12.4.1	Preparing for an Audit	353
	12.4.2	Conducting an Audit	355
	12.4.3	Documenting an Audit	357
12.5	Measures of Success		358
	References		359

Section IV   Practical Approaches for Achieving Process Safety Excellence

13. Develop Personal Capability in Process Safety 363
13.1	Introduction		364
13.2	Process Safety Roles and Responsibilities		364
13.3	How to Be Effective in Process Safety Roles		367
	13.3.1	Developing Technical Process Safety Knowledge	368
	13.3.2	Influencing the Organization	370
	13.3.3	Thinking and Communicating Independently	372
13.4	Developing a Learning Plan		373
	References		374

14. Commit to a Safe Future 377
    14.1   Maintaining an Effective Process Safety Program......................377
    14.2   Looking to the Future of Process Safety 379
    References 382

Epilogue 383
Index 385

Acknowledgments

We appreciate the support of our families, as well as many colleagues who provided comments at various stages in the development of this book.

Disclaimer

The views expressed in this book are those of the authors, based on their knowledge and experience. They do not necessarily reflect the views of the current or former employers of the authors. The authors have made a diligent effort to properly reference the work of others throughout this book.

The approaches discussed in this book are intended to be introductory and should be supplemented with additional material when conducting process safety program activities. Neither the authors, the current or former employers of the authors, nor the publisher is responsible for any liability whatsoever arising from the use of this book, in whole or in part, by any person, company, or entity of any type. The user of this book assumes complete and total responsibility for whatever consequences arise from or are related to the use of this book, in whole or in part.

About the Authors

James A. (Jim) Klein is a senior process safety consultant with 36 years of experience. He is currently conducting process hazard and safety culture assessments, compliance audits, and other process safety services. Before this, he was a senior process safety competency consultant and PSM co-lead for North America Operations at DuPont, with experience in process safety, engineering, and research. At DuPont, he led global teams for chemical reactivity, consequence analysis, process safety training, and operational discipline. He has more than 50 publications, conference presentations, and university talks and has participated in several Center for Chemical Process Safety (CCPS) book projects, including as the leader of *Conduct of Operations and Operational Discipline* and as a member for the *Risk Based Process Safety* and other projects. He has a BS degree in chemical engineering from MIT, Cambridge, Massachusetts, an MS degree in chemical engineering from Drexel University, Philadelphia, Pennsylvania, and an MS degree in management of technology from the University of Minnesota, Minneapolis, Minnesota.

Bruce K. Vaughen has more than two and a half decades of experience in process safety, including engineering, research, consulting, and teaching experience with DuPont, DuPont Teijin Films, Cabot Corporation, and BakerRisk, and as a visiting assistant professor at Rose-Hulman Institute of Technology, Terre Haute, Indiana. His roles have included leading global process safety management (PSM) efforts, updating and developing corporate PSM standards, and developing PSM training and workshops. He is the principal author of two Center for Chemical Process Safety (CCPS) guideline books (*Siting and Layout of Facilities; Integrating Management Systems and Metrics to Improve Process Safety Performance*) and has developed training modules for the American Institute of Chemical Engineers (AIChE) Safety and Chemical Engineering Education (SAChE) committee. He holds a BS degree in chemical engineering from the University of Michigan, Ann Arbor, Michigan, and MS and PhD degrees in chemical engineering from Vanderbilt University, Nashville, Tennessee, and is a registered professional engineer.

The authors can be contacted at pskcpacrc@gmail.com

List of Abbreviations

ALARP	As Low As Reasonably Practicable
BPCS	Basic Process Control System
CA	Consequence Analysis
CCPS	Center for Chemical Process Safety
COL	Conservation of Life
CSB	Chemical Safety Board (US)
DHA	Dust Hazards Assessment
EPA	Environmental Protection Agency (US)
EU	European Union
FMEA	Failure Mode and Effect Analysis
FS	Facility Siting
FTA	Fault Tree Analysis
GHS	Globally Harmonized System of Classification and Labeling of Chemicals
HAZOP	Hazards and Operability Study
HEE	Hazardous Event Evaluation
HF	Human Factors
HIRA	Hazards Identification and Risk Assessment
HSE	Health and Safety Executive (UK)
IHA	Intrinsic Hazards Assessment
ISP	Inherently Safer Processes
ITPM	Inspection, Testing and Preventative Maintenance
KPI	Key Performance Indicators
LOPA	Layer of Protection Analysis
MOC	Management of Change
OD	Operational Discipline
OSHA	Occupational Safety and Health Administration (US)
PDCA	Plan, Do, Check, Act
PHA	Process Hazards Analysis
PHRA	Process Hazards and Risk Analysis
PPE	Personal Protective Equipment
PSI	Process Safety Information
PSM	Process Safety Management
QRA	Quantitative Risk Assessment
RAGAGEP	Recognized and Generally Accepted Good Engineering Practices
RBPS	Risk Based Process Safety
RCA	Root Cause Analysis
SIS	Safety Instrumented System
SWP	Safe Work Practices

Glossary

Operational Discipline: following system and procedure requirements correctly every time

Process: parts of a facility that contain process hazards due to the presence of hazardous materials or conditions, which may include large or small manufacturing facilities, research pilot plants, laboratories, and universities

Process hazard: intrinsic properties of materials and processes that can either directly or through release of energy lead to harmful effects, such as toxicity, flammability, explosivity, and reactivity

Process risk: a measure of potential harm associated with process activities and events, based on potential consequences (e.g., injury, property damage, environmental) and frequency (e.g., years)

Process safety: application of management systems to ensure that process hazards are identified, evaluated, and managed to reduce the risk of potentially hazardous incidents and injuries

Process safety elements: distinct process safety activities that may have been established through regulatory or industry standards and guidelines

Process safety incident: a combination of conditions and events that result in toxic releases, fires, explosions, or runaway reactions that harm people, damage the environment, and cause property and business losses

Process safety systems: systems integrating the process safety elements to ensure that process hazards are identified, evaluated, and understood, and that appropriate risk management controls are provided to achieve and maintain safe and reliable operations

Safety culture: the normal way things are done at a facility, company, or organization, reflecting expected organizational values, beliefs, and behaviors, that set the priority, commitment, and resource levels for safety programs and performance

Preface: A Different Approach

> It does not do to leave a live dragon out of your calculations, if you live near him.
>
> **J.R.R. Tolkien**

Copyright James A. Klein, used by permission.

Although dragons are not in the workplace, if you are reading this book, it is likely that you do have process hazards. And if you do, they should not be ignored or treated complacently, because process hazards, like dragons, have the potential to cause great harm at any time. Rather, process hazards must be identified, evaluated, eliminated (if possible), and rigorously controlled. This requires both awareness and action. You must know if process hazards are present in your workplace, and, if they are, constant action is required to manage the risks associated with these hazards.

This book provides the key concepts and practical approaches needed to get started on or to improve your efforts in managing process hazards. Like the knight with the dragon above, this book takes a somewhat different approach. The traditional element-based approach to process safety is enhanced by a comprehensive process safety program that provides the primary goals and requirements of an integrated management system framework, with an emphasis on how different traditional elements work together. Lack of awareness or complacency about the process hazards in your workplace, as with dragons, can result in death and destruction. An appropriate focus on effective process safety programs, as described in this book, can help make sure potentially catastrophic incidents do not happen.

Section I

Key Concepts
of Process Safety

1

Introduction to Process Safety

There is an old saying that if you think safety is expensive, try an accident. Accidents cost a lot of money, not only in damage to plant and claims to injuries, but also in the loss of the company's reputation.

Trevor Kletz

1.1 Introduction

The path to sustained, excellent process safety performance requires a continuous journey with constant effort by everyone in an organization when process hazards are present. "Everyone," for example, probably includes you if you are reading this book, but more generally:

- senior and middle managers, who manage people and provide resources
- technical personnel, who design and evaluate processes and equipment
- operating personnel, who operate equipment
- mechanical personnel, who maintain reliable equipment
- many other personnel, who support research, production, and safety.

Process safety programs go beyond traditional personal safety, which helps ensure that people perform their work tasks safely by, for example, wearing their safety glasses, hard hat, and the correct gloves, when required. Rather, effective process safety programs address the process hazards that may lead to toxic releases, fires, explosions, and runaway reactions, potentially resulting in events with catastrophic consequences. If your facility has process hazards, then you have an important role:

- in some cases, by contributing to the design and implementation of effective new or improved process safety systems
- in all cases, by helping to ensure the rigorous day-to-day application of the existing process safety system requirements.

By becoming knowledgeable about your organization's process safety goals and systems, you can help achieve safe and reliable operations, contributing to both excellent safety performance and high-quality research or manufacture.

This book provides key concepts, practical approaches, and tools for establishing and maintaining effective process safety programs to successfully identify, evaluate, and manage process hazards. The commitment to process safety should always be zero injuries and zero severe incidents. This means:

- both you and your coworkers complete your work safely every day
- your organization safely uses hazardous materials and processes
- your customers and society safely benefit from your efforts.

Use of this book to understand and apply key process safety concepts and systems can help achieve these goals. Most likely, your facility already has some process safety practices and systems in place. This book will help you understand why these process safety systems have been implemented, what they are intended to achieve, how you can apply them daily to achieve safe and reliable operations, and possibly, how you can improve them, as needed.

Unfortunately, there have been many catastrophic incidents when process safety programs and systems were ineffectively implemented or maintained, or the need for process safety programs was never recognized. The consequences of these incidents have been very high, including

- fatalities and life-changing injuries
- significant environment harm
- catastrophic property damage
- lost business and employment opportunities.

The Deepwater Horizon Gulf Oil Spill in 2010, for example, resulted in 11 fatalities, 17 injuries, a major oil spill reported to have caused over $60 billion in environmental and economic harm, and destruction of a $590 million oil rig, as shown in Figure 1.1 [1]. Investigation of this tremendous loss of life and property pointed to significant gaps in process safety systems and their applications [1–2]. Understanding and applying the principles and practical methodologies discussed in this book can help make sure that severe hazardous incidents do not occur at your facility by ensuring that process hazards are properly identified, evaluated, and managed. Many additional examples of process incidents are discussed throughout this book.

1.1.1 Purpose and Scope

The purpose of this book is to take a different approach introducing key concepts and practical approaches to process safety for someone who needs to safely store, handle, and use potentially hazardous materials and processes, similar to

FIGURE 1.1
The Deepwater Horizon Gulf Oil Spill, 2010 [1].

what might be included in an introductory process safety training course. It is intended to help readers understand, work with, consistently apply, and possibly improve the existing process safety systems that impact their daily work, helping to contribute to safe, reliable performance and to improved personal productivity and effectiveness. Process safety systems are introduced with basic goals and requirements, real-world examples, practical methodologies, and more detailed case studies to help introduce and apply basic process safety principles and systems. If you work with process hazards, you need to understand and apply the information in this book as a starting point. If you do not currently have a fully implemented process safety program, this book will help you assess potential gaps to develop improvement plans. If you are mainly concerned about regulatory compliance, this book should help you both understand the purpose of many requirements and practice effective approaches for meeting the requirements (e.g., see Chapter 3 for a brief discussion on the applicable regulations).

Process safety consists of many comprehensive, integrated systems that describe how to identify, evaluate, and manage process hazards. This book is organized to help readers:

- understand the key concepts of effective process safety programs (Chapters 1 through 4)
- apply process safety systems using practical concepts, approaches, and tools to design safe processes (Chapters 5 through 7) and implement and maintain daily focus on process safety (Chapters 8 through 12)
- develop personal capability in process safety and achieve and sustain process safety excellence (Chapters 13–14).

The intent is to introduce process safety systems in a way that can be helpful in designing and implementing a new or improved process safety program at different types of facilities, but also to help readers understand the purpose and use of overall process safety system requirements. The practical approaches provided in this book should help you apply these concepts in your daily work. Most of us work at facilities that have already developed and implemented a process safety program based on the process hazards that are present, so knowing how to design and implement new process safety systems is often not necessary. Understanding *why* process safety systems have been designed in a particular way, understanding *what* they are intended to achieve, and understanding *how* they should function day-to-day are essential to ensure continued safe and reliable operations. By reading this book, you should be well prepared to:

- understand and apply key concepts and methodologies to support effective process safety systems
- evaluate potential gaps between the approaches presented in this book and the current practices in your facility
- contribute to maintaining or improving your facility's process safety program.

This book provides an integrated framework for process safety systems based on desired activities and results rather than organizing them by specific "elements" that comprise the systems, which has been typical in the past [3–5]. In effect, this approach combines several traditional process safety elements into systems with common goals and requirements. For example, Chapter 8 combines "traditional" operating procedures, training, and contractor safety elements together to ensure that people working with process hazards are adequately trained and knowledgeable about process safety. Similarly, Chapter 10 combines "traditional" management of change and prestart-up safety review elements to ensure that process and equipment changes are made and reviewed safely before use. The approach taken in this book, based on a practical understanding of how these traditional process safety systems work together to achieve safe and reliable operations, is discussed in more detail in Chapter 3.

Even if not implemented in this way, the basic understanding of how elements work together in systems is fundamental to helping ensure the effectiveness of process safety programs.

1.1.2 Application

Regardless of global location, if hazardous materials and processes are present, process safety is applicable in all parts of a facility's life cycle, from its design, construction, operation, and maintenance stages, to its eventual shutdown and dismantlement. If a facility has process hazards, with the potential for toxic releases, fires, explosions, or uncontrolled reactions, the implementation of a process safety program is necessary to protect employees and the communities near the facility against injuries and potentially catastrophic incidents. Application of process safety systems is therefore important for safe and reliable operations at:

- companies, both large and small
- research facilities, including universities.

For simplicity, the wide range of operations that have process hazards will be referred to in this book as a process:

> *Process*: Parts of a facility that contain process hazards due to the presence of hazardous materials or conditions, which may include large or small manufacturing facilities, research pilot plants, laboratories, and universities.

The type of process and the quantity of hazardous materials being used may impact the type of process safety program that is implemented. Research facilities and universities, for example, may choose to implement process safety programs appropriate for their scale and level of hazard. However, if process hazards are present, they must always be effectively managed through an appropriate process safety program. Many incidents have occurred in processes where failures to identify, evaluate, and manage process hazards have resulted in fatalities and serious injuries, severe environmental impact, significant property damage, and loss of production capability. Many examples of serious incidents will be discussed throughout this book to help reinforce the need to understand design and implementation of process safety systems, what process safety systems intend to achieve, and how these systems should function day-to-day for safe and reliable operation.

In many countries, regulations exist that require implementation of a process safety program. Although it is necessary to meet these regulations, it is also sometimes necessary to "go beyond" them in order to manage hazardous materials and processes effectively. The key is to identify, evaluate, and understand the risks associated with process hazards and then to ensure that they are managed appropriately. In some cases, this may require additional process safety systems not required by regulations, or application of process safety systems to facilities not covered specifically by regulations. Process hazards exist in many different industries, such as chemical, energy, nuclear, aerospace, electronics, food, pharmaceutical, and mining. Proper identification of process hazards and implementation of appropriate risk management strategies, regardless of the industry, will help ensure the successful application of the process safety systems described in this book.

Many process safety books are written for process safety specialists who may be developing or implementing new process safety systems. This book is intended to be different. Although it should still be of interest to process safety specialists, this book is intended for a broader audience that works with process safety systems as part of their daily activities, including knowing and following their facility's requirements on managing hazardous materials and processes. Working with these process safety systems, understanding why they are important, and contributing to overall safety and reliability are part of their key responsibilities. Hence, the audiences for this book include:

- **Executives, managers, and other supervisors**, who must implement and enforce process safety system requirements, and who must provide sufficient and capable resources to maintain their desired performance levels for safe and reliable operations.

- **Engineers, chemists, and others with specific technical roles**, who must work with these process safety systems as they continuously improve products, equipment, and processes through their research, design, and support of safe and reliable operations.

- **Operators, mechanics, electricians, contractors, and others directly supporting operations**, who must work every day with these hazardous processes and process safety systems to manufacture the products and maintain the equipment for safe and reliable operations.

- **Researchers and technicians**, who usually work in laboratories or pilot facilities, and who must understand how to work safely and also how their "small-scale" process hazards have to translate to "large-scale" operations for safe and reliable operations.

- **Professors, instructors, and students**, who may supervise or conduct experiments or small-scale projects, who may design processes involving process hazards, and who need to develop capability in process safety as they conduct their research and/or prepare for industrial or research careers.

In effect, there is a need for *everyone* who works with or will potentially work with hazardous materials and processes, regardless of the scale, to have an awareness and understanding of process safety.

An example that reinforces this view that everyone potentially working with process hazards must have, at minimum, an awareness of process safety is provided by the Texas City refinery incident in 2005, where 15 workers were killed after a large flammable release and explosion occurred [6,7]. Recommendations from investigations following this incident include:

BP should develop and implement a system to ensure that its executive management, its refining line management above the refinery level, and all U.S. refining personnel, including managers, supervisors, workers, and contractors, possess an appropriate level of process safety knowledge and expertise [6].

The material discussed in this book is both fundamental and appropriate for raising the awareness and capabilities of a broad audience on process safety goals, systems, and implementation for managing process hazards at a wide range of facilities. Process safety is a comprehensive, integrated set of management systems that work together to ensure that process hazards are identified, evaluated, and managed appropriately and effectively. Gaps in awareness or in implementation of effective systems when managing hazardous materials and processes, can be, and have, proven deadly. See Chapter 13 for additional discussion on how to develop personal capability in process safety.

1.2 Key Concepts

1.2.1 What Is Process Safety?

Managing potentially hazardous materials, energies, and processes to prevent serious incidents and injuries is an essential part of process safety programs, requiring the continuous dedication and commitment of everyone in an organization. Unfortunately, there have been many examples of catastrophes caused by not properly recognizing, evaluating, and managing process hazards, as shown in Table 1.1. The results far too often are deadly exposures to hazardous materials, fires, explosions, runaway reactions, and other serious events that lead to fatalities, property damage, and significant

TABLE 1.1

Some Major Process Safety Incidents [Modified from 1,8,10,11]

Process Incident	Year	Fatalities
Flixborough	1974	28
Bhopal	1984	2500+
Chernobyl	1986	31+
Piper Alpha	1988	167
Pasadena	1989	24
Toulouse	2001	30
Texas City	2005	15
Deepwater Horizon (Macondo Well)	2010	11
Amuay Refinery	2012	47
West Fertilizer	2013	14

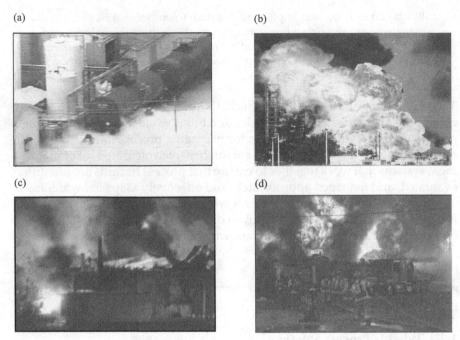

FIGURE 1.2
Examples of consequences from process hazards. (a) Toxicity—toxic releases, (b) flammability—explosion and fire, (c) combustible dust—explosion and fire, and (d) reactivity—explosion and fire [Modified from www.csb.gov].

environmental harm, as shown in Figure 1.2. The Bhopal incident in 1984, for example, involved the unintended reaction and large release of toxic methyl isocyanate (MIC) that caused thousands of fatalities and injuries, some of which may have passed from one generation to the next [8]. More recently, in addition to the large flammable release and explosion at the refinery in Texas City in 2005 [6–7], serious incidents have occurred in an oil storage depot in the UK in 2008 [9], on an oil rig in the Gulf of Mexico in 2010 [1–2], in a Venezuelan refinery in 2012 [10], and in a fertilizer warehouse in Texas [11], among many others.

Effective process safety programs can help prevent these types of incidents. Industry has benefited time after time from having implemented effective process safety programs to properly control the risks associated with process hazards, preventing serious incidents or mitigating the potential consequences to avoid more severe incidents. The basis for process safety programs, as distinguished from traditional occupational safety and health programs (see Section 1.2.2), is the presence of process hazards in a manufacturing, research, or related process:

TABLE 1.2

Process Hazards

Materials	Processes
• Toxicity, including low oxygen content • Flammability • Explosivity/energetics • Dust combustibility • Chemical reactivity • Corrosivity/acidity	• High/low temperature • High/low pressure • Mechanical • Rotating equipment

Process hazard: Intrinsic properties of materials and processes that can either directly or through release of energy lead to harmful effects, such as toxicity, flammability, explosivity, and reactivity.

Process hazards, as shown in Table 1.2, typically include both material hazards, such as toxicity and flammability, and processing hazards, such as high temperature or pressure, that if not controlled or managed properly through effective process safety programs can potentially lead to significant incidents. Regardless of the size of the facility or process involved, process hazards must always be identified if present and managed appropriately to avoid injuries and potentially serious incidents. For example, very serious consequences are possible even when laboratory equipment or small quantities of hazardous materials are used, so the presence of any process hazards will determine the need for an appropriate process safety program. It is important to note that process hazards may be present at any or all stages of a facility in its life cycle: for example, the incident at Bhopal occurred when the plant was no longer in operation, but hazardous materials were still present at the facility.

Process safety is a comprehensive, integrated set of management systems designed to help manage the risk associated with process hazards with the primary goal of preventing process-related incidents and injuries:

Process safety: Application of management systems to ensure that process hazards are identified, evaluated, and managed to reduce the risk of potentially hazardous incidents and injuries.

Process safety programs are designed to lower the process risk involved in the storage, handling, and use of hazardous materials and processes:

Process risk: A measure of potential harm associated with process activities and events, based on potential consequences (e.g., injury, property damage, environmental) and frequency (e.g., years).

Risk is a function of the potential consequences, such as fatalities, cost, or some other term ("fatalities/event"), multiplied by the potential frequency, usually expressed in years ("events/year"), to give units such as "fatalities/year," as shown in Equation 1.1:

$$\text{Risk} = \text{Frequency} \times \text{Consequence} \quad (1.1)$$

Obviously, the goal is to reduce process risks by evaluating and implementing different risk management strategies to reduce the consequences and/or frequency of potentially hazardous events (see additional discussion in Chapter 3). Your business may have established its own criteria for determining if risk is "acceptable" or "tolerable," and many countries have also established regulatory requirements that, at minimum, must be met [12]. These include documented "ALARP" justifications, where risk is reduced to "as low as reasonably practicable," given current available technologies (see Chapter 3). Typical criteria for the risk of a fatality, for example, are 10^{-5} or 10^{-6}, which would be 1 fatality in 100,000 or 1,000,000 years. This level is much lower than the risk of fatality for more common causes, such as car accidents.

To achieve effective process safety performance and consistent management of process risk, effective process safety programs must consist of three interrelated foundations as shown in Figure 1.3. These are discussed in more detail in Chapters 2 through 4 and include:

- **Safety culture/leadership:** A process safety program will only be as effective as required and supported by the underlying safety culture and leadership. If an organization has a weak safety culture, process safety efforts are probably doomed due to conflicting priorities, such as production schedules or lowering operating costs. A strong safety culture, based on a deep commitment to core values on safety, health, and environmental issues, is reflected by an organization's policies, goals, metrics, and day-to-day decision making that support establishing and maintaining strong process safety systems. Safety priorities are recognized as inherently necessary for completing any task the right way and do not conflict with other important tasks. Management leadership and personal commitment, as heavily influenced by the existing safety culture, are visible to everyone in the organization, helping to both sustain and improve the safety culture. The actions of all levels of management, from senior management

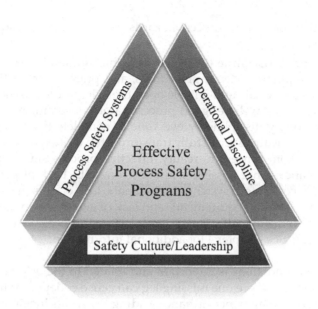

FIGURE 1.3
The foundations of an effective process safety program.

to first line supervisors, must support and reinforce strong process safety programs and accountability:

- beginning with forming and fostering a strong safety culture
- establishing firm policies that set high expectations for excellent safety performance
- continuing with day-to-day decisions to provide needed resources and support for safety-related activities.

See Chapter 2 for additional information.

- **Process safety systems:** A comprehensive process safety program includes integrated management systems to provide a process safety framework to ensure that process hazards and risks are properly identified, evaluated, and controlled. In many cases, a good starting point for developing strong process safety systems is existing regulations, if available, but regulations should generally be considered as a starting point only, representing a minimum essential level of practice. In some cases, significant process hazards may not be covered by regulations, requiring a detailed analysis of process hazards and risks to ensure that an appropriate process safety program is implemented. Additional requirements that go beyond regulatory requirements in order to better meet organizational needs and goals for managing process safety may be needed to achieve adequate risk reduction of your process hazards. See Chapter 3 for additional information.

- **Operational discipline (OD):** OD describes human behavior in following required systems, procedures, and practices correctly every time to achieve safe and reliable operations. Regardless of how comprehensive and well-designed process safety programs may be, it is the day-to-day discipline by everyone to complete their daily work tasks correctly that successfully transforms the written process safety program from concept to reality. Effective process safety program performance is not possible without strong OD. Lack of knowledge about process safety system requirements, lack of personal commitment in following requirements, and lack of awareness in daily work will contribute to ineffective systems and poor process safety performance [13]. See Chapter 4 for additional information.

Each of these foundations is important and related to the other parts, like a three-legged stool, where one missing leg can cause collapse, or in this case, cause poor process safety performance leading to serious incidents and injuries. For example, a strong safety culture and leadership depends on effective process safety systems and OD, but process safety systems and OD in turn are influenced by the safety culture and leadership.

The scope and depth of process safety programs may vary from process to process, depending on the process hazards and risks [4], recognizing that system consistency at any one site or organization is an important factor. For example, some highly hazardous materials or processes may be heavily regulated and may have very detailed process safety requirements to ensure safe usage. However, other materials or processes may have low to moderate process hazards, allowing for more flexible process safety requirements. Hence, proper evaluation of your process hazards and their associated risks is always necessary to ensure that appropriate process safety programs have been implemented, as described in Chapters 6 and 7. Regulatory requirements may also impact how process safety programs are designed and implemented, as discussed in Chapter 3.

1.2.2 Process Safety Is Not "Personal Safety"

Process safety programs are typically different than "traditional" occupational safety and health programs, though there may be significant overlap. Occupational safety and health is primarily concerned with personal workplace hazards that can cause personal injuries. These include tool and machinery safety, slips-trips-and-falls, electrical safety, moving equipment and vehicles, ergonomics, etc., as shown in Table 1.3. Common safety metrics include personal injury rates, such as medical treatment or restricted/ lost time injury rates. Based on these measures, the chemical industry is a relatively safe place to work, as shown in Table 1.4, where the lost workday

TABLE 1.3

Safety and Process Safety

Safety	Process Safety
• Driving safety • Slips, trips, and falls • Tool and machine safety • Lifting and ergonomics • Electrical safety • Moving equipment • Noise • Nonionizing radiation	• Process hazard assessment • Process risk management

TABLE 1.4

Industry Injury Rates, 2014 [Modified from 29]

Industry	Total Recordable Injury Rate	Lost Workday Injury Rate
Soft drink manufacturing	7.5	2.1
Motor vehicle manufacturing	6.9	1.6
Food manufacturing	5.1	1.3
All manufacturing	4.0	1.0
All private industry	3.2	1.0
Paper manufacturing	2.9	0.9
Mining	2.8	1.4
Basic chemical manufacturing	2.1	0.6
Petroleum refineries	0.7	0.2
Petrochemical manufacturing	0.5	<0.1

Injury rate = (number of injuries and illnesses × 200,000)/employee hours worked.

injury rate for basic chemical manufacturing (0.6), petroleum refineries (0.2), and petrochemical manufacturing (<0.1) are well below the rate for all private industry (1.0). Safety programs target identification, elimination, and management of workplace hazards through worker awareness, skill training, and implementation of workplace guards and controls, such as safety procedures and personal protective equipment.

Typically the types of hazards and the scale involved in traditional safety programs differ significantly from process safety programs, which are based on process hazards that can lead to potentially catastrophic incidents involving large numbers of fatalities or serious injuries, major financial loss, and significant environmental harm. There is some common ground in application, such as with confined space entry, personal protective equipment, etc., and in overall safety culture and leadership practices. Due to the hazard and scale differences, though, process safety programs must be managed as separate but related programs with different goals, systems, and metrics to help ensure process safety program effectiveness. For example, the U.S. Chemical Safety Board concluded in the BP Texas City refinery explosion in 2005 that:

> BP's approach to safety largely focused on personal safety rather than on addressing major hazards. BP Group and the Texas City officials almost exclusively focused on, measured, and rewarded reductions in injury rates and days away from work rather than the improved performance of its process safety systems [7].

Similarly, programs designed to manage safety during transport of hazardous materials contain many common features with process safety programs, but usually involve separate requirements, partly due to regulatory differences. Management of transportation and distribution risks is therefore beyond the scope of this book.

1.3 Why Is Process Safety Important?

Many companies may implement process safety programs solely to meet regulatory requirements. While that is important, it can be ineffective since regulations may not adequately manage the process risks present at a facility. Process safety must be implemented at every point in the facility's life cycle as process hazards may be present that must be managed to prevent or mitigate serious injuries and incidents. In many cases, these process hazards may be covered by appropriate regulations, and in some cases, they may not be covered or may not be adequately controlled based on regulations alone. Effective process safety programs, in addition to meeting regulatory requirements, are fundamental to any business activity involving process hazards:

> Safe operations and sustainable success in business cannot be separated. Failure to manage process safety can never deliver good performance in the long term, and the consequences of getting control of major hazards wrong are extremely costly... Major accidents may not just impact your bottom line profitability—they could completely wipe it out [14].

Think for a minute what it would be like to have a major process incident at your facility. The emergency response during the event would be stressful and scary, even if you are not in the direct area of the event. People who you probably know might be injured or even killed. The facility might be heavily damaged. Would it be rebuilt, or would jobs be lost? People in the local community may be angry and may not support continued operation of the facility. Investigations to determine the cause of the incident may continue for many months, and the findings may be difficult to accept. Lawsuits may be filed. Your career may be impacted in many ways. The future of the facility

TABLE 1.5

Potential Consequences of Process Incidents [Modified from 14]

- Harm to people, including loss of life and serious injury
- Environmental damage—for example, air, water, and land contamination
- Damage to business efficiency from disruption of production and loss of customers or suppliers
- Potentially huge costs—both direct (asset replacement or repair, lawsuits, fines) and indirect (increased insurance premiums, loss of shareholder confidence resulting in falling share value)
- Negative impacts on the local economy
- Long-term damage to an organization's reputation from adverse publicity, legal action, and harm to the company "brand"
- Cessation of the company as a viable, ongoing entity

and the organization may be uncertain. You will always remember the day of the incident and wonder if it could have been prevented. A major process incident can have devastating and lasting consequences:

> A large process safety accident can happen in an instant, often the result of some small bit of carelessness or a push to finish a task on time. Yet an instant is all it takes to claim lives and to change the lives of family, friends, and coworkers forever. An instant is all it takes to change a corporation's reputation. An instant is all it takes to impact public perceptions of an entire industry or of a profession... This is all part of a broader change in how we as an industry have to view process safety. No longer should it be just one important aspect of how we manage our operations. It must be central to what we do [15].

A summary of potential consequences due to a major process incident is provided in Table 1.5. After the incident, you will probably think about how a more effective process safety program may have helped. Focus on the benefits of a process safety program at your facility in advance—now—and help make sure this type of incident never occurs at your facility.

A study of the 100 largest property damage losses in the hydrocarbon industry since 1972 shows that the average property damage cost was over $300 million [16]. The total costs of these incidents, including other factors such as business loss, litigation costs, fines, etc., are much higher. The property damage for the Deepwater Horizon Gulf Oil Spill in 2010, for example, was $590 million, but the total costs for this incident including incident response, environmental clean-up, litigation, and other factors likely exceeded $60 billion. And, of course, as shown in Table 1.1, 11 people were killed by the initial explosion and fire. As shown in Table 1.6, the largest property damage incidents involve many different countries and demonstrate the global need for application of process safety [16].

TABLE 1.6

Global Nature of Large Process-Related Incidents
[Modified from 16]

Country	Property Loss ($ Mn, 2013)	Year
Algeria	$500+	2004
Argentina	$500+	2013
Aruba	$250+	2001
Australia	$750+	1998
Bangladesh	$100+	1991
Belgium	$250+	1975
Brazil	$750+	2001
Canada	$250+	2011
France	$500+	2001
Germany	$250+	2005
India	$250+	2005
Indonesia	$100+	2004
Italy	$100+	2008
Japan	$500+	2011
Kuwait	$750+	2000
Lithuania	$150+	2006
Malaysia	$250+	1997
Mexico	$100+	1996
Morocco	$100+	2002
Norway	$750+	1991
Russia	$100+	1994
Saudi Arabia	$100+	1987
Singapore	$100+	2011
Thailand	$100+	2012
United Kingdom	$1500+	1988
United States	$1000+	1989
Venezuela	$250+	2012

1.3.1 Business Impact

Implementing and maintaining a strong process safety program to manage process risk is fundamental for sustained, strong business performance. Business benefits typically result in the following areas, as summarized in Table 1.7:

- **Corporate reputation and sustainability** – Having a corporate reputation for being a safe, responsible citizen in the surrounding community and beyond can be invaluable, supporting and enhancing relationships with employees, community neighbors, customers, suppliers, regulators, and others. One serious process incident can damage these relationships, leading to regulatory investigations

TABLE 1.7

How Business Benefits from Process Safety [Modified from 30]

Reductions in Potential Consequences	Other Benefits
• Fatalities and injuries • Property damage costs • Business interruption costs • Loss of market share • Litigation costs • Incident investigation costs • Regulatory penalties • Regulatory attention	• Increased productivity • Lower production costs • Lower maintenance costs • Lower capital budget • Lower insurance premiums • Improved quality

and fines, loss of production capacity, interrupted product supply, reduced market share, environmental damage, unsupportive neighbors, and unhappy employees. Ultimately, the sustainability of companies to stay in business will be impacted by the effectiveness of their process safety program in addition to many other factors.

- **Employee protection** – Ensuring a safe workplace is essential for attracting and keeping a qualified workplace and maintaining high levels of employee engagement. Effective process safety programs are fundamental to avoiding fatalities and serious injuries, and potentially, high employee turnover rates with associated higher costs for training.

- **World class manufacturing** – Production capability resulting from achieving safe and reliable operations is essential to high product quality, asset productivity, and low capital, operating, and maintenance costs.

- **Reduced costs** – Incident costs, such as property damage, emergency response and recovery, incident investigation, litigation, regulatory penalties, higher insurance premiums, etc., can be avoided, contributing to improved financial performance.

1.3.2 Code of Ethics

Ensuring strong professional process safety focus and performance is a fundamental responsibility of engineers and related occupations:

> It is our responsibility in the process industry to operate safely and responsibly. It is your responsibility as a practicing chemical engineer or as a leader in the process industry to ensure that process safety is integral to every aspect of what you and your company do. It is quite simply a moral and ethical obligation [15].

In recognition of this, many engineering societies have developed a code of ethics for their members that include safety as a key component. In the

United States, for example, the American Institute of Chemical Engineers requires that, among other things, members shall:

- Hold paramount the safety, health and welfare of the public and protect the environment in performance of their professional duties.
- Formally advise their employers or clients (and consider further disclosure, if warranted) if they perceive that a consequence of their duties will adversely affect the present or future health or safety of their colleagues or the public [17].

Similar statements and beliefs may be found in various organizations for their members globally.

1.3.3 Conservation of Life

Process safety can be thought of as the application of "conservation of life" (COL)—the prevention of serious human injury—as a fundamental principle of engineering similar to the conservation of energy and the conservation of mass [18–19]. COL when applied at its most basic level can be simply viewed as "people in = people out," where everyone who shows up for work goes home safely at the end of the day. COL reflects the need for fundamental awareness and application of process safety and product sustainability concepts in engineering education, design, and practice and can be used by universities as a concept and unifying theme for increasing awareness, application, and integration of process safety throughout the undergraduate engineering curriculum. The application of COL principles will help you achieve "the goal is zero" with respect to injuries, incidents, and environmental and social impact. Awareness and use of these principles can help everyone understand their important roles as engineers in helping make achievement of this goal a reality once they enter the workforce.

1.4 Case Studies

Not all process incidents are catastrophic, of course, but many have the possibility of causing significant injury, property and environmental damage, and business impact. The following incidents provide examples for large companies, small companies, and universities.

1.4.1 Large Companies

- **Bayer CropScience, 2008, U.S.** [20] – On startup of the methomyl process, used to make insecticides, a runaway reaction occurred inside a 4500-gallon pressure vessel that exploded killing two workers and injuring eight. Several roads were closed and 40,000 people in the surrounding areas were asked to shelter-in-place as a precaution. The U.S. Chemical Safety Board (CSB) report concluded that written startup procedures were not followed, safety devices were bypassed, and many other factors contributed to the incident. Fortunately, a nearby tank containing a highly toxic material, methyl isocyanate (MIC), was not damaged by the explosion or the incident may have been worse. The methomyl process was not operating, and $80 million was invested to improve the safety of MIC storage and handling, although the company decided to shutdown the use of MIC at the facility before startup [21].

- **Imperial Sugar, 2008, U.S.** [22] – This sugar refinery processed raw cane sugar into granulated sugar. Conveyers and elevators were used to transport sugar through the plant and to large storage silos. An initial dust explosion occurred in a conveyer near the storage silos that dispersed additional sugar dust into the air, leading to secondary dust explosions and fires throughout the facility. Fourteen people were killed, 36 were injured, and the facility was severely damaged. CSB concluded that equipment design and housekeeping practices allowed sugar dust to accumulate that led to the explosions.

- **Bhopal, 1984, India** [8] – A pesticide plant in India was not operating, but a large release of highly toxic methyl isocyanate (MIC) occurred from storage that resulted in many fatalities and injuries in the nearby community in the middle of the night. Water was introduced into a storage tank containing MIC that led to a runaway reaction and subsequent venting of MIC from the tank. Many safeguards at the site, including tank refrigeration, instrumentation, scrubber, and process flare, were shut down or not operating.

- **Piper Alpha, 1988, United Kingdom** [8] – Piper Alpha was a large oil platform in the North Sea. After the accidental release of flammable hydrocarbons, a series of explosions and fires occurred that led to total destruction of the platform, 167 fatalities, and a property loss of $1.5 billion. The investigation of the catastrophe determined that a primary cause of the event was a faulty work permit system that separated the work permits for a condensate pump and its associated relief valve, allowing the pump to be inadvertently started up with the relief valve still out-of-service.

1.4.2 Small Companies

- **T2 Laboratories, 2007, U.S.** [23] – T2 was a small company employing 12 people. On the date of the incident, T2 was making its 175th batch of a gasoline additive called MCMT. A runaway reaction occurred due to failure of the reactor cooling system, leading to an explosion equivalent to 1400 lbs of TNT. Four employees were killed, 32 people were injured, and significant property damage occurred. The CSB report concluded that inadequate hazard recognition and evaluation of the reaction hazards contributed to inadequate design of the reactor cooling and emergency relief systems.

- **Concept Sciences, 1999, U.S.** [24] – At the time of the incident, Concept Sciences had 20 employees with 10 assigned in a new production facility. When making the first commercial batch of a 50% hydroxylamine (HA) solution for use in the semiconductor industry, a runaway reaction occurred, leading to an explosion equivalent to ~700 lbs of TNT that killed 5 people and injured 14. The production facility was severely damaged, resulting in $3.5–4 million in property damage. CSB concluded that inadequate process design and operation information was developed and that inadequate hazard analysis was completed on the reactive hazards of concentrated HA solutions.

1.4.3 Universities

- **Texas Tech University, 2010, U.S.** [25,26] – Two graduate students were synthesizing a nickel hydrazine perchlorate (NHP) derivative. In order to make more material for testing, batch size was increased from 50–300 mg to approximately 10 g. On the smaller scale, NHP was observed to be stable when wet with water or hexane, so one of the graduate students was mixing the large batch of NHP with hexane in a mortar and pestle to break up clumps of material. During mixing at one point, the material detonated, leading to three lost fingers, burns on the hands and face, and an injured eye. CSB concluded that the hazards of the NHP material were not adequately assessed, evaluation of the increase in batch size was not evaluated, and that safety accountability and oversight at the university were insufficient.

- **UCLA, 2008, U.S.** [26] – A research assistant was working with tert-butyl lithium, a pyrophoric material that ignites spontaneously when exposed to air. When this material was being transferred in a syringe, the syringe came apart and ignited, seriously burning the assistant who later died. Investigations determined that the assistant allegedly had not been instructed on safe lab procedures working with this material and was not wearing a lab coat or flame-retardant clothing. The state of California fined the university, and criminal charges were eventually settled after several years [27–28].

1.5 Key Questions

Table 1.8 lists some basic questions that most people working in facilities with process hazards should be able to answer. They are not exhaustive, and more detail will be provided in the remainder of this book. If you are not sure of the answers, these are good areas for starting to learn more as quickly as possible. In addition, senior leadership, especially in large companies must have a special focus on process safety program implementation at their facilities. For example, an investigation finding following the Texas City refinery explosion in 2005 focused on senior leadership: Although safety culture has likely been a factor in all major process incidents, the BP Texas City incident [17,54], discussed later in this chapter, was one of the first that emphasized safety culture:

> The board of directors of BP, BP's executive management (including its Group Chief Executive), and other members of BP's corporate management must provide effective leadership on and establish appropriate goals for process safety. Those individuals must demonstrate their commitment to process safety by articulating a clear message on the importance of process safety and matching that message both with the policies they adopt and the actions they take [6].

Senior leadership are encouraged to use a self-assessment checklist, containing sections on leadership and culture, risk awareness, safety information, competence, and action, that is available from the OECD website [14] to consider the quality of process safety implementation at their facilities.

TABLE 1.8

Basic Questions on Application of Process Safety [Modified from 14 and 31]

- Do you know what process hazards and risks are present at your facility? How are they being evaluated and managed?
- Are comprehensive material, process hazards, and process design documentation readily available as part of the process technology/process safety information?
- What practices do you have in place to ensure safe, reliable design and operation of your facility? Has this information been well documented?
- Do you know what major process incidents can occur? What can initiate an incident or near miss and how is it detected? What are the possible consequences? Who would be impacted?
- Do you know what safeguards are available to prevent or mitigate an incident? How confident are you that current process safety systems and safeguards will function correctly?
- Do you know how to respond to an incident?
- What practices do you have in place to learn about potential problems before they occur, such as audits, metrics, and learning from previous incidents?
- What practices do you have to ensure a knowledgeable, well-trained workforce?
- What practices do you have to maintain equipment integrity and reliability?
- What practices do you have in place to evaluate and manage potential equipment and processing changes before they are made?

References

1. U. S. Chemical Safety Board. 2014. *Explosion and Fire at the Macondo Well*. Report No. 2010-10-I-OS2013-02-I-TX. www.csb.gov.
2. BP. 2010. *Deepwater Horizon Accident Investigation Report*. www.bp.com/en/global/corporate/gulf-of-mexico-restoration/deepwater-horizon-accident-and-response.html.
3. Center for Chemical Process Safety. 1994. *Guidelines for Implementing Process Safety Management Systems*. Wiley-AIChE, New York.
4. Center for Chemical Process Safety. 2007. *Guidelines for Risk Based Process Safety*. Wiley-AIChE, New York.
5. U.S. OSHA. 1992. 29 CFR 1910.119: *Process Safety Management of Highly Hazardous Chemicals*. www.osha.gov.
6. Baker, James A., Frank L. Bowman, Glenn Erwin, Slade Gorton, Dennis Hendershot, Nancy Leveson, Sharon Priest, Isadore Rosenthal, Paul V. Tebo, Douglas A. Wiegmann, and L. Duane Wilson. 2007. *The Report of BP US Refineries Independent Safety Review Panel*. www.bp.com/bakerpanelreport.
7. U.S. Chemical Safety Board. 2007. *Refinery Explosion and Fire*. Report No. 2005-04-I-TX. www.csb.gov.
8. Atherton, John and Frederic Gil. 2008. *Incidents That Define Process Safety*. Wiley-AIChE, Hoboken, NJ.
9. UK HSE. 2011. *Buncefield: Why Did It Happen?* Competent Authority Strategic Management Group (CASMG). www.hse.gov.uk.
10. Venuzuala. 2013. *PDVSA, Evento Clase A Refineria de Amuay*. Gobierno Boivariano de Venuzuela, Ministerio dei Poder Popular de Petroleo y Minera. www.pdvsa.com/interface.sp/database/fichero/publicacion/8264/1632.PDF.
11. U.S. Chemical Safety Board. 2016. *West Fertilizer Company Fire and Explosion*. Report No. 2013-02-I-TX. www.csb.gov.
12. Center for Chemical Process Safety. 2009. *Guidelines for Developing Quantitative Safety Risk Criteria*. Wiley-AIChE, Hoboken, NJ.
13. Klein, James A. and Bruce K. Vaughen. 2008. A revised model for operational discipline. *Process Safety Progress* 27:58–65.
14. OECD. 2012. *Corporate Governance for Process Safety: OECD Guidance for Senior Leaders in High Hazard Industries*. www.oecd.org.
15. Dolan, Michael J. 2012. The intersection of process safety and corporate responsibility. *Chemical Engineering Progress* 108(6):24–27.
16. Marsh Energy Practice. 2014. *The 100 Largest Losses 1974–2013: Large Property Damage Losses in the Hydrocarbon Industries*, 23rd ed. www.marsh.com.
17. AIChE. No date. *Code of Ethics*. www.aiche.org/about/code-ethics.
18. Klein, James A. and Richard A. Davis. 2011. Conservation of life as a unifying theme for process safety in chemical engineering education. *Chemical Engineering Education* 45:126–130.
19. Davis, Richard and James A. Klein. 2012. Implementing conservation of life across the curriculum. *Chemical Engineering Education* 46:157–64.
20. U.S. Chemical Safety Board. 2011. *Pesticide Chemical Runaway Reaction Pressure Vessel Explosion*. Report No. 2008-08-I-WV. www.csb.gov.
21. Moore, David A. 2011. *Challenges In Applying Inherent Safety for Chemical Security*. 7th Global Congress on Process Safety.

22. U.S. Chemical Safety Board. 2009. *Sugar Dust Explosion and Fire*. Report No. 2008-05-I-GA. www.csb.gov.
23. U.S. Chemical Safety Board. 2009. *T2 Laboratories, Inc. Runaway Reaction*. Report No. 2008-03-I-FL. www.csb.gov.
24. U.S. Chemical Safety Board. 2002. *The Explosion at Concept Sciences: Hazards of Hydroxlamine Case Study*. Report No. 1999-13-C-PA. www.csb.gov.
25. U.S. Chemical Safety Board. 2010. *Texas Tech University Lab Explosion Case Study*. Report No. 2010-05-I-TX. www.csb.gov.
26. C&EN. 2012. 'Systemic Failures' Cited in UCLA Lab Fatality. www.cen.acs.org/articles/90/web/2012/01/Systemic-FailuresCited-UCLA-Lab-Fatality.html.
27. Blum, Deborah. 2012. *Bad Chemistry*. www.wired.com/wiredscience/2012/07/sangji-chemistry-death-ucla/.
28. AP. 2014. *Settlement in Case of UCLA Chemist in Fatal Fire*. bigstory.ap.org/article/settlement-case-ucla-chemist-fatal-fire.
29. Bureau of Labor Statistics. 2014. *Incidence Rates of nonfatal occupational injuries and illnesses by industry and case types, 2014*. www.bls.gov.
30. Center for Chemical Process Safety. 2006. *The Business Case for Process Safety*, 2nd ed. www.aiche.org/ccps/about/business-case.
31. Johnson, Robert W., Steven W. Rudy, and Stephen D. Unwin. 2003. *Essential Practices for Managing Chemical Reactivity Hazards*. Wiley-AIChE, Hoboken, NJ.

2

Process Safety Culture, Leadership, and Performance

Sustained great results depend upon building a culture of self-disciplined people who take disciplined action.

Jim Collins

Why Process Safety Culture, Leadership and Performance Are Important

When a decision has to be made about getting overdue production sent to a customer or dealing with a related safety problem first, how does (and should) a supervisor decide what to do? When a problem occurs in the middle of the night with limited supervision, how do workers decide how to deal with the problem? Is it reported the next morning? When new products or experiments are being designed, how do potential process, environmental, and product hazards get identified and managed? Collectively, these are issues related to safety culture and leadership that help define how an organization approaches problems and issues related to safety. Is safety a core value with high priority, or is it an afterthought subject to other organizational priorities? Safety culture influences the daily behaviors of leadership and workers, who either reinforce and improve the safety culture over time or allow it to degrade. Safety culture and leadership are the foundation of any process safety effort, where competing priorities, such as productivity or cost reduction, must be balanced and appropriately prioritized to allow successful implementation and operation of effective process safety programs.

A process safety program cannot be effective without
a strong safety culture and continuing daily focus by
leadership and employees to achieve appropriate priority
for safety as fundamental for business success.

2.1 Introduction

In the early days of the safety movement in the early 1900s, an appropriate industrial focus on safety was championed, introducing concepts such as incident prevention and "safety first" to promote both corporate responsibility and individual worker safety. At this early stage, a focus on industrial safety organizations and leadership, as shown in Table 2.1, can be found to include many elements of what is considered as safety culture today. For example:

> Virtual accident elimination is approached only … [w]hen the new concepts have infiltrated the minds of all units in the industrial establishment and have become a guiding force educing fixed habits of safe thought and action [1].

Important concepts of safety culture, such as shared values, attitudes, and practices promoting safe behaviors and performance, have always been recognized as fundamental to implementing and improving effective safety programs, even if they have not always been called "safety culture." The challenge of creating and sustaining a positive safety culture, however, is a challenge that continues today.

The importance of safety culture and associated human factors became more formally recognized following nuclear plant incidents at Three Mile Island in 1979 and Chernobyl in 1986. The Nuclear Regulatory Commission (NRC) concluded that "the principal deficiencies … are not hardware problems, they are management problems [2]." Following Chernobyl, the International Nuclear Safety Group (INSAG) [3] investigation found that "Formal procedures, properly reviewed and approved, must be

TABLE 2.1

1926 Fundamental Requirements for Successful Safety Organization [Modified from 1]

- Proper attitude on the part of the company and plant executives and willingness to take their part as leaders in the [safety] work.
- Continued demonstration of their attitude by personal interest, by personal example, by precept, and by action.
- Application of the principle of "safety first" in settling questions in which safety and production appear to conflict.
- To make safety a part of every employee's duties and not permit it to become the function of a single person, group, or class.
- … the accident situation shall be ascertained and used for guidance in the prosecution of preventive work.

Challenger 1986   Columbia 2003

FIGURE 2.1
The NASA space shuttle incidents [7, 9].

supplemented by the creation and maintenance of a 'nuclear safety culture'."
The three primary aspects of safety culture, according to INSAG, were:

- The environment created by local management
- The attitudes of individuals at all levels
- The actual safety experience at the plant.

Development and assessment of safety culture were extensively explored [3–6].

Safety culture was also a major focus of investigation at NASA following the Challenger shuttle explosion in 1986 and again following the loss of the Columbia shuttle in 2003, as shown in Figure 2.1. The Rogers Commission Report [7] on the Challenger explosion found a serious flaw in the decision-making process at NASA, partly based on ineffective communication of potentially serious concerns about the launch and that "NASA appeared to be requiring a contractor to prove that it was not safe to launch, rather than proving it was safe." Subsequent investigation [8] concluded that problems in NASA's safety culture contributed to the explosion, including:

- a production culture, where competition and scarcity for resources effectively reduced priority for safety relative to cost considerations and meeting flight schedules
- normalization of deviation, where more and more risk was accepted with NASA in an incremental fashion

- structural secrecy, where organizational obstacles contributed to poor communication and ineffective decision making on critical safety issues.

Similarly, when Columbia was lost during reentry in 2003, the Columbia Accident Investigation Board [9] concluded that "Cultural traits and organizational practices detrimental to safety were allowed to develop..." A Center for Chemical Process Safety (CCPS) review of the Columbia accident [10–12] identified several factors that are important when evaluating safety culture, as shown in Table 2.2. These and other factors will be discussed later in this chapter. Similarly, the importance of safety culture as a factor in process incidents, such as Flixborough and Piper Alpha, has been discussed elsewhere [12]. More recently, the BP Texas City explosion in 2005 [13] was examined in depth for organizational and cultural factors and is discussed as a case study in this chapter.

Most companies now have the technical knowledge and capabilities to safely identify and manage process hazards and risks. As improved guidance and tools for implementing process safety programs have been developed, the focus on underlying causes of major incidents has shifted in many cases to understanding and improving organizational and cultural causes. Certainly, many, if not most, incidents in the petrochemical and other industries have always had some degree of cultural causes, but improved technical capabilities have made—or should have made—other types of causes less frequent. Kletz [14], for example, expressed in the early 1990s that "new" incidents rarely occur; rather, the same kinds of incidents are repeated and therefore should be preventable. More recently, Hendershot [15] stated:

> We know how to improve process safety performance. Our biggest challenge is not technical, it is cultural. We need to do what we already know how to do, we need to do it well, and we need to do it everywhere and all of the time.

TABLE 2.2

Safety Culture Themes from the Columbia Shuttle Explosion [Modified from 10–12]

- Maintaining a sense of vulnerability
- Combating normalization of deviation
- Establishing an imperative for safety
- Performing valid, timely hazard/risk assessments
- Ensuring open and frank communications
- Learning and advancing the culture

2.2 Key Concepts

2.2.1 What Is Safety Culture?

Safety culture is part of the overall organizational culture in an organization that encompasses all the ways that things are or should be done. If process risks are present, then safety culture, as needed to support an effective process safety program, must be a core part of the culture:

> What's important ... what the public and the shareholders expect ... what we are called to do ... is to elevate process safety to a central role in our operations and a critical component of corporate social responsibility [16].

Although safety culture has likely been a factor in all major process incidents, the BP Texas City incident [17,54], discussed later in this chapter, was one of the first where subsequent investigations emphasized safety culture:

> Although we necessarily direct our report to BP, we intend it for a broader audience. We are under no illusion that deficiencies in process safety culture, management, or corporate oversight are limited to BP. Other companies and their stakeholders can benefit from our work. We urge these companies to regularly and thoroughly evaluate their safety culture, the performance of their process safety management systems, and their corporate safety oversight for possible improvements [18].

Differences in safety culture can have dramatic effects on safety performance, including injuries and significant incidents. Another dramatic example was provided during the 2011 Japanese tsunami that led to the nuclear meltdown at the Fukushima power plant. Although the Fukushima power plant suffered this catastrophic incident, the nearby Onagawa power plant was able to shut down safely with relatively minor damage. Some of this difference has been attributed to differing safety cultures between the owners of the two power plants [19].

It is simply not possible to talk about process safety without considering the impact of the safety culture and the leadership at a company, organization, or university. Safety culture and leadership are therefore the foundation of effective process safety program implementation and performance, as shown in Figure 1.3. Together, they can either doom process safety efforts and performance based on conflicting priorities, such as cost reduction and/or production schedules, or they can support recognition that good safety performance is an inherent part of world class manufacturing capability that contributes to good business and financial performance. Think about the safety culture at your facility: do you need to prove that something is unsafe before a task is stopped, or do you need to ensure that something is safe before proceeding? The answer says a lot about your culture and the probable

TABLE 2.3

Definitions of Safety Culture

- Safety culture is that assembly of characteristics and attitudes in organizations and individuals which establishes that, as an overriding priority, nuclear plant safety issues receive the attention warranted by their significance. INSAG 1991 [3]
- The safety culture of an organization is the product of individual and group values, attitudes, perceptions, competencies, and patterns of behavior that determine the commitment to, and the style and proficiency of, an organization's health and safety management. HSE 2002 [21]
- Organizational culture refers to the basic values, norms, beliefs, and practices that characterize the functioning of a particular institution. At the most basic level, organizational culture defines the assumptions that employees make as they carry out their work; it defines 'how things are done'. An organization's culture is a powerful force that persists through reorganizations and the departure of key personnel. CAIB 2003 [9]
- The combination of group values and behaviors that determine the manner in which process safety is managed. Kadri and Jones 2006 [10]
- The combination of group values and behaviors that determines the manner in which process safety is managed. A sound process safety culture refers to attitudes and behaviors that support the goal of safer process operations. CCPS 2007 [22,23]
- Individual culture is the tendency in each of us to want to do the right thing in the right way at the right time, ALL the time—even when/if no one is looking. Arendt 2008 [24]
- See also [25].

effectiveness of process safety systems for preventing serious incidents and for achieving strong process safety performance.

Safety culture is part of the overall organizational culture at a facility, company, or organization. In general, organizational culture is described by [20]:

- Group and individual shared values (what is important)
- Beliefs and attitudes (how things work)
- Competencies and behaviors (how things are done).

Safety culture has been defined in many ways, as shown in Table 2.3, although a good definition is:

Safety culture: The normal way things are done at a facility, company, or organization, reflecting expected organizational values, beliefs, and behaviors, that set the priority, commitment, and resource levels for safety programs and performance.

Safety culture has also simply been defined by many as "how the organization behaves when no one is watching [15,24]." The Center for Chemical Process Safety (CCPS) has also defined "committed culture" as part of its Vision 20/20 effort [26]:

In a Committed Culture, executives involve themselves personally, managers and supervisors drive excellent execution every day, and all employees maintain a sense of vigilance and vulnerability.

Safety cultures are often referred to as good/bad or stronger/weaker. In this book, we prefer stronger/weaker. Ultimately, the safety culture is an accumulation of many different factors, such as the essential features discussed in the next section. Some will be stronger, and some will be weaker, leading to a balance that defines how positive the safety culture is related to supporting effective safety programs and performance. A positive or stronger safety culture gets good results over time. A weaker safety culture may get good results for a short period, but over a long period of time, performance will normally degrade. A contrast of stronger/weaker safety culture behaviors is shown in Table 2.4.

Safety cultures may have different levels of maturity [5,27–29], ranging from early hazard awareness to interdependent teams responsible for safety. Safety culture is also not stagnant, leading either to purposeful improvement or unintended degradation. As a result, developing and

TABLE 2.4

Safety Culture Behaviors

Stronger Safety Culture Examples	Weaker Safety Culture Examples
• Leadership provides a clear safety vision	• Leadership does not provide a clear vision
• Resources are provided for safety activities	• Safety activities are "unfunded mandates"
• Strong personal "felt" leadership	• Leadership rarely seen in informal ways
• Balances production pressures with safety	• Pushes production as top priority
• Determines system level root causes	• Blames human error
• Engages employees	• Commands employees
• Authorizes standard work practices	• Encourages getting the job done
• Challenges incorrect practices	• Ignores or accepts incorrect practices
• Work must be proved safe to proceed	• Work must be proven unsafe to not proceed
• Provides and maintains documentation	• Rubberstamps documentation needs
• Recognizes safety contributions	• Recognizes "saving the batch" or shortcuts
• Measure key safety performance indicators	• Measures mainly production indicators
• Communicate openly and extensively	• Communicates infrequently
• Involves high level safety personnel	• Decides without key safety personnel
• Focuses on continuous improvement	• Focuses on maintaining the status quo
• Learns and remembers	• Filters information that does not fit
• Maintains strong hazard awareness	• Becomes complacent
• Manages changes carefully	• Allows safety margins to degrade
• Strong systems for identifying and completing safety-related problems	• Slow completion of safety-related problems if even identified
• People like working here and feel safe (low turnover)	• People often do not feel safe and leave (high turnover)
• Ultimately, good and/or improving safety performance (injuries, incidents, and metrics)	• Ultimately, poor and/or declining safety performance (injuries, incidents, and metrics)

sustaining a good safety culture must be continually reinforced by leadership, as discussed in this chapter.

2.2.2 Characteristics of Safety Cultures

No consistent set of factors, used to evaluate the strength of a safety culture, have been adopted, though the 12 essential features defined by CCPS [23], as shown in Table 2.5, are often used. A discussion of the essential features is provided below:

- **Establish process safety as a core value** – Core values are deeply held beliefs that are beyond compromise. By establishing process safety as a core value, in vision and mission statements, by clear and constant communication, and by implementing cultural rituals that reinforce desired beliefs and behaviors (e.g., beginning all meetings with discussion of safety), high value for effective process safety programs can be realized.

- **Provide strong leadership** – Strong leadership for process safety is based on understanding and valuing process safety, sharing personal commitment with others by displaying desired behaviors, providing resources, involving and supporting safety personnel, and consistently considering risk management in day-to-day decision making.

- **Establish and enforce high standards of performance** – Provide clear and consistent expectations, including in annual individual performance reviews, to follow safety systems and operating procedures without tolerating intentional shortcuts or other violations of requirements.

- **Document the process safety culture emphasis and approach** – Document safety culture core values, expectations, responsibilities, and accountabilities, including mechanisms for periodically evaluating and sustaining a strong culture.

- **Maintain a sense of vulnerability** – Provide systems and training to develop awareness and respect for process hazards and potential

TABLE 2.5

CCPS Essential Features of Safety Culture [Modified from 22 and 23]

• Establish process safety as a core value	• Defer to expertise
• Provide strong leadership	• Ensure open and effective communications
• Establish and enforce high standards of performance	• Establish a questioning/learning environment
• Document the process safety culture emphasis and approach	• Foster mutual trust
• Maintain a sense of vulnerability	• Provide timely response to process safety issues and concerns
• Empower individuals to successfully fulfill their safety responsibilities	• Provide continuous monitoring of performance

process incidents to prevent complacency and to ensure appropriate sensitivity to operations, including recognition of possible warning signs, effective incident investigations, and records of historical incidents.

- **Empower individuals to successfully fulfill their safety responsibilities** – Ensure personnel are trained in all aspects of their roles and provided with appropriate resources so they can complete work correctly and safely—and stop work if concerned about safety if needed.

- **Defer to expertise** – Create leadership positions where knowledgeable safety personnel have access to and credible input for decision-making processes, involving other safety professionals as appropriate.

- **Ensure open and effective communications** – Communicate consistently and clearly on process safety goals, activities, and accomplishments, and provide systems for reporting of safety-related issues requiring timely response.

- **Establish a questioning/learning environment** – Provide risk management systems to identify process hazards and help prevent process incidents, including mechanisms for learning from experience, ensuring input from all personnel, and maintaining critical knowledge.

- **Foster mutual trust** – Create an environment where personnel are comfortable participating in activities, communicating with leadership and each other honestly, reporting mistakes, and making decisions made without fear, based on consistent management principles and behavior.

- **Provide timely response to process safety issues and concerns** – Provide systems for reporting process safety concerns, following up and completing action items in a timely manner, and communicating the resolutions to demonstrate consistent application of process safety principles and to avoid credibility problems.

- **Provide continuous monitoring of performance** – Develop key performance indicators for process safety and safety culture, periodically review and evaluate them to identify continuous improvement opportunities, and share results with affected personnel.

Training on safety culture should be provided to leadership, who are expected to lead safety activities, and to all personnel so that they are building awareness, understanding, and commitment to ensuring stronger safety culture and process safety performance. Although some differences in safety culture in a large organization may be expected due to differing hazards, products, sites, businesses, or geographic locations, core aspects of safety culture and safety programs can be established universally.

The foundations of any safety culture are effective risk management systems (know the hazards) and maintaining a sense of vulnerability (continue to know and control the hazards). If the hazards are not identified, understood, and managed, then an appropriate and stronger safety culture cannot be developed and effective process safety systems cannot be implemented. Any degree of complacency about these hazards, whether due to ignorance, familiarity, lack of incidents, or other causes, reduces the priority for ensuring continued excellent process safety performance. Previous strong safety performance, especially when a serious injury or incident has not happened for many years, is often a factor in complacency, associated with a false belief that success is routine [30] rather than something requiring continued focus and diligence.

An early example of complacency issues occurred in the Bhopal 1984 incident, where many process safeguards were taken out-of-service in the mistaken belief that hazardous events could not occur in a process plant that was shutdown [17,31]. More recently, the Baker Panel [18] investigation following the BP Texas City explosion in 2005 found that "... apparent complacency toward serious process safety risks existed at each of BP's U.S. refineries." Further, the Baker Panel stated:

> Preventing process accidents requires vigilance. The passing of time without a process accident is not necessarily an indication that all is well and may contribute to a dangerous and growing sense of complacency. When people lose an appreciation of how their safety systems were intended to work, safety systems and controls can deteriorate, lessons can be forgotten, and hazards and deviations from safe operating procedures can be accepted. Workers and supervisors can increasingly rely on how things were done before, rather than rely on sound engineering principles and other controls. People can forget to be afraid [18].

Although a return to the early days of the chemical industry, as described below, is not desirable or necessary, a deep and healthy respect for the hazards in the workplace is always necessary:

> Like a Damoclean sword,* hanging over the head of every powder man, was the dread of explosions. It was a nameless terror, waking men at night to prowl at odd hours about the yards, observing things that would have drawn scarcely a glance from less anxious eyes—random sparks from a chimney, wind direction, faintest indication of heavy skies that might bring thunder and lightning [32].

Some additional approaches for keeping a sense of vulnerability have been described [33].

If a sense of vulnerability is not maintained, the culture and systems will ultimately degrade due to complacency and the normalization of small changes over time leading to reduced safety margins and performance.

---

* The sword suspended by a single horsehair over Damocles as he sat on the throne, so that King Dionysius could demonstrate the inherent risks associated with being king.

A continuing sense of vulnerability reinforces the safety culture and leads to high operational discipline in following process safety systems, the foundations for effective process safety programs as shown in Figure 1.3.

2.2.3 Employee Participation

It is a requirement of the U.S. OSHA PSM regulation that employers provide a written plan of action regarding employee participation in process safety. Typically, the employee participation plan describes the ways that:

- employees are involved in the implementation of all process safety systems
- employees are consulted on the ongoing conduct and development of process safety systems, with emphasis on process hazard and risk analysis studies
- employees are provided access to process safety system documentation, with emphasis on process hazard and risk analysis.

Employee participation plans are normally documented separately or as part of the process safety documentation that describes the scope of the process safety program, such as the boundaries of covered regulatory processes at a facility. The plan should be reviewed periodically to ensure that it is current and its requirements are being met.

2.3 Evaluating and Improving Safety Cultures

Improving weaker safety cultures to stronger safety cultures, or strong to stronger, requires methodologies for evaluating the current state of the culture and identifying potential improvement opportunities. There is no one best safety culture; rather, a stronger safety culture is one that supports setting and achieving of goals that help ensure effective process safety program implementation and performance in appropriate priority balance with other organizational goals. Changing safety culture can be very difficult:

> The current organizational and safety cultures of an organization will have evolved over its life, and will have been influenced by the beliefs and values of the founders, their basic cultural assumptions and the experience of the organization to the present point in time. During this evolution, the main cultural components will have become embedded in the organization. Once the cultural elements have stabilized, further change becomes more complicated. It now involves having to unlearn beliefs, attitudes, values and basic assumptions, as well as learning new

ones. People resist change because such unlearning is uncomfortable and anxiety producing [34].

Some type of safety culture exists in an organization—stronger or weaker—and this obviously serves as the starting point for any improvement effort. Changing the culture for the better, rather than starting from scratch, is generally the only practical goal.

To get started, it usually makes sense to identify and evaluate the safety issues that are present, and then to consider how the current culture impacts these issues, and how it can be more effective, if needed. Employee concerns, organizational strengths and weaknesses, motivations and barriers for change, and external factors such as regulatory requirements, industry standards, and benchmarking can be important considerations. If small changes can be made to achieve the desired goals with only minimal impact on the culture, this reduced path of resistance may be the most successful path. Cultures are often like a spring that when pushed against, pushes back; so planning major cultural changes must be approached carefully. In some cases, such as catastrophic incidents, financial crises, or acquisitions, a strong driving force that deeply impacts the organization may exist that eases the path for significant changes. Of course, if change can occur without these possibly tumultuous events, the organization can respond with planning and purpose to mitigate resistance to change. Incremental changes may be just as or more effective than step changes. Collins [35] likens this approach as similar to a flywheel that requires a great deal of effort to start turning, but then slowly builds momentum until it starts to turn on its own effort.

Evaluating something as comprehensive and complex as a safety culture can be a difficult task, so multiple tools or approaches are commonly used [34,36,37], including:

- **Performance data** – process metrics and other key performance indicators provide "hard" data on how the organization is performing (e.g., number of incidents and number of overdue action items). Quantitative data are hard to argue with. It may suggest that the safety culture is supporting excellent performance; it may suggest that results are terrible. As discussed in Chapter 12, careful thought should be given to ensuring that appropriate metrics are developed that provide insights into process safety systems but also on safety culture and operational discipline characteristics. In particular, use of root cause analysis [38] and other methods, such as Six Sigma, to identify safety culture issues in injuries and incidents can help identify underlying problems.

- **Surveys** – Computer surveys provide semi-quantification of what is primarily subjective data from possibly large numbers of people in a relatively short time. A wide range of questions can be asked to ensure the survey is comprehensive and yet also targeted to the specific essential features of safety cultures that were discussed in the previous section. Questions can be based on yes/no, level of agreement, or free text entries. The level of agreement as "% yes" or "% in agreement" (e.g., strongly agree or agree) can be scored to provide measures of the responses, with special focus on essential features or questions that score highest or lowest. Surveys can also be filtered to allow comparison of results from subgroups in an organization, such as management, technical, operations and maintenance. Some typical survey questions are shown in Table 2.6.

- **Interviews** – Interviews allow discussion with employees by experienced facilitators who can probe more deeply into certain topic areas of interest. Sometimes, an interview may go in directions that are unanticipated but that can provide great insights into potential issues at a site. Generally, one-on-one interviews are conducted to help ensure anonymity, but sometimes group discussions can lead to richer discussions where participants build on each other's thoughts and stimulate new ideas. Often, a combination of single and group interviews can be a practical approach for obtaining the most information in the limited time that is usually available. Planning should ensure that a wide range of interviews for different roles in the organization are conducted.

- **Observation** – Experienced facilitators can also simply observe the work activities of many people in the organization to see how work is commonly done. The actual work process—how work is planned, how work is conducted, the results that are obtained, and how work is evaluated—can reveal both strengths and weaknesses of the safety culture that can confirm or weaken results obtained from other methods. In some cases, data collected as part of behavior-based safety (BBS) programs may also be useful.

TABLE 2.6

Typical Process Safety Culture Survey Questions [Modified from 4 and 24]

- Plant management is sincerely interested in feedback on process safety from employees
- My coworkers routinely report all process safety issues, including near misses
- We do not shortcut safety in order to maintain or improve production
- Management focuses on correcting issue rather than "blaming someone"
- We never reward someone for taking a safety shortcut even if the result is positive
- Our supervisor never asks us to perform an unsafe operation or task
- My coworkers never "work around" a safety problem rather than report it
- Our process safety training is very good
- Equipment problems are repaired as soon as possible
- We have the authority to stop work if we think it is not safe
- Our recognition program includes process safety
- Our leaders talk about process safety frequently
- We have clear goals on process safety established every year
- Well-written and current procedures are provided for work activities
- We have good procedures for reviewing job safety

Note: These questions are usually answered on a scale from strongly agree to strongly disagree.

Some or all of these approaches are used to conduct full evaluations of the safety culture periodically (e.g., annually or every few years) to help measure progress of improvement activities and to see if new issues have developed. In some cases, it may also be beneficial to more frequently conduct partial evaluations using only a few of the methods described above or in targeted areas or work areas [39], such as those that scored poorly previously or are actively being worked on, in order to see if better results are being obtained or if revisions need to be made. Results of the evaluations and suggested improvements should be reviewed with all affected employees for awareness and to solicit additional input and commitment on potential activities. Use of group meetings to allow for discussion can often prove beneficial versus communication via e-mail or other more impersonal approaches.

The results of a safety culture evaluation usually include both "hard" and "soft" data that help develop a picture of the stronger and weaker scoring essential features, which were discussed in the previous section. A culture may have strong leadership, but have lost to a degree its sense of vulnerability. Workers may be empowered to complete their work, but not really trust that they can do so without repercussions. The opportunities for improving the culture can be presented as shown in Figure 2.2, where the results can represent either the essential features or specific recommendations (actions 1–12). Some may be easy to improve but may not provide great benefit, such as action 4. Some may provide great benefit but may be difficult to implement and sustain, such as action 8. Some possible actions for improving the essential features are shown in Table 2.7. Additional suggestions are also available [36,37].

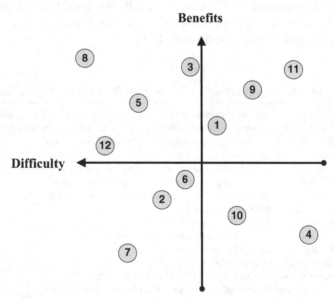

FIGURE 2.2
A safety culture improvement grid.

TABLE 2.7

Improving Safety Culture

Essential Feature	Example Improvement Activities
Establish process safety as a core value	• Clearly communicate and reinforce the importance of process safety at all opportunities • Set annual goals for process safety
Provide strong leadership	• Emphasize process safety in formal and informal activities, such as safety meetings and field audits • Provide training on process safety fundamentals and leadership to line managers
Establish and enforce high standards of performance	• Provide clear expectations and goals to ensure accountability in work activities, including process safety, as part of individual performance reviews • Establish a procedure-based organization focused on high levels of operational discipline
Document the process safety culture emphasis and approach	• Provide resources to implement and sustain effective process safety programs • Document process safety policies, requirements, and guidance to ensure that personnel know how to do their work correctly and safely
Maintain a sense of vulnerability	• Maintain an effective incident investigation program with broad sharing of incidents • Provide detailed training on process hazards and actual and potential hazardous events to maintain awareness
Empower individuals to successfully fulfill their safety responsibilities	• Provide sufficient process safety support personnel and resources to assist individuals with meeting process safety requirements • Ensure that "safety is part of the job" when planning work activities, resources, and schedule
Defer to expertise	• Provide staff level positions in process safety equal to operating positions to provide input and credibility • Involve process safety professionals in risk management decisions
Ensure open and effective communications	• Provide systems for identifying and following up on process safety issues and/or improvements • Communicate process safety program goals, activities, and metrics to affected personnel
Establish a questioning/ learning environment	• Encourage involvement of all personnel, especially with differing opinions, to avoid groupthink • Train personnel on warning signs related to possible operating and process safety problems
Foster mutual trust	• Ensure open, credible, and consistent leadership • Engage personnel to help develop, commit to, and be highly involved in sustaining a strong process safety program
Provide timely response to process safety issues and concerns	• Maintain and review metrics to track timeliness of action item closure • Communicate metrics and closure actions to affected personnel
Provide continuous monitoring of performance	• Maintain, review, and share key performance indicators at an appropriate frequency • Identify improvement opportunities and provide resources to ensure change

TABLE 2.8

Effective Change Processes [Modified from 40]

1. Establish a sense of urgency (gaps/priorities)
2. Create a guiding coalition (leadership)
3. Develop a vision and strategy (goals/plans)
4. Communicate the vision change (engagement)
5. Empower employees for action (resources)
6. Generate short-term wins (monitor)
7. Consolidate gains and produce more change (iterate)
8. Anchor new approaches in the culture (sustain)

TABLE 2.9

Supporting Cultural Change [Modified from 34]

- Provide a compelling positive vision so that employees believe that they and the organization will be better off if they adopt the new way of thinking and working. Senior management must be committed to the vision and communicate it to others
- Give employees formal training in the new ways of thinking and working
- Involve employees in designing their own optimal learning processes, thus recognizing that everyone learns slightly differently
- Provide opportunities to practice the new ways and give feedback, so that people can make mistakes and learn from them without disrupting the organization
- Give informal training to groups so that new norms and assumptions can be built collectively. A person should not feel deviant in engaging with the new learning
- Provide positive role models so that people can see the new behavior and attitudes in others with whom they can identify
- Form support groups so that problems associated with the new learning can be discussed, and so that people can speak openly about difficulties with others who may be experiencing similar difficulties
- Ensure that systems and structures are consistent with the new way of thinking, for example, reward and discipline systems

Much has been written on how to manage major change efforts within organizations. Kotter's [40] approach, as shown in Table 2.8, provides a typical model of leading change. Careful application of these steps can help ensure the success of a safety culture improvement effort. The International Atomic Energy Agency (IAEA) [34] has also provided some activities to support safety culture change, as shown in Table 2.9, to help the organization understand why changes are being made, what the benefits are, how long is required, what kind of results are being obtained, and what involvement and follow-through from everyone is needed. Building awareness, commitment, and involvement of the organization increases the likelihood of an improved safety culture and better process safety performance. The challenges to leadership are many, so a focus on strong leadership is also essential, as discussed in the next section.

2.4 Process Safety Leadership

Although strong leadership is an essential feature of safety culture, it is also instructive to consider process safety leadership separately, especially if the safety culture and leadership are not in sync, as shown in Figure 2.3. The four quadrants shown in this figure are:

- **Sustaining** (strong leadership and strong safety culture) – this organization (company, division, facility, or department) should normally be able to achieve and sustain excellent process safety performance with continued focus.

- **Conflicting** (strong leadership and weak safety culture) – this organization probably does not have excellent process safety performance, especially over longer time periods, but new or revitalized leadership is engaged and working to improve. The existing safety culture, however, resists change and the resulting conflict of status quo versus change may lead to uncertainties on whether the leadership will succeed in changing the safety culture and achieving improved performance. Ultimately, leadership may turn around a poor culture if successful.

- **Conflicting** (weak leadership and strong safety culture) – this organization probably has good process safety performance due to the strong safety culture, but future performance may be in jeopardy due to leadership most likely with other priorities. Again, the resulting conflict between the safety culture and leadership may lead to

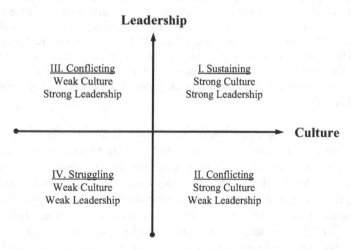

FIGURE 2.3
A culture and leadership grid.

uncertainties on "who wins," and safety performance may degrade. Ultimately, the culture may win and the leader will depart, or the leader will win and the culture will be severely damaged.

- **Struggling** (weak leadership, weak safety culture) – this organization likely has poor process safety performance, especially over the longer term, and possibly may not even recognize the process hazards that may be present and the risks associated with them. Most likely, performance will remain poor or degrading until a major reason for change occurs, or, ultimately, business failure occurs.

When conflict between safety culture and leadership occurs, organizational change based on an external event (e.g., merger, regulatory, or legal actions) or internal event (significant injury or incident) may unfortunately be needed to implement a more effective process safety program and achieve improved performance, as discussed further in the next section. This may require personnel change in key roles from senior leadership to frontline supervision to ensure that the best leaders are in place to achieve desirable organizational safety goals.

The fate of leaders often depends on the actions of senior leadership, including the organization's board of directors. The Baker Panel [18] investigation following the BP Texas City incident found:

> The Panel believes that leadership from the top of the company, starting with the Board and going down, is essential. In the Panel's opinion, it is imperative that BP's leadership set the process safety "tone at the top" of the organization and establish appropriate expectations regarding process safety performance. Based on its review, the Panel believes that BP has not provided effective process safety leadership and has not adequately established process safety as a core value across all its five U.S. refineries. While BP has an aspirational goal of "no accidents, no harm to people," BP has not provided effective leadership in making certain its management and U.S. refining workforce understand what is expected of them regarding process safety performance [18].

Good leaders need support, resources, and time to effectively manage change. Poor leaders need to be developed or may need to be replaced. Both actions require that senior leadership, starting with the board of directors (or equivalent) understand the process risks associated with their businesses, ensuring that effective process safety programs are implemented and providing strong leadership and resources for sustaining programs and performance [41–43]. Leaders must also consider all stakeholders, including communities, stockholders, employees, industry associations, and regulators, as part of their responsibilities. Communication with and development

TABLE 2.10

Factors that Influence Leadership Capability [Modified from 45 and 46]

• Committed	• Personality
• Ethical	• Courageous
• Visionary	• Emotional intelligence
• Engagement	• Sense of vulnerability
• Courage	• Communication
• Values	• Collaboration
• Influence	• Accountability
• Visibility	• Change agent
• Responsive	• Persistent
• Consistent	• Experience

of relationships with these stakeholders in good times can be very beneficial if problems occur later. OECD [44] has provided guidance for senior leaders in high-hazard industries, including a set of self-assessment questions for senior leaders (see Table 1.8).

The business and safety literature provides a wealth of guidance on how to be good leaders and managers, including the leadership characteristics shown in Table 2.10. Although it is impractical to review the literature in detail in this book, the following characteristics stand out in importance for leaders who are able to achieve excellent safety and business performance:

- **Commitment** – leaders are personally committed to achieving excellent performance and link safety to other organizational priorities through a clear vision of desired performance and clear goals for specific activities in support of improvement.

- **Focus** – leaders ensure that they spend time communicating about safety, reviewing programs, measuring performance, providing guidance and resources, and establishing accountability for results.

- **Involvement** – leaders provide "felt" leadership personally through interacting with and listening to personnel, leading by example, and building trust and employee engagement through visibility and consistent action.

Certainly, many other leadership qualities are important [45,47,48], including personality, credibility, and influence as listed in Table 2.10. Ultimately, it is the personal commitment, focus, and involvement of leaders to safety success as a starting point for excellent safety leadership. How to develop capability and leadership as a process safety professional is discussed in Chapter 13.

2.5 Process Safety Performance

For most companies, PSM has been implemented for at least 20 years, due to the 1992 U.S. OSHA PSM Regulation in the United States, the 1982 Seveso Directive in Europe, and other regulations around the world [49]. As a result, guidance on design and implementation of process safety programs has been available for many years, and generally speaking, most companies now have the technical knowledge and capabilities to safely identify and manage process hazards and risks [20,23]. Often, the result has been literally thousands of program requirements based on regulations, Recognized and Generally Accepted Good Engineering Practices (RAGAGEP), and other guidance. Once implemented, most companies have worked to continuously improve their process safety programs, driven by factors shown in Table 2.11. Yet, despite continuous improvement of program requirements [49–52], industry continues to have performance problems, sometimes resulting in serious injuries and catastrophic incidents.

What do we mean by "performance?" Dictionary definitions refer to executing actions or something accomplished. Process safety performance therefore relates to executing program requirements and systems with the intent of achieving program objectives. The primary objective of process safety programs is ensuring safe and reliable processes, resulting in no injuries and no major incidents. Many leading and lagging performance indicators have been developed to measure program performance, as discussed in Chapter 12, which have proven useful in identifying important improvement opportunities. The related concept of "effective" process safety programs is used in this book and has also been described by Arendt [50], where effective is defined as a function of efficiency (related to resources and cost) and performance (related to reduced injuries and incidents). We focus only on the performance part of this equation, recognizing that, although issues of resources and costs are always present and are important considerations in any practical program and improvement effort, they are generally outside the scope of this book.

This book describes the foundations of effective process safety programs as shown in Figure 1.3, with the intent to help achieve improved process safety

TABLE 2.11

Driving Forces for Improving Process Safety Performance [Modified from 45 and 46]

- Near miss and incident learning, including significant incidents as well as incident trends
- Poor or degrading leading and lagging process safety metrics, including audits
- Costs associated with poor performance (e.g., lost production, repair costs, poor quality)
- New regulations and good engineering practices (e.g., RAGAGEP)
- New technology, including new hazards as well as new applications (e.g., biological)
- Benchmarking with other companies
- Corporate restructuring, acquisitions, and mergers
- Stakeholder relations (e.g., communities, stockholders, employees, and regulators)

TABLE 2.12

Some Potential Causes of Poor Process Safety Performance
[Modified from 45, 49, and 53]

- Weak safety culture and/or leadership
- Lack of senior leadership operational experience, focus, and commitment
- Compliance-based mentality rather than risk-based mentality
- Poor interpretation/implementation of regulations and good engineering practices (i.e., RAGAGEP)
- Poor hazard recognition/identification
- Poor process safety system design
- Poor operational discipline
- Cost/staffing/resource pressures
- Poor management of mergers and acquisitions
- Lack of experienced, knowledgeable management and technical personnel
- Poorly designed feedback/measurement systems
- Complacency/no sense of vulnerability resulting from past good performance

performance. Some possible causes of poor performance, based on these foundations, are shown in Table 2.12. Ultimately, of course, each organization, in fact each specific facility, must evaluate their own operations, metrics, and performance to identify the specific issues that are important for them.

While all aspects of an effective process safety program are important, a model [53] for process safety performance is shown in Figure 2.4. In this model, the main elements are:

- **World class process safety principles and programs** – Organizations must be aware of external regulations, good engineering practices, literature, and organizations that provide direction on how effective process safety programs can be implemented. This information, in many ways, provides the guidance on what needs to be accomplished, how it should be accomplished, and how well other companies are doing in achieving excellent process safety performance. Knowledge of external information and activities is essential for leveraging organizational capabilities in process safety. The Center for Chemical Process Safety (CCPS), among other organizations, has provided comprehensive guidance on risk-based process safety [23] and a vision for process safety programs in the future [26].

- **Organizational safety culture** – As discussed in this chapter, the organization culture underlies and influences all things that are done at the organization, for good or bad, depending on how strong the culture is.

- **Leadership commitment and focus** – Also as discussed in this chapter, the organization's leadership must be committed and focused on process safety in order to ensure that the organization provides sufficient resources and capability in process safety based on established policies and specific goals.

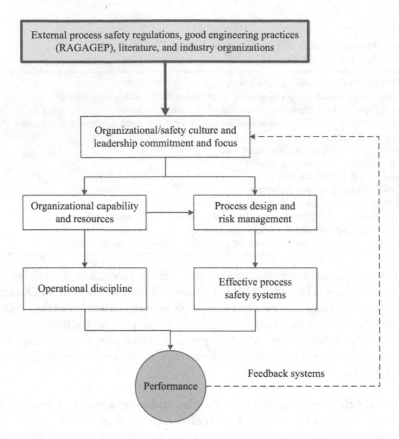

FIGURE 2.4
A process safety performance model [Modified from 53].

- **Capable organization and resources** – The foundation of safety culture and leadership must provide resources that are sufficient, targeted and purposeful, and trained and competent to ensure that the organization develops capability in process safety. Specialty resources, such as consulting services in any aspects of process safety, may be necessary to supplement the internal capabilities that it is reasonable for an organization to develop, based on process hazards that are present, organization size, and other factors.

- **Appropriate design and risk management** – Hazard recognition, evaluation, and control are the technical starting points for designing and implementing effective process safety systems for managing process risks. If process hazards are not identified, then the scope of process safety systems must necessarily be deficient, increasing the risk of catastrophic incidents.

- **Effective process safety systems** – The process safety systems and their requirements are discussed in Chapter 3 and later in this book. Systems must be (1) designed appropriately based on process hazards and regulatory requirements, (2) implemented effectively based on appropriate design and risk management, and (3) improved continuously based on actual performance.

- **Operational discipline (OD)** – As discussed in Chapter 4, an effective OD program must ensure that personnel have the knowledge, commitment, and awareness to complete their individual work tasks correctly and safely every time. A lapse in OD, through lack of knowledge or a shortcut, even one time, can undermine the requirements of well-designed process safety systems, leading to poor performance and possibly catastrophic incidents.

- **Feedback systems** – As discussed in Chapter 12, sensitivity to operations, based on leading and lagging metrics that are periodically reviewed, is essential for identifying possible warning signs of performance issues and improvement opportunities. Development of learning organizations to trend and interpret metrics and to maintain important design and operational documentation is a key to ensuring that essential process safety information is not lost over time.

The results of excellent process safety performance were highlighted in Table 1.7.

Effective process safety programs and excellent process safety performance are the goals of this book. We hope that awareness of and action on the key concepts and practical applications discussed in the remainder of this book will contribute to achieving these goals.

2.6 Case Study

In 2005, a major fire and explosion at the BP Texas City refinery resulted in 15 fatalities, 180 injuries, and $1.5 billion in damages. A large flammable release occurred while the isomerization unit at the refinery was being started up. This incident has been extensively investigated [17,18,54] and was one of the first petrochemical incidents (versus nuclear or space shuttle incidents) where safety culture issues were emphasized. Performance issues at the site led to a safety culture assessment in 2004, where 1080 employees were surveyed and 112 employees were interviewed. This assessment [54] reported that:

- Production pressures impact managers "where it appears as though they must compromise safety."
- "Production and budget compliance gets recognized and rewarded before anything else at Texas City."
- "The pressure for production, time pressure, and understaffing are the major causes of accidents at Texas City."

- "The quantity and quality of training at Texas City is inadequate... compromising other protection-critical competence."
- "Many [people] reported errors due to a lack of time for job analysis, lack of adequate staffing, a lack of supervisor staffing, or a lack of resident knowledge of the unit in the supervisory staff."
- Many employees also reported "feeling blamed when they had gotten hurt or they felt investigations were too quick to stop at operator error as the root cause." There was a "culture of casual compliance."
- Serious hazards in the operating units from a number of mechanical integrity issues: "There is an exceptional degree of fear of catastrophic incidents at Texas City."
- Leadership turnover and organizational transition; the creation and dismantling of the South Houston site "made management of protection very difficult."
- The strong safety commitment by the Business Unit Leader "is undermined by the lack of resources to address severe hazards that persist," and "for most people, there are many unsafe conditions that prove cost cutting and production are more important than protection. Poor equipment conditions are made worse in the view of many people by a lack of resources for inspection, auditing, training, and staffing for anything besides 'normal operating conditions.'"
- Texas City was at a "high risk" for the "check the box" mentality. This included going through the motions of checking boxes and inattention to the risk after the check off. "Critical events (breaches, failures or breakdowns of a critical control measure), are generally not attended to."

Results of the culture assessment were accepted by refinery management, but the incident occurred soon after. The U.S. Chemical Safety Board [54] report concluded that "Senior executives ... did not provide effective safety culture leadership and oversight to prevent catastrophic accidents." Several comments from the Baker Panel investigation following this incident have already been included in this chapter. Among its findings, the Baker Panel concluded:

> BP has not instilled a common, unifying process safety culture among its U.S. refineries. Each refinery has its own separate and distinct process safety culture. While some refineries are far more effective than others in promoting process safety, significant process safety culture issues exist at all five U.S. refineries, not just Texas City. Although the five refineries do not share a unified process safety culture, each exhibits some similar weaknesses. The Panel found instances of a lack of operating discipline, toleration of serious deviations from safe operating practices, and apparent complacency toward serious process safety risks at each refinery [18].

Some of the process safety system findings associated with this incident are discussed in Chapter 3.

The U.S. Chemical Safety Board has issued a case study on process safety culture [55], based on their investigations of two process incidents at the Tesoro Martinez refinery. In both incidents, workers were injured by exposure to sulfuric acid releases. Noting that process safety culture "contributed to a pattern of sulfuric acid exposure incidents" at the refinery, the case study also concludes that:

> Safety culture assessments ... can provide critical insight into cultural weaknesses that may contribute to process safety incidents. Simply identifying safety culture deficiencies, however, is not sufficient. Effective continual improvement programs are necessary to address safety culture weaknesses in order to prevent significant accidents, worker injuries, and the potential for community impact.

References

1. DeBlois, Lewis A. 1926. *Industrial Safety Organization for Executive and Engineer.* McGraw-Hill.
2. Sorensen, J.N. 2002. Safety culture: a survey of the state-of-the-art. *Reliability Engineering and System Safety.* 75:189–204.
3. International Nuclear Safety Advisory Group. 1991. *Safety Culture.* Safety Series No. 75-INSAG-4. International Atomic Energy Agency, Vienna, Austria.
4. International Atomic Energy Agency. 1996. ASCOT *Guidelines: Guidelines for Organizational Self-Assessment of Safety Culture and for Reviews by the Assessment of Safety Culture in Organizations Team.* IAEA-TECDOC-860. International Atomic Energy Agency, Vienna, Austria.
5. International Atomic Energy Agency. 1998. *Developing Safety Culture in Nuclear Activities.* Safety Reports Series No. 11. International Atomic Energy Agency, Vienna, Austria.
6. International Atomic Energy Agency. 2002. *Self-Assessment of Safety Culture in Nuclear Installations: Highlights and Good Practices.* International Atomic Energy Agency, Vienna, Austria.
7. Report of the Presidential Commission on the Space Shuttle Challenger Accident. 1986. history.nasa.gov/rogersrep/genindex.htm.
8. Vaughan, Diane. 1997. *The Challenger Launch Decision: Risky Technology, Culture, and Deviance at NASA.* University of Chicago Press, London.
9. Report of Columbia Accident Investigation Board. 2003. www.nasa.gov/columbia/home/CAIB_Vol1.html.
10. Kadri, Shakeel H. and David W. Jones. 2006. Nurturing a strong process safety culture. *Process Safety Progress.* 25:16–20.
11. Center for Chemical Process Safety. 2005. *Self Evaluation Tool: Key Lessons from the Columbia Shuttle Disaster (adapted to the process industries).* www.aiche.org/ccps/topics/elements-process-safety/commitment-process-safety/process-safety-culture/building-safety-culture-tool-kit/process-safety-culture/key-lessons-columbia-disaster.

12. Center for Chemical Process Safety. 2005. *Building Process Safety Culture ToolKit.* www.aiche.org/ccps/topics/elements-process-safety/commitment-process-safety/process-safety-culture/building-safety-culture-tool-kit.

13. U.S. Chemical Safety Board. 2007. *Refinery Explosion and Fire.* Report No. 2005-04-I-TX. www.csb.gov.

14. Kletz, T.A. 1993. *Lessons From Disaster: How Organizations Have No Memory and Accidents Recur.* Gulf Professional Publishing.

15. Hendershot, Dennis C. 2012. Process safety management—you can't get it right without a good safety culture. *Process Safety Progress.* 31:2–5.

16. Dolan, Michael J. 2012. The intersection of process safety and corporate responsibility. *Chemical Engineering Progress.* 108(6):24–27.

17. Atherton, John and Frederic Gil. 2008. *Incidents That Define Process Safety.* Wiley-AIChE.

18. Baker, James A., Frank L. Bowman, Glenn Erwin, Slade Gorton, Dennis Hendershot, Nancy Leveson, Sharon Priest, Isadore Rosenthal, Paul V. Tebo, Douglas A. Wiegmann, and L. Duane Wilson. 2007. *The Report of BP US Refineries Independent Safety Review Panel.* www.bp.com/bakerpanelreport.

19. Meshkati, Najmedin. 2014. *Onagawa: The Japanese Nuclear Power Plant that Didn't Melt Down on 3/11.* thebulletin.org/onagawa-japanese-nuclear-power-plant-didn%E2%80%99t-melt-down-311.

20. Center for Chemical Process Safety. 2016. *Guidelines for Implementing Process Safety Management.* Wiley-AIChE.

21. Collins, A.M. and Gadd, S. 2002. *Safety Culture: A Review of the Literature.* UK Health & Safety Laboratory. HSL/2002/25.

22. Frank, W.L. 2007. *Process Safety Culture in the CCPS Risk Based Process Safety Model.* Process Safety Progress. 26:203–208.

23. Center for Chemical Process Safety. 2007. *Guidelines for Risk Based Process Safety.* Wiley-AIChE, New York.

24. Arendt, Steve. 2008. *Connecting Process Safety Performance Outcomes to Process Safety Cultural Root Causes.* 1st Latin American Process Safety Conference: Building Culture and Competency, Buenos Aires.

25. Choudhry, Rafiq M., Dongping Fang, and Sherif Mohamed. 2007. The nature of safety culture: A survey of the state-of-the-art. *Safety Science.* 45:993–1012.

26. Center for Chemical Process Safety. 2014. *Vision 20/20.* www.aiche.org/ccps/about/vision-2020.

27. Fleming, Mark. 2001. *Safety Culture Maturity Model.* UK Health & Safety Laboratory. Offshore Technology Report 2000/049.

28. DuPont. Bradley Model. www.dupont.com/products-and-services/consulting-services-process-technologies/brands/sustainable-solutions/sub-brands/operational-risk-management/uses-and-applications/bradley-curve.html.

29. Parker, Dianne, Matthew Lawrie, and Patrick Hudson. 2005. A framework for understanding the development of organizational culture. *Safety Science.* 44:551–562.

30. B and W Pantex. 2008. *High Reliability Organizations.* www.pantex.com/safety/HRO/index.htmCached-Similar.

31. LaPierre, Dominique and Javier Moro. 2002. *Five Past Midnight in Bhopal.* Warner Books, New York.

32. Klein, James A. 2009. Two centuries of process safety at DuPont. *Process Safety Progress*. 28:114–122.

33. Sanders, Roy E. 2013. Keep a sense of vulnerability: For safety's sake. *Process Safety Progress*. 32:119–121.

34. International Atomic Energy Agency. 2002. *Safety Culture in Nuclear Installations: Guidance for Use in the Enhancement of Safety Culture*. IAEA-TECDOC-1329. International Atomic Energy Agency, Vienna, Austria.

35. Collins, Jim. 2001. *Good To Great: Why Some Companies Make the Leap ... And Others Don't*. HarperBusiness, New York.

36. Arendt, Steve, Rick Curtis, Lelio DePaiva SaFreitas, and Ron Henderson. 2009. *Practical Steps to Improving Process Safety/HSE Culture*. 2nd Annual CCPS Latin American Process Safety Conference and Expo, Sao Paulo, Brazil.

37. Curtis, Rick, Herzi Marouni, Ken Hanchey, San Burnett, and Steve Arendt. 2009. *Practical Steps to Improve Process Safety Culture—Case Studies and Lessons from the Process Industry*. 2nd Annual CCPS Latin American Process Safety Conference and Expo, Sao Paulo, Brazil.

38. Sutton, Ian. 2008. Use root cause analysis to understand and improve process safety culture. *Process Safety Progress*. 27:274–279.

39. Reiman, T. and P. Odewald. 2004. Measuring maintenance culture and maintenance core task with CULTURE-questionnaire—a case study in the power industry. *Safety Science*. 42:859–889.

40. Kotter, John P. 2012. *Leading Change*. Harvard Business Review Press, Boston, MA.

41. Karthikeyan, I. 2015. Moving process safety into the board room. *Chemical Engineering Progress*. 111(9):43–46.

42. AIChE. 2016. Leadership Q. and A: Taking a leading role in process safety. *Chemical Engineering Progress*. 112(1):19–23.

43. UK HSE. 2003. *The Role of Managerial Leadership in Determining Workplace Safety Outcomes*. Research Report 044. www.hse.gov.uk.

44. OECD. 2012. *Corporate Governance for Process Safety: OECD Guidance for Senior Leaders in High Hazard Industries*. www.oecd.org.

45. Arendt, Steve. 2013. *Process Safety Culture, Leadership, and Operational Discipline*. 26th Annual TCC/ACIT EHS Seminar. Galveston, TX.

46. Killimett, Patrick. 2006. Organizational factors that influence safety. *Process Safety Progress*. 25:94–97.

47. Oregon OSHA (USA). No date. *Keys to Effective Safety Leadership*. OR-OSHA 118 0203. www.cbs.state.or.us/osha/educate/materials/Safety-Leadership-163/1-163i.pdf.

48. Oregon OSHA (USA). No date. *Safety and the Supervisor*. orosha.org/educate/materials/Safety-and-the-Supervisor-160/1-160i.pdf.

49. Klein, James A. and Seshu Dharmavaram. 2012. Improving the performance of established PSM programs. *Process Safety Progress*. 31:261–265.

50. Arendt, Steve. 2006. Continuously improving PSM effectiveness—a practical roadmap. *Process Safety Progress*. 25: 86–93.

51. McCavit, Jack, Scott Berger and Louisa Nara. 2014. Characteristics of companies with great process safety performance. *Process Safety Progress*. 33:131–135.

52. Hanchey, Ken and James R. Thompson. 2011. The challenge to implement and maintain an effective PSM program. *Process Safety Progress*. 30:319–322.

53. Klein, James A. and James R. Thompson. 2015. *Improving Process Safety Performance*. 11th Global Process Safety Conference. Austin, TX.
54. U.S. Chemical Safety Board. 2007. *Refinery Explosion and Fire*. Report No. 2005-04-I-TX. www.csb.gov.
55. U.S. Chemical Safety Board. 2016. *Tesoro Martinez Refinery: Process Safety Culture Case Study*. Report No. 2014-02-I-CA. www.csb.gov.

3

Process Safety Systems

These systems are interdependent; a significant change in one may upset the equilibrium of the whole.

Stephen Covey

Why Process Safety Systems Are Important

People need to know what to do. They have assigned tasks to accomplish during their workdays, and it is important, of course, that their work gets done correctly and safely. If they work at a facility where significant process hazards are present, they or their coworkers could be seriously injured or killed if something is done incorrectly or goes wrong. Management systems must be provided to describe what requirements and specific practices have been implemented, why they have been implemented, where they must be followed, who is involved or affected, and how specific work activities are to be done. Management systems exist for business, financial, quality, and many other purposes to ensure that effective practices are provided for achieving organizational goals and for meeting regulatory/legal requirements. When process hazards and risks are present, such as from the use of hazardous materials with toxic or flammable properties, effective process safety programs must be implemented that include appropriate and well-designed process safety systems. Process safety systems provide the detailed requirements of the process safety program to ensure that process hazards have been identified and evaluated before being first introduced into the workplace, and that process risks are successfully controlled... every minute of every day.

Process safety systems must be implemented to ensure that process hazards are identified, evaluated, and managed by providing specific requirements and practices for daily work.

3.1 Introduction

Management systems are "formally established and documented activities that are designed to produce specific goals in a consistent manner on a sustainable basis [1]." Process safety systems, the second foundation of effective process safety programs, are intended to ensure that process hazards are identified, evaluated, and understood, and that appropriate risk management controls are provided to achieve and maintain safe and reliable operations. Process safety systems must be:

- **Designed appropriately,** meeting regulatory requirements at minimum and based on the types and levels of process hazards and risks that are present
- **Implemented effectively,** ensuring that process hazards and risks are identified, controlled, and managed consistently and continuously, that sufficient and capable resources are available, and that appropriate management controls and measures are provided for monitoring performance and achieving organizational goals
- **Improved continuously,** learning from experience (such as incidents), technical and organizational change, and management review to improve performance and better achieve organizational goals.

Table 3.1 provides the scope of subjects commonly included in effective management systems.

TABLE 3.1

Scope of Effective Management Systems
[Modified from 1]

- Purpose
- Resource needs
- Roles and responsibilities
- Support requirements and tools
- Qualifications and training
- Schedule and timing
- Work activities
- Expected results
- Document control and revisions
- Management controls and measures
- Audit requirements
- Regulatory/legal issues

Process safety systems have traditionally been based on process safety elements provided in:

- **Regulations,** such as the 14 elements in the U.S. OSHA process safety management standard [2]
- **Industry guidance,** such as the 20 elements using a risk-based process safety approach developed by the Center for Chemical Process Safety (CCPS) [3].

In addition, Recognized and Generally Accepted Good Engineering Practices (RAGAGEP), such as RP-754 (Process Safety Performance Indicators for the Refining and Petrochemical Industries) developed by the American Petroleum Institute, NFPA 30 (Flammable and Combustible Liquids Code) developed by the National Fire Protection Association, and many, many more codes, standards, and best practices provide additional requirements that must be incorporated into process safety systems, as appropriate. As discussed in Chapter 1, a good starting point for developing effective process safety systems is meeting the requirements of existing regulations, which provide a minimum essential level of practice. Additional systems or requirements that go beyond regulations may also be needed to successfully manage the process hazards and risks present at a facility and to meet specific RAGAGEP requirements.

This chapter provides a framework for effective process safety systems and includes additional discussion on management of process risks.

3.2 Key Concepts

3.2.1 Process Risk

An effective process safety program is designed to reduce the risk of catastrophic and other potentially hazardous process incidents to acceptable levels, as defined by process safety system requirements. Process risk (harm/year) was introduced in Chapter 1 as the product of frequency (events/year) and consequences (harm/event), as shown in Equation 1.1. Harm is usually described in terms of human health effects (e.g., fatalities/year) but also may be provided in terms of financial (e.g., dollars/year), environmental (e.g., permit violations/year), or other consequences. Frequency is sometimes described as likelihood with the same units as frequency, or in some cases, as a probability; although since probability is dimensionless, its use should be minimized. In terms of process safety, the risk of high consequence or severity events should typically be a small value, such as 1×10^{-5} or 1×10^{-6}, where the risk of a fatality, for example, would be 1 fatality in 100,000 or 1,000,000 years. If the risk of a

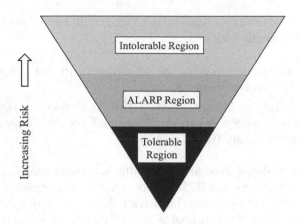

FIGURE 3.1
Guidance for risk reduction ranges—"As Low As Reasonably Practicable" (ALARP) [Modified from 5].

fatality is 1×10^{-2} (1 fatality in a 100 years), additional safeguards must be provided to reduce the risk to a more acceptable level as discussed in this chapter.

Generally, if process hazards are present, zero risk for a hazardous event cannot exist. Risk can be very low, as described above, such as 1 fatality in 1,000,000 years, but some level of risk is present if the process hazard is present. Reduction of risk to an acceptable level is therefore an important activity of an effective process safety program. "Acceptable" is often better described as "tolerable," indicating the risk is low enough that no further risk reduction is required. Criteria for defining tolerable risk levels are based on regulatory, industry, or most commonly, specific organization guidelines [4]. As shown in Figure 3.1, these criteria usually present a target level of risk where risk becomes tolerable, and in many applications, achieving tolerable risk levels is possible. In some cases, regulations in some countries and/or industry practice may also recognize the concept of ALARP, which is an acronym for "As Low As Reasonably Practicable" [5]. ALARP represents a level of risk which may be considered higher than desired because any further realistic risk reduction efforts and benefits are ultimately impractical because of cost or other factors. An example might be the handling of a hazardous material where further risk reduction efforts could preclude use of the material at all because of the high costs or impractical features associated with additional safeguards.

Methods exist to rigorously calculate risk [6], as discussed in Chapter 7, though this is beyond the scope of this book. The use of a qualitative risk matrix, as shown in Figure 3.2, though, is a common and cost-effective approach for simply and clearly estimating and discussing process risk levels. An organization typically has its own risk matrix, with the frequency (likelihood) and consequence (severity) shown as the two axes. A risk matrix commonly has three to six levels for each axis, based on corporate risk

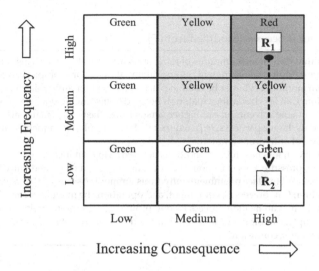

FIGURE 3.2
A qualitative risk matrix.

criteria defined in corporate risk management or hazard analysis standards. In Figure 3.2, a simple matrix is shown where three levels (low, medium, and high) are provided for both the frequency and consequences. Possible definitions for these values could be:

- **Frequency:** Low ($<10^{-4}$ years), Medium (10^{-2} to 10^{-4} years), and High ($>10^{-2}$ years)
- **Consequences:** Low (minor injury), Medium (major injury), and High (fatality)

Risk estimates are made individually for specific event scenarios, usually based on the predicted worst case consequences and the actual frequency, as discussed further in Chapter 7. Different risk levels are typically indicated by showing the boxes as different colors, such as green, yellow, and red. The risk level indicates if the resulting risk, based on the estimates of the frequency and consequences, is tolerable. If the risk level is green, for example, no risk reduction is required. If the risk level is yellow, risk reduction may be desirable but not required, and if the risk level is red, recommendations for additional safeguards as discussed in the following section may be required to reduce the risk estimate to a lower level, such as the green risk level.

3.2.2 Managing Process Risk

Management of process risk necessarily requires that either the frequency or consequence of a potentially hazardous event be reduced if the risk level is not considered tolerable or ALARP. Though a range of possible hazardous events are always possible (if not likely) if process hazards are present, *permanent*

TABLE 3.2

Inherently Safer Processes [Modified from 7]

- **Minimization:** Can smaller amounts of hazardous materials be used? This can include reactors or other process equipment, piping, materials, etc. For example, can a 60-kilogram ammonia cylinder be used in place of a 2000-liter ammonia storage tank?
- **Substitution:** Can less hazardous materials be used? This can include use of water in place of flammable solvents, use of higher versus lower flash point materials, use of less toxic materials, larger particle size powders, etc. For example, can aqueous ammonia be used in place of anhydrous ammonia?
- **Moderation:** Can less hazardous conditions be used? This can include lower temperature, pressure, concentration, etc. For example, can refrigerated anhydrous ammonia be used in place of ambient anhydrous ammonia?
- **Simplification:** Can processes, equipment, and operations be made less complex or error-prone? This can include process layout, piping configurations, valve accessibility, etc. For example, can piping headers be simplified and clearly labeled to help prevent potential operator confusion?

---

elimination or reduction of worst case consequences is usually based on Inherently Safer Process (ISP) approaches [7]. ISP concepts were advocated by Trevor Kletz based on the simple concept that what you do not have, cannot leak [8]. In other words, if process hazards are eliminated or reduced, the process is inherently safer than if the hazards were still present, and further risk reduction efforts could similarly be greatly reduced or eliminated. Several of the common methods of ISP are shown in Table 3.2. Application of ISP is usually most beneficial early in the design of a facility in order to avoid significant investment in risk reduction engineering and additional design requirements, and once built, significant redesign is often impractical. ISP is therefore an important consideration in the early stages of project management, though it should also be reviewed as part of periodic process hazard reviews, as discussed in Chapter 7. ISP is sometimes also referred to as inherently safer technology (IST) or inherently safer design (ISD).

If process hazards are present, risk reduction strategies must be used to reduce the frequency and/or consequences of possible hazardous events, including a range from relatively minor to worst case consequences. These strategies are often called the hierarchy of controls, which are shown in Table 3.3. If application of ISP approaches is possible, especially early in the design as discussed above, this is the most effective strategy for managing risk. If, however, application of ISP is impractical, such as when a particular hazardous material must be used for a desired chemical reaction, passive and/or active engineering controls must normally be used to reduce risk to tolerable levels. If effective engineering controls are not possible or are not sufficient in reducing risk to tolerable levels, administrative controls that rely on systems, procedures, and training (i.e., human action) must be used. Examples of administrative controls include preventative maintenance, checklists, and use of personal protective equipment (PPE), which are intended to provide human

TABLE 3.3

Risk Management Strategies [Modified from 7]

- **Inherent:** Permanently eliminate or reduce process hazards as part of the process design. This includes minimization, substitution, moderation, and simplification, as discussed in Table 3.2.
- **Passive engineering:** Reduces event frequency or consequences without requiring specific activation. This includes improved design, such as a higher vessel design pressure, containment for releases, such as catch tanks and diking, and isolation, such as protective barricades.
- **Active engineering:** Specific activation and function is required to reduce event frequency or consequences. This includes pressure relief devices, such as relief valves or rupture disks, safety instrumented systems (SIS), such as automatic dilution systems, and other automatic engineering controls, such as sprinklers for fires.
- **Administrative:** Systems and procedures that rely on human action. This includes basic process safety systems, operating procedures, safe work practices, maintenance tests and inspections, emergency response, and use of personal protective equipment.

---

controls to prevent or mitigate potentially hazardous events. The selection and design of protection layers based on these concepts are discussed further in Chapters 5 and 7.

A typical pressure profile for an out-of-control or runaway reaction is shown in Figure 3.3, which provides an example of how protection layers may be designed. In this example, a runaway reaction is characterized by the unintended, sudden, and rapid increase in pressure where the duration may be short or long, depending on the chemistry and operating conditions. A worst case runaway reaction may exceed the design parameters of the

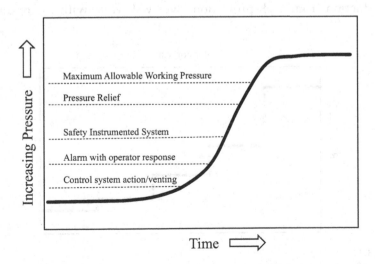

FIGURE 3.3
Example of protection layers for increasing pressure.

reactor, resulting in the catastrophic release of energy and hazardous materials. Process control systems, instrumentation, and procedures are typically provided to make the desired reaction products safely without occurrence of significant process deviations.

As shown in Figure 3.3, if a runaway reaction occurs, the control system will initially respond as programmed to the high-pressure deviation. If this does not control the pressure, pressure may start to vent through a reactor vent, and if the pressure gets high enough, a high-pressure alarm may sound. If operator response to the alarm, an administrative control, is not sufficient, engineering controls, such as safety instrumented systems (SIS), may be automatically activated to control the reaction, such as through the addition of a reaction inhibitor. A pressure relief valve or rupture disk may also be provided to relieve the pressure from the reactor to atmosphere, a catch tank, or a flare to prevent possible rupture of the reactor if the maximum allowable working pressure of the reactor is exceeded. If the action of any of these protection layers, or combination of protection layers, is successful, the pressure should return to normal or at least a catastrophic event should be prevented.

This approach of providing multiple protection layers is essential for reducing and managing the risk of catastrophic process events. If the risk of an event is high, as shown in the qualitative risk matrix by R_1 in Figure 3.2, the addition of one or more protection layers is necessary to reduce the risk of the evaluated event to a tolerable level, as shown by R_2 in the figure. Since the worst case consequence is typically shown in the qualitative risk matrix, as discussed in Chapter 7, the addition of protection layers typically reduces the frequency of the event until a tolerable risk level is achieved. As shown in Figure 3.4, the magnitude of the risk reduction from each protection layer will vary, with the number of

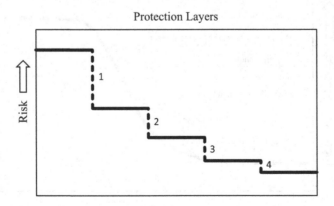

FIGURE 3.4
Addition of protection layers to reduce risk.

required protection layers determined ultimately by the magnitude of risk reduction needed to achieve a tolerable level, as shown in Figure 3.2. Similarly, not all safeguards will be effective protection layers for reducing risk, since the specificity, quality, and reliability of the safeguard may be insufficient to impact risk reduction significantly, or the cost or other factors may be prohibitive. The selection and evaluation of protection layers are discussed further in Chapters 5 and 7.

An example of a sequence of protection layer action is shown in Figure 3.5. In this case, a sequence possibly leading to a potentially hazardous event has been initiated, such as an equipment failure leading to the runaway reaction shown in Figure 3.3. In this sequence, the first protection layer will either work to stop or mitigate the event, or it will fail allowing the event to continue. If it is successful, the resulting Event 1 shown in Figure 3.5 may have minor consequences, although this will be dependent on the specific event sequence and hazards. For example, action of this first protection layer may prevent a loss of containment of a hazardous material, or it may limit the release to a very small amount. Since a layered protection layer approach is used, failure of the first protection layer may not be catastrophic because additional protection layers are provided as shown in Figure 3.5. Ultimately, the effectiveness of the protection layers will determine a range of possible events, such as the five events shown in the figure, where the consequences may vary from minor to catastrophic. The number, design, and

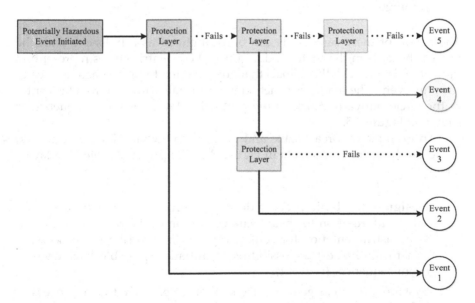

FIGURE 3.5
An example of protection layer effectiveness.

reliability of protection layers are evaluated in the process hazards and risk analysis (PHRA), as discussed in Chapter 7. The success or failure of protection layers is also an important consideration in incident investigations, as discussed in Chapter 11.

3.2.3 Protection Layer Model

Many frameworks are possible to describe the multiple protection layers that can be provided to reduce and manage process risk. The following protection layer goals should be considered:

- **Prevent** – ensure safe and reliable operations with appropriate design and systems
- **Detect** – monitor and alarm for process deviations, resulting in automatic and/or human responses
- **Control** – automatic systems to stop or control hazardous process deviations
- **Relieve** – discharge pressure to prevent catastrophic equipment failure
- **Contain** – capture releases in dikes or other catch systems to minimize impacted areas
- **Respond** – emergency response, involving both on- and off-site resources

The action of these safeguards can be thought of as a bulls-eye with the process hazards in the center and additional concentric circles representing protection layers with these goals that are intended to protect against a catastrophic event. The sequence of action will normally proceed from the center of the circle outward, much as the protection layers acted in sequence as shown in Figure 3.5.

An example of a reasonable and practical protection layer framework, based on these goals, is shown in Figure 3.6. The primary protection layers include:

1. **Design** – the basic process design incorporates known process safety information to ensure safe operation of the process, to provide instrumentation for controlling and monitoring the process, and to minimize the possibility of initiating possible hazardous event sequences (prevent).
2. **Systems** – the management systems that are part of effective process safety programs, including risk analysis, process integrity, management of change, and emergency planning, that provide initial and daily application of process safety concepts (prevent).

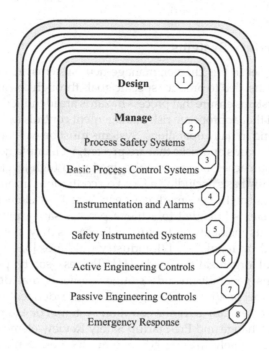

FIGURE 3.6
An example of protection layer hierarchy.

3. **Automation** – the basic process control systems, used for safe daily operations (prevent, detect).

4. **Alarms** – the instrumentation and warnings that indicate deviation from expected operating parameters has occurred, prompting for automatic and/or human response (detect).

5. **Safety Instrumented Systems (SIS)** – highly reliable safety instrumented systems that are designed to respond to and stop specific process deviations (control).

6. **Active Controls** – pressure relief devices, sprinkler systems, and other active engineering controls, such as a flare, that mitigate hazardous events (control, relieve).

7. **Passive Controls** – dikes, catch tanks, some fire protection safeguards, and other passive engineering controls that contain or mitigate hazardous events (contain).

8. **Emergency Response** – emergency response by trained internal and/or external resources (respond).

Process safety systems are introduced in Section 3.3, and various barrier methods for displaying and evaluating specific protection layers are discussed in Section 3.4.

3.3 Process Safety Systems

As discussed earlier in this chapter, management systems are designed to prescribe the actions needed to produce specific goals through consistent execution. Process safety systems ensure that process hazards are identified, evaluated, and understood, and that appropriate risk management controls are provided in the process design and in daily operations. Systems must be well designed to manage the process hazards and risks that are present, well implemented with practical, effective approaches, and well maintained with high levels of operational discipline on a daily basis, including a focus on continuous improvement.

Most process safety systems have been based on elements that define management systems intended to achieve process safety program goals for preventing serious incidents and injuries. Process safety systems have been developed based on regulatory [2], industry guidance [3,9,10], and in some cases, individual organization needs [11]. Process safety programs based on these systems generally include similar elements that define the scope and activity of the programs. Often, elements are defined separately, even though they are essentially part of a common system. For example, traditional Management of Change and Pre-Startup Safety Review elements comprise an overall system for safely making equipment changes. Similarly, Emergency Planning and Response and Incident Investigation elements comprise a system for anticipating possible incidents, for responding to them if they occur, and for investigating them to learn what happened to help prevent further incidents.

The process safety systems described in this book reorganize the traditional element-based approaches to provide an integrated management system framework to meet the primary goals of effective process safety programs. The process safety systems are summarized in Table 3.4, based on practical approaches for both designing safe processes and for implementing process safety on a continuing basis. Process safety systems for designing safe processes include:

- **Design safe processes** – Design safe processes to ensure that appropriate technology and safeguards are incorporated to control process risks and achieve safe and reliable operations.

- **Identify and assess process hazards** – Identify and assess process hazards to ensure that process risks can be properly evaluated and managed.

- **Evaluate and manage process risks** – Evaluate and manage process risks to ensure safe process design and that process safety systems are properly designed, implemented, and maintained.

Since many processes are constantly changing to improve technology and cost effectiveness, designing safe processes often occurs continuously, as needed, but is separated here as these systems provide the foundation for

TABLE 3.4

Process Safety Systems

Chapter	Process Safety System	Purpose
Part II – Practical Approaches for Designing Safe Processes		
5	Design Safe Processes	Design safe processes to ensure that appropriate technology and safeguards are incorporated to control process risks and achieve safe and reliable operations.
6	Identify and Assess Process Hazards	Identify and assess process hazards to ensure that process risks can be properly evaluated and managed.
7	Evaluate and Manage Process Risks	Evaluate and manage process risks to ensure safe process design and that process safety systems are properly designed, implemented, and maintained.
Part III – Practical Approaches for Implementing Process Safety		
8	Operate Safe Processes	Operate safe processes by ensuring that personnel are provided appropriate procedures and training to develop capability in operations and process safety.
9	Maintain Process Integrity and Reliability	Maintain process integrity and reliability by ensuring that appropriate tests and inspections, preventative maintenance, and quality assurance programs are conducted.
10	Change Processes Safely	Change processes safely by ensuring that process, equipment, system, and organizational changes are evaluated and authorized and that operational readiness reviews are conducted.
11	Manage Incident Response and Investigation	Manage incident response and investigation by ensuring that personnel are trained how to respond during emergencies and that incidents are thoroughly investigated.
12	Monitor Process Safety Program Effectiveness	Monitor process safety program effectiveness by ensuring that appropriate metrics are selected, evaluated, and used to measure and improve performance.

the other process safety systems that are focused on daily safe and reliable operation of the process. Process safety systems for daily implementation of process safety include:

- **Operate safe processes** – Operate safe processes by ensuring that personnel are provided appropriate procedures and training to develop capability in operations and process safety.
- **Maintain process integrity and reliability** – Maintain process integrity and reliability by ensuring that appropriate tests and inspections, preventative maintenance, and quality assurance programs are conducted.
- **Change processes safely** – Change processes safely by ensuring that process, equipment, system, and organizational changes are evaluated and authorized and that operational readiness reviews are conducted.

- **Manage incident response and investigation** – Manage incident response and investigation by ensuring that personnel are trained how to respond during emergencies and that incidents are thoroughly investigated.
- **Monitor process safety program effectiveness** – Monitor process safety program effectiveness by ensuring that appropriate metrics are selected, evaluated, and used to measure and improve performance.

Tables 3.5, 3.6, and 3.7 show how the process safety system framework in this book relates to the U.S. OSHA PSM Regulation, the EU Seveso III Articles, and the CCPS risk-based process safety elements, respectively. For facilities in the United States, there are also some state and local regulations, such as in California, Delaware, and New Jersey, that may have different requirements. Note that the U.S. OSHA and U.S. EPA process safety regulations were being reviewed under Executive Order 13650, "Improving Chemical Facility Safety and Security," that was issued on August 1, 2013. Some changes to these regulations may occur as a result after the time this book was written. Thus, it is essential that you understand the current requirements for your location.

TABLE 3.5

U.S. OSHA Process Safety Management Elements

	Paragraph / Element in the US OSHA PSM	Chapter / Process Safety System in this Book
c	Employee participation	2 Safety Culture and Leadership
	Process Safety Information (PSI)	
d	• Hazards of Materials	6 Identify and Assess Process Hazards
	• Process design basis • Equipment design basis	5 Design Safe Processes
e	Process hazards analysis (PHA)	7 Evaluate and Manage Process Risks
f	Operating procedures	8 Operate Safe Processes
g	Training	8 Operate Safe Processes
h	Contractors	8 Operate Safe Processes
	Mechanical integrity (MI)	
i	• Equipment integrity and quality assurance	9 Maintain Process Integrity and Reliability
	• Maintenance training	8 Operate Safe Processes
j	Pre-startup safety reviews (PSSR)	10 Change Processes Safely
k	Hot work	8 Operate Safe Processes
l	Management of Change (MOC)	10 Change Processes Safely
m	Incident investigation	11 Manage Incident Response and Investigation
n	Emergency planning and response	11 Manage Incident Response and Investigation
o	Compliance audits	12 Monitor Process Safety Program Effectiveness
p	Trade secrets	Not applicable

TABLE 3.6

EU Seveso III Articles

Article from the Seveso III Directive		Chapter / Process Safety System in this Book	
5	General obligations of the operator		
7	Major-accident prevention policy See Annex III Below	2	Safety Culture and Leadership
9	Safety report, paragraph (1)(a)		
13	Information on safety measures		
6	Notification	10	Change Processes Safely
7	Major-accident prevention policy	11	Manage Incident Response and Investigation
	See Annex III below	8	Operate Safe Processes
8	Domino effect	5	Design Safe Processes
		7	Evaluate and Manage Process Risks
9	Safety report, paragraph (1)(b)	6	Identify and Assess Process Hazards
	Safety report, paragraph (1)(b)	7	Evaluate and Manage Process Risks
	Safety report, paragraph (1)(c)	5	Design Safe Processes
	Safety report, paragraph (1)(c)	9	Maintain Process Integrity and Reliability
	Safety report, paragraph (1)(d)	11	Manage Incident Response and Investigation
	Safety report, paragraph (1)(e)	10	Change Processes Safely
	Details noted in Annex II	6	Identify and Assess Process Hazards
	See Annex III below	12	Monitor Process Safety Program Effectiveness
	See Annex III below	10	Change Processes Safely
10	Modification of an installation, an establishment or a storage facility	10	Change Processes Safely
11	Emergency plans See Annex IV below	11	Manage Incident Response and Investigation
12	Land-use planning	5	Design Safe Processes
14	Information to be supplied by the operator following a major accident	11	Manage Incident Response and Investigation
15	Information to be supplied by the Member States to the Commission	11	Manage Incident Response and Investigation
18	Inspections	8	Operate Safe Processes
II	Minimum data and information to be considered in the safety report specified in Article 9	6	Identify and Assess Process Hazards
III	Principles Referred To In Article 7 And Information Referred To In Article 9 On The Management System And The Organization Of The Establishment With A View To The Prevention Of Major Accidents	8	Monitor Process Safety Program Effectiveness
		12	Monitor Process Safety Program Effectiveness
IV	Data And Information To Be Included In The Emergency Plans Specified Under Article 11	11	Manage Incident Response and Investigation

TABLE 3.7

CCPS RBPS Elements [Modified from 3]

Chapter / Risk Based Process Safety (RBPS)		Chapter / Process Safety System in this Book	
3	Process Safety Culture		
6	Workforce Involvement	2	Safety Culture and Leadership
7	Stakeholder Outreach		
17	Conduct of Operations	4	Operational Discipline
4	Compliance with Standards	5	Design Safe Processes
5	Process Safety Competency	8	Operate Safe Processes
		13	Develop Personal Capability
8	Process Knowledge and Management	5	Design Safe Processes
		6	Identify and Assess Process Hazards
9	Hazard Identification and Risk Analysis	6	Identify and Assess Process Hazards
		7	Evaluate and Manage Process Risks
10	Operating Procedures	8	Operate Safe Processes
11	Safe Work Practices		
12	Asset Integrity and Reliability	9	Maintain Process Integrity and Reliability
13	Contractor Management	8	Operate Safe Processes
14	Training and Performance Assurance		
15	Management of Change	10	Change Processes Safely
16	Operational Readiness		
18	Emergency Management	11	Manage Incident Response and Investigation
19	Incident Investigation		
20	Measurement and Metrics		
21	Auditing	12	Monitor Process Safety Program Effectiveness
22	Management Review and Continuous Improvement		

Figure 3.7 shows how these process safety systems can be organized in a traditional Plan, Do, Check, Act (PDCA) framework as follows:

- The **Plan** stage involves designing safe processes and effective process safety systems.
- The **Do** stage involves the daily application of the process safety systems to ensure consistent implementation.
- The **Check** stage involves the use of process metrics and audits as feedback on process and system performance to identify opportunities for improvement as needed.
- The **Act** stage involves the improvement of the process and/or process safety systems using an effective management of change system, based on system performance, technology changes, and productivity improvements.

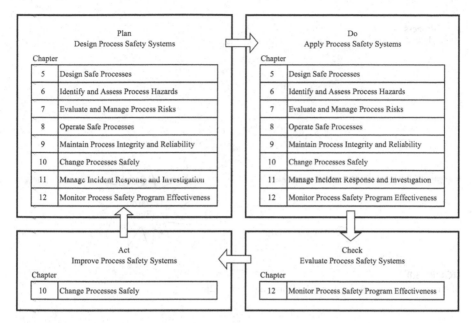

FIGURE 3.7
The process safety system framework.

See Parts II and III for additional information on key concepts and practical approaches for process safety systems.

3.4 Barrier Protection Layer Models

Use of multiple protection layers for reduction of process risk, as discussed earlier in this chapter, is an application of barrier model concepts [12–15]. Use of barrier models is based on the observation that many incidents result from the failure of a series of safeguards that are intended to prevent or mitigate the incident. Quoting Benjamin Franklin:

> For want of a nail the shoe was lost.
> For want of a shoe the horse was lost.
> For want of a horse the rider was lost.
> For want of a rider the message was lost.
> For want of a message the battle was lost.
> For want of a battle the kingdom was lost.
> And all for the want of a horseshoe nail.

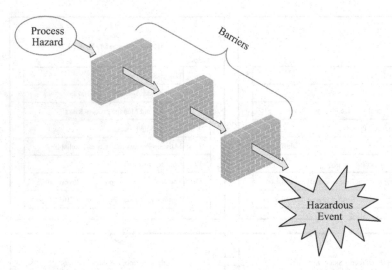

FIGURE 3.8
The barrier protection layer model [Modified from 12].

A simple, linear image of the barrier model approach is shown in Figure 3.8. When process hazards are present, the risk of a serious incident is possible, and appropriate protection layers are provided to reduce the risk. If one or more of these protection layers is successful in controlling the hazard, then the incident, or at least the worst case incident, is avoided. Barrier models can be useful in support of hazard analysis, training, emergency planning, incident investigations, preventative maintenance, and other process safety system activities. Failure of a barrier may also indicate that the initial design, maintenance, or other factors were faulty. The lack of a barrier when one should have been provided is often an area of focus in incident investigations, as discussed in Chapter 11. Concepts of barrier specificity, reliability, redundancy, diversity, separation, and other factors are generally beyond the scope of this book, though they must be considered in process hazards and risk analysis, as discussed in Chapter 7.

The two most commonly encountered barrier models in process safety applications—the Swiss Cheese Model and Bow Tie diagrams—are discussed in the following sections. Leveson [16] has also developed an alternative to sequential barrier models based on systems engineering approaches. Additional discussion on applying these models to both simple and complex systems in incident investigations can be found in Chapter 11.

3.4.1 Swiss Cheese Model

The Swiss Cheese Model was introduced by Reason [17] to show that incidents occur due to "holes" in protection layers, as shown in Figure 3.9, that look like slices of Swiss cheese. When multiple protection layers are provided,

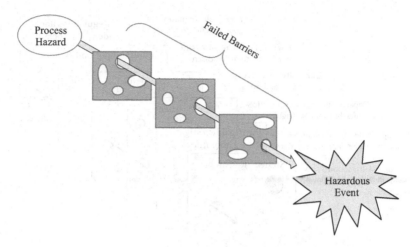

FIGURE 3.9
The Swiss Cheese Model [Modified from 17].

then holes must be present in each protection layer, and the holes must line up, to allow the incident to propagate to the worst case event. Holes are the result of active failures and/or latent conditions. Active failures are generally unsafe acts or human errors that are relatively immediate to the timing of the incident. Examples of active failures include not following a procedure, accidentally turning a valve in the wrong direction, or failing to add the correct amount of a material. Latent failures generally occur at some point, perhaps even many years, before the incident. Examples of latent failures include design errors, ineffective management of change, or poor preventative maintenance.

The Swiss Cheese Model is often used to qualitatively show the failures of protection layers when incidents have occurred or as a visual tool for hazard analysis or training [18,19]. Figure 3.10 shows a Swiss Cheese Model for the Deepwater Horizon incident described in previous chapters [20]. The flammability hazard of reservoir hydrocarbons is shown on the left of the figure, and multiple protection layers failed, resulting in the release and ignition of the hydrocarbons, and ultimately the fire and spill catastrophic event is shown on

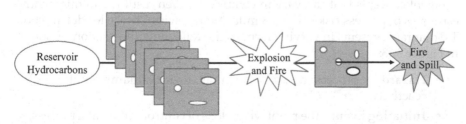

FIGURE 3.10
BP Deepwater Horizon incident [Modified from 20].

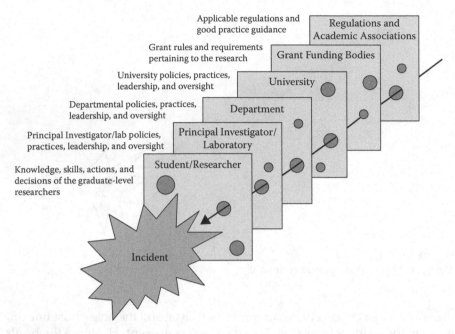

Applicable regulations and
good practice guidance

Regulations and
Academic Associations

Grant rules and requirements
pertaining to the research

Grant Funding Bodies

University policies, practices,
leadership, and oversight

University

Departmental policies, practices,
leadership, and oversight

Department

Principal Investigator/lab policies,
practices, leadership, and oversight

Principal Investigator/
Laboratory

Knowledge, skills, actions, and
decisions of the graduate-level
researchers

Student/Researcher

Incident

FIGURE 3.11
The Swiss Cheese Model for the Texas Tech University Laboratory Incident [21].

the right side of the figure. The U.S. Chemical Safety Board has also used the
Swiss Cheese Model, as shown in Figures 3.11 and 3.12 [21,22].

The Swiss Cheese Model has also been adapted for evaluation of human
error in aviation and other incidents [23], where specific Swiss cheese layers for
organizational influences, unsafe supervision, preconditions for unsafe acts,
and unsafe acts are defined. Specific failure categories (e.g., inadequate supervision) and subcategories (e.g., failed to track performance) were classified to
assist with incident investigation and to provide a framework for trending of
incident causes.

3.4.2 Bow Tie Diagrams

The Bow Tie diagram provides an excellent visual presentation of risk management concepts that can vary in complexity, even resulting in quantitative
estimates of process risk [24–27]. Similar to the Swiss Cheese Model, the Bow
Tie diagram presents in varying formats the following information, as shown
in Figure 3.13:

- **Hazard** – process hazards, such as toxicity, flammability, and
  reactivity.
- **Initiating Event** – the point where loss of control occurs in a process,
  such as the loss of refrigeration in an ammonia process or the mis-
  charge of a reactant in a chemical reaction.

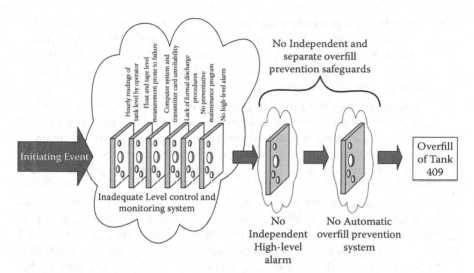

FIGURE 3.12
The Swiss Cheese Model for the Caribbean Petroleum Company Incident [22].

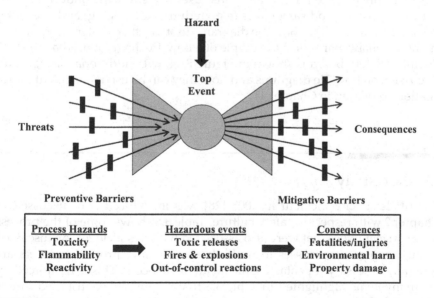

FIGURE 3.13
The Bow Tie diagram.

- **Controls** – the preventive and mitigative barriers that are provided to manage process risk. Preventive barriers are generally designed to prevent the loss of control, and mitigative barriers are designed to either bring the process back into control or to mitigate the potential consequences of the event.
- **Consequences** – the range of potentially hazardous events resulting from loss of control, such as toxic releases and exposures, fires, explosions, and out-of-control reactions. Depending on the severity of events and the success or failure of the barriers, a range of different consequences may be possible, as shown in Figure 3.5.

Since various failure pathways can lead to the loss of control, and various consequences may be possible, the visual presentation of these varying possibilities can take on the appearance of a Bow Tie, as shown in Figure 3.13. The Bow Tie diagram clearly shows defined engineering and administrative controls as barriers in the various pathways. Note that generally, only highly effective and reliable controls would be shown, depending on the intended use. Training applications, for example, may wish to show multiple barriers to reinforce the needs for awareness and reliability, whereas quantitative risk management applications will normally only show and use barriers meeting specific, stringent criteria. It is also often useful to show the underlying barrier failure modes and safeguards (e.g., redundancy, testing, and preventative maintenance) on the Bow Tie diagrams to strengthen systems required to maintain reliable barriers. An example of a Bow Tie diagram used by the U.S. Chemical Safety Board is shown in Figure 3.14, with additional background discussion on Bow Tie diagrams and how they can be used in incident investigations provided in Chapter 11.

3.5 Case Study

The BP Texas city incident in 2005 [28] was introduced and discussed in Chapter 2 with respect to safety culture. Table 3.8 shows some of the process safety system issues that were also associated with this incident. As discussed in this chapter, weaknesses in multiple systems and protection layers are generally necessary for catastrophic incidents to occur. This is reinforced by the many items highlighted in Table 3.8 that potentially contributed to this incident, based on subsequent investigations [29,30]. Concepts of operational discipline, as discussed in the next chapter, are also associated with many of these issues [31].

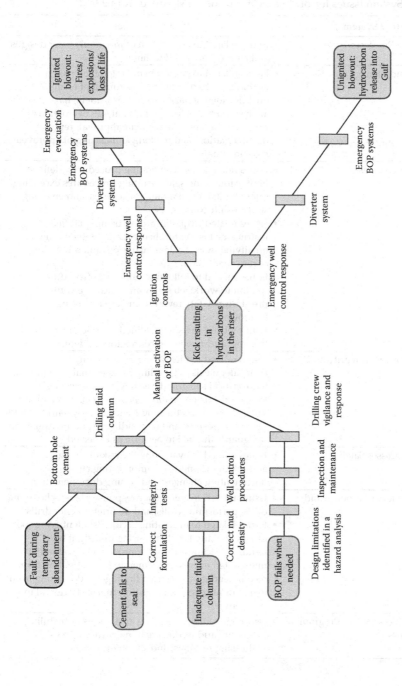

FIGURE 3.14

A Bow Tie diagram for the Deepwater Horizon Incident [28].

TABLE 3.8

Potential System Issues for BP Texas City Incident [Modified from 29–31]

Process Safety System	Potential Issues
Design Safe Processes	• Issues related to application of inherently safer designs and current safeguard technology
Evaluate and Manage Process Risks	• Issues related to definition of "tolerable" risk; risk evaluation; chartering and resourcing of teams; qualifications of hazards analysis leaders; hazards analysis methods (HAZOP; What/If); evaluation of protective layers; management approval; report documentation; and tracking and closing of approved recommendations
Operate Safe Processes	• Issues related to safe operating procedures (defined limits); following safe operating procedures exceeding safety limits); maintaining procedures; and ensuring shift-to-shift consistency • Issues related to qualifications; using qualified instructors; testing for theoretical understanding; qualifying by skills demonstration; and scheduling of refresher training • Issues related to following safe work procedures (permit to work; job safety analyses); performing pre-startup safety reviews; and documenting handovers • Issues related to using qualified contractors; contractor training; and reviewing contractor accidents
Maintain Process Integrity and Reliability	• Issues related to identifying critical equipment and their safeguards; scheduling planned maintenance on these critical safeguards; testing and inspecting safeguards; using qualified people to inspect and test safeguards; evaluating and trending safeguard test and inspection results; and responding to and investigating safeguards that fail to perform as expected
Change Processes Safely	• Issues related to evaluating the risk of changes; authorizing changes; communicating changes; documenting changes; and closing documentation
Manage Incident Response and Investigation	• Issues related to emergency response plans; scheduling drills; performing drills; and de-briefing after drills • Issues related to evaluating accidents; identifying root causes; looking for blame; using hindsight bias; trending leading indicators; proactively evaluating unexpected events; implementing investigation team recommendations; sharing findings; and learning from incidents at other locations (internal and external to company)
Monitor Process Safety Program Effectiveness	• Issues related to scheduling area audits; scheduling procedure audits; documenting audit findings; addressing findings; and sharing findings

References

1. Center for Chemical Process Safety. 2016. *Guidelines for Implementing Process Safety Management.* Wiley-AIChE.
2. U.S. OSHA. 1992. 29 CFR 1910.119: *Process Safety Management of Highly Hazardous Chemicals.* www.osha.gov.
3. Center for Chemical Process Safety. 2007. *Guidelines for Risk Based Process Safety.* Wiley-AIChE, New York.
4. Center for Chemical Process Safety. 2009. *Guidelines for Developing Quantitative Safety Risk Criteria.* Wiley-AIChE, New York.
5. HSE. *ALARP "At a Glance."* www.hse.gov.uk/risk/theory/alarpglance.htm.
6. Center for Chemical Process Safety. 2001. *Layer of Protection Analysis: Simplified Process Risk Assessment.* AIChE, New York.
7. Center for Chemical Process Safety. 2009. *Inherently Safer Chemical Process. A Life Cycle Approach,* 2nd Edn. Wiley-AIChE, Hoboken NJ.
8. Kletz, T. A. 1978. What you don't have, can't leak. Chemistry and Industry 287–292.
9. American Petroleum Institute. www.api.org/oil-and-natural-gas/health-and-safety/process-safety/process-safety-standards.
10. Energy Institute. www.energyinst.org/technical/PSM.
11. Klein, James A. 2009. Two centuries of process safety at DuPont. *Process Safety Progress* 28:114–122.
12. Hollnagel, E. 1999. *Accident Analysis and Barrier Functions.* Institute for Energy Technology, Halden, Norway.
13. Sklet, Snorre. 2006. Safety Barriers: Definition, classification, and performance. *Journal of Loss Prevention in the Process Industries* 19:494–506.
14. Jan Duijm, Nijs. 2009. Safety-barrier diagrams as a safety management tool. *Reliability Engineering and System Safety* 94:332–341.
15. Petroleum Safety Authority. 2013. *Principles for Barrier Management in the Petroleum Industry.*
16. Leveson, Nancy. 2011. *Engineering a Safer World: Systems Thinking Applied to Safety.* MIT Press, Cambridge, MA.
17. Reason, James. 1997. *Managing the Risks of Organizational Accidents.* Ashgate Publishing, Limited.
18. Eurocontrol. 2006. *Revisting the Swiss Cheese Model of Accidents.*
19. Broadribb, Michael P., Bill Boyle, and Steven J. Tanzi. 2009. Cheddar or Swiss? How strong are your barriers? *Process Safety Progress* 28:367–272.
20. BP. 2010. *Deepwater Horizon Accident Investigation Report.* www.bp.com/en/global/corporate/gulf-of-mexico-restoration/deepwater-horizon-accident-and-response.html.
21. U.S. Chemical Safety Board. 2010. *Texas Tech University Lab Explosion Case Study.* Report No. 2010-05-I-TX. www.csb.gov.
22. U.S. Chemical Safety Board. 2015. *Caribbean Petroleum Tank Terminal Explosion and Multiple Tank Fires.* Report No. 2010.02.I.PR. www.csb.gov.
23. Wiegmann, Douglas A. and Scott A. Shappell. 2003. *A Human Error Approach to Aviation Accident Analysis.* Ashgate Publishing, Limited.
24. www.abs-group.com/What-We-Do/Safety-Risk-and-Compliance/Risk-Management/THESIS-BowTie-Risk-Management-Software/.

25. Cockshott, J. E. 2005. Probability bow-ties: A transparent risk management tool. *Process Safety and Environmental Protection* 83:307–316.
26. Khakzad, Nima, Faisal Khan, and Paul Amyotte. 2012. Dynamic risk analysis using bow-tie approach. *Reliability Engineering & System Safety* 104:36–44.
27. Lewis, Steve, and Kris Smith. 2010. *Lessons learned from real world application of the bow-tie method.* 6th Global Congress of Process Safety. San Antonio, TX.
28. U.S. Chemical Safety Board. 2014. *Explosion and Fire at the Macondo Well.* Report No. 2010-10-I-OS. www.csb.gov.
29. Baker, James A., Frank L. Bowman, Glenn Erwin, Slade Gorton, Dennis Hendershot, Nancy Leveson, Sharon Priest, Isadore Rosenthal, Paul V. Tebo, Douglas A. Wiegmann, and L. Duane Wilson. 2007. *The Report of BP U.S. Refineries Independent Safety Review Panel.* www.bp.com/bakerpanelreport.
30. U.S. Chemical Safety Board. 2007. *Refinery Explosion and Fire.* Report No. 2005-04-I-TX. www.csb.gov.
31. Vaughen, Bruce K. and James A. Klein. 2012. What you don't manage will leak: A tribute to Trevor Kletz. *Process Safety and Environmental Protection* 90:411–418.

4

Operational Discipline

You may believe that the accidents could not happen at your plant because you have systems to prevent them. Are you sure that they are always followed, everywhere, all the time?

Trevor Kletz

Why Operational Discipline Is Important

People make mistakes. Although there is nothing surprising about that, it is extremely important when people are working in facilities that have significant process hazards. Some of the reasons for mistakes include:

- Human fallibility, capability, complacency, and commitment
- Training issues, including procedure quality and training effectiveness
- Workplace environment, including accessibility of information and distractions
- Familiarity with the work being done and the time since it was last done
- Fitness-for-duty considerations, such as alcohol, drugs, stress, and fatigue
- Urgency for completing a task quickly.

What is the probability that a routine task will be completed correctly every day for a year? Using a common error rate of 1 mistake in 100 chances (0.99 success rate) for completing the task correctly each day, this probability can be calculated as

$$\text{Probability of success}(\text{year}) = (0.99 \text{ for day } 1) \times (0.99 \text{ for day } 2)\ldots$$

$$= (0.99)^{365} = 2.6\%$$

A mistake should therefore be expected at some point over the year. Even if a mistake is not made in one year, there is a good chance it will occur in the few years after that. If this work task includes a high-risk activity involving significant process hazards, the results could potentially be catastrophic. Studies have shown that human error may account for a high percentage of incidents that occur in the chemical industry [2]. Operational discipline (OD), similar to "conduct of operations," must be a key part of an effective process safety program to both anticipate that mistakes are likely to occur and to minimize the potential for human error [3–7] through specific worker programs and multiple protection layers, as discussed in Chapter 3.

A focus on operational discipline can help ensure that human error is minimized and system requirements are rigorously followed, preventing serious injuries and incidents and improving overall safety and business performance.

4.1 Introduction

A simple look at site metrics may tell the story well enough. People are getting hurt. Incidents are occurring too frequently, and the number of close calls is causing concern. Quality and other aspects of site operations may not be performing well either. And yet, comprehensive and detailed systems and procedures, as discussed in Chapter 3, have been implemented that are meant to avoid these problems. For some reason, performance is not matching up to expectations for strong results. Continued focus on improving procedures may help, but improvement should only be incremental since the procedures are pretty good already. Assuming that site leadership is committed to safety and has already developed a good safety culture, as discussed in Chapter 2, a focus on operational discipline may provide the best opportunity for improving performance.

Operational discipline (OD) is defined as:

Operational discipline: Following system and procedure requirements correctly every time.

OD is used to describe human behavior in following required systems, procedures, and practices to achieve high quality, reliable, and safe processes [8,9]. If an organization has good discipline in implementing and following effective managing systems and requirements, a high level of OD can be achieved and many operating benefits, such as lower cost, higher productivity, and reduced incidents and injuries, can result. If, however, a low level of OD occurs, due to, for example, poorly documented procedures, ineffective training, shortcuts, complacency, no sense of vulnerability, and other factors, good operating and safety performance will be difficult to achieve. A focus on OD requires a day-to-day commitment by the organization's leadership to ensure appropriate systems, procedures, and other requirements have been established by workers at all levels to consistently follow and meet these requirements. Ensuring day-to-day focus on OD as part of a process safety program is a continuing challenge, but one needed to achieve desired performance.

4.2 Key Concepts

4.2.1 Benefits of Strong OD

An effective OD program emphasizes completing all tasks correctly and safely every time, regardless of the role of individuals in the organization— OD is for everyone, not just for operators. Certainly some tasks have more process safety impact than others, but the benefits of an effective program include [10,11]:

- Process hazards and risks are identified, evaluated, and managed
- Equipment and facilities are properly designed, operated, and maintained
- Management systems are well designed, implemented, and executed
- Operating problems and incidents are consistently investigated and corrected.

In addition to being a foundation of effective process safety programs, OD efforts should improve overall operational and business performance through higher productivity, higher quality, reduced waste, and lower costs, as shown in Table 4.1. The focus of OD programs on completing work tasks correctly is fundamental for any organization that hopes to remain competitive or to achieve world-class manufacturing operations.

OD programs help ensure that well-designed process safety systems are successfully implemented and help achieve consistent, desired, day-to-day performance. The safety triangle, as shown in Figure 4.1, illustrates qualitatively how significant consequences, such as serious injuries or catastrophic incidents are often the result of, or predicted by, a larger number of smaller, undesirable, unsafe acts, or behaviours. A focus on minimizing or

TABLE 4.1

Benefits of Operational Discipline (OD) Programs [Modified from 8 and 13]

- **Operations**—improved efficiencies and costs in support of achieving world class manufacturing goals.
- **Safety, Health, and Environmental (SHE)**—prevention or reduction of workplace injuries, occupational exposures, fires, environmental releases, and associated costs.
- **Productivity**—reduction in unscheduled shutdowns, poor process utilization, inefficient use of manpower, downtime related to incidents, and associated costs.
- **Quality and Waste**—reductions in off-specification product, rework and waste costs, low yields, poor quality, and customer complaints.

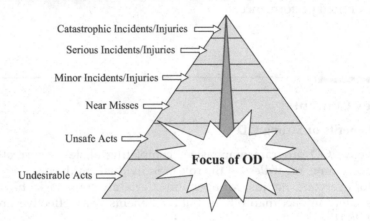

FIGURE 4.1
The Safety Triangle [Modified from 9].

eliminating problems at the bottom of the triangle helps to prevent the more serious events at the top. Thus, an effective OD program works to reduce the number of unsafe acts and undesirable behaviors in the organization, as seen at the base of the safety triangle, helping to prevent more serious incidents and injuries. This is especially important for many high-risk activities, such as working with highly toxic materials or electrical work. In these cases, the base of the safety triangle can be very narrow, meaning that even one unsafe act or mistake can directly lead to a serious injury or fatality [12].

Another useful way to consider the impact of effective OD programs on process safety is to qualitatively modify the risk equation presented in Equation 1.1 as [9]:

$$\text{Risk} = \frac{\text{Frequency} \times \text{Consequence}}{f\,(\text{Operational Discipline})} \tag{4.1}$$

As discussed in Chapter 3, risk is normally expressed as:

- the frequency of a possible hazardous event, often determined by the effectiveness of process safety systems and multiple protection layers

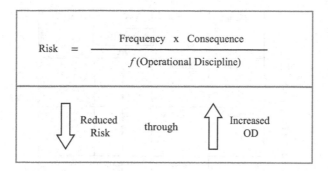

FIGURE 4.2
The Effect of OD on Risk [Modified from 9].

- the potential consequences of the event, characterized by the inherent substance and process hazards.

OD, expressed simply in fractional form (e.g., 0.5 for 50% OD) for illustration, has been added to the denominator of this equation, suggesting that the actual risk is higher than the normally calculated risk if OD problems are present. As OD performance increases, meaning that process safety systems are consistently followed and safeguards are well designed and maintained, risk decreases. If a high level of OD is not achieved, then the risk of specific work tasks and the overall risk of injury or a serious incident increase. Although the relationship is more complex, this simple approach illustrates that a poor OD level of 50%, however measured, suggests that risk is doubled. These relationships are shown in Figure 4.2, where the impact of OD is shown as a function rather than any particular mathematical relationship.

The impact of OD on risk can also be illustrated in a qualitative risk matrix (see Chapter 3), as shown in Figure 4.3. Poor OD most often increases the potential frequency of a hazardous event, although it may, in some cases, increase the potential consequences. This increases the risk from level R_1, the perceived risk where OD is not considered, to a higher level, the actual risk, depending on the extent of the frequency increase. The resulting higher risk may be considered unacceptable in a process hazards and risk analysis (see Chapter 7) due to the greater likelihood of serious incidents and injuries, but might not be fully evaluated if OD factors (which may impact the initiating frequency of the event and/or the reliability of safeguards) are not accounted for. An example of this is the Bhopal incident in 1984, discussed previously, where multiple safeguards designed to prevent or mitigate a large release of methyl isocyanate failed to work or were not available. Many people died in the surrounding community due to OD and other failures at the facility, when a traditional risk study at the time may have concluded that adequate protection layers were provided if properly maintained.

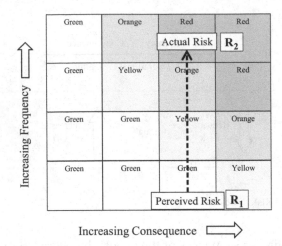

FIGURE 4.3
The Impact of OD on Risk [Modified from 13].

Many protection layers are typically provided to help prevent injury or significant incidents associated with process hazards, as discussed in Chapter 3. Yet, when an injury or incident occurs, all of the safeguards have failed for one reason or another. OD is often one of the major reasons. This was illustrated with the Swiss Cheese Model, which shows that many protection layers typically must fail before a catastrophic incident can occur. Often, when the underlying reasons for failure of the safeguards are evaluated, the "holes" are found to be related to OD as well as other factors.

4.2.2 OD Program Characteristics

An effective OD program includes a focus on both [9,13,14]:

- **Organizational OD** – for site or business leadership to create the programs and work environment and to provide appropriate resources that support desired behaviors
- **Personal OD** – for individual workers at all levels of the organization to know what they need to do their work correctly and safely every time.

Both organizational and personal OD are defined by different characteristics that provide the foundation for describing, evaluating, and improving OD. These characteristics, which together are the basis for effective OD programs, are discussed in the following sections. Ultimately, the purpose of the organizational OD characteristics is to provide the programs and resources to achieve strong personal OD performance, as shown in Figure 4.4.

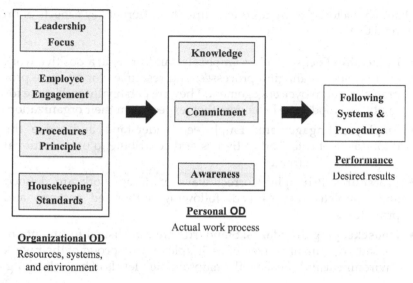

FIGURE 4.4
OD Characteristics [Modified from 9, 13].

4.2.2.1 Organizational OD

Organizational OD helps define what corporate, business, and site leadership should do to promote and achieve high levels of OD. Good OD will not happen on its own, but rather it is something that requires continued management attention to:

- demonstrate personal attention and dedication
- develop effective processes to facilitate employee understanding, involvement, and follow-through
- provide appropriate resources
- implement, standardize, maintain, and improve appropriate systems.
- support, evaluate, and monitor performance.

The many benefits of OD in helping achieve world class manufacturing operations, reduce risk, and prevent injuries provide a basis for engaging leadership to support and maintain effective OD programs. OD helps leadership:

- set and support high expectations for safe and reliable operations
- provide and maintain comprehensive, integrated process safety managing systems and procedures
- foster and sustain a strong safety culture.

Culture and leadership are discussed in Chapter 2.

Four characteristics are used to define the different key aspects of organizational OD:

- **Leadership Focus:** Leaders emphasize and provide a positive work environment, managing processes, and resources for effective programs and employee engagement. They are personally passionate for safety and model the behaviors they expect from their organization.
- **Employee Engagement:** Employees understand and value the importance of safe work activities and contribute to organizational programs and activities.
- **Procedures Principle:** Correct ways of doing work are defined and completed as planned, following authorized systems and procedures.
- **Housekeeping Standards:** Standards are established for maintaining safe equipment and facilities. Employees are proud of their work environment and consistently maintain high levels of housekeeping.

The foundation of any OD effort is Leadership Focus, as shown in Figure 4.5, without which priority and support for OD improvement cannot exist. OD programs must also promote strong Employee Engagement, where high levels of uninterested and uninvolved employees will limit the organization's ability to achieve strong performance. An important part of Leadership Focus is providing the culture and work environment to engage employees and involve them in the safety effort as well as to understand their part in achieving a high level of OD, regardless of their position in the organization. The most visible results of Leadership Focus and Employee Engagement

FIGURE 4.5
Organizational OD Characteristics [Modified from 9].

TABLE 4.2

Organizational OD Characteristics [Modified from 9]

Leadership Focus	Procedures Principle
Leaders are passionate for SHE[a] and model the behavior they expect from others.	*Work is completed as planned, following authorized and current procedures.*
<ul><li>Visibly demonstrate personal priority for SHE as a core value in their decisions.</li><li>Clearly document, maintain up-to-date, and communicate SHE goals, standards, and systems.</li><li>Monitor SHE performance via metrics, audits, and personal involvement to drive continuous improvement.</li><li>Provide sufficient and capable resources to sustain 'The Goal is Zero' SHE performance.</li><li>Develop and support processes to facilitate employee involvement and effective teams.</li><li>Recognize and celebrate good SHE performance.</li></ul>	<ul><li>Procedures are documented and readily available for all appropriate SHE activities.</li><li>Clear expectations exist for following procedures and for not taking shortcuts.</li><li>Procedures are periodically reviewed and authorized to keep them current, including employee participation.</li><li>All changes, tests, and deviations are reviewed and authorized before use.</li><li>Training and field audits are conducted to ensure procedures are understood and followed.</li></ul>
Employee Engagement	**Housekeeping Standards**
Employees are active and enthusiastic about participating in SHE activities.	*Employees are proud of their workplace and maintain consistently high levels of housekeeping.*
<ul><li>Know and share the organization's SHE core values and goals.</li><li>Volunteer and are active in SHE activities and teams.</li><li>Provide feedback and suggestions for improvement.</li><li>Show pride in being part of the organization.</li></ul>	<ul><li>Clear expectations are established for maintaining good housekeeping.</li><li>Standards for equipment and area housekeeping are documented and clearly communicated.</li><li>Audits are conducted to monitor and help improve housekeeping.</li></ul>

[a] Safety, Health, and Environmental.

are employees following approved systems and procedures (Procedures Principle) and maintaining equipment and work areas in good operating condition (Housekeeping Standards). Additional information on the organizational OD characteristics is shown in Table 4.2.

4.2.2.2 Personal OD

Although ultimately organizational leadership is accountable for results, achieving high levels of OD, and subsequently achieving excellent operational and safety performance, requires the active involvement of all employees in completing every job task correctly and safely every time.

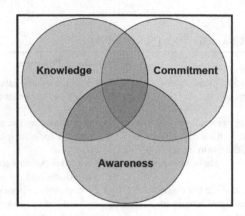

FIGURE 4.6
Personal OD Characteristics [9].

A strong organizational OD focus by corporate and facility leadership, as discussed previously, provides the systems, procedures, tools, expectations, resources, and support for personal OD programs for workers at all levels in the organization. Efforts to improve personal OD must necessarily be viewed as part of or overlapping with other safety and operational management programs intended to reduce human error [10,14].

Three characteristics are used to define the different key aspects of the personal OD program:

- **Knowledge:** I understand how to do my work task correctly and safely
- **Commitment:** I commit to do my tasks the right way, every time
- **Awareness:** I anticipate potential problems and recognize unusual situations.

Everyone needs to know how to do a work task, needs to plan to actually do it that way with no deviations or shortcuts, and needs to anticipate and be prepared for what can go wrong. All three personal OD characteristics are essential for an individual to work with a high level of OD, as shown in Figure 4.6. If, for example, an employee knows what to do and plans to do it correctly, but does not have a high level of Awareness, it is more likely that some part of the task may not be completed correctly when things do not go quite as planned. The result may be a quality problem, a serious incident, or an injury. Similarly, if an individual is not committed to completing a task correctly and safely every time, shortcuts may be taken or distractions may occur that lead to a lack of focus and an injury or incident may occur. These types of unwanted behaviors can ultimately lead to a reduced sense of vulnerability and complacency about workplace hazards, potentially leading to further injuries and incidents. Additional information on the personal OD characteristics is shown in Table 4.3.

TABLE 4.3

Personal OD Characteristics [Modified from 9]

Knowledge

I understand how to do my tasks safely and correctly.

- I know the correct way to do my job task, based on procedures, training, and other SHE system requirements.
- I understand why the job task is being done in a certain way, what needs to be accomplished, and how it should be done.
- I ensure that my equipment, tools, and PPE, if needed, are in good condition.
- I make sure my co-workers also know how to do their job task safely.

Commitment

I commit to do my tasks the right way every time.

- I take personal responsibility for properly understanding my job task and making sure it can be completed safely every time.
- I plan to follow procedures carefully without shortcuts.
- I trust that procedures have been developed for a purpose, but suggest changes if they don't make sense.
- I focus on the task at hand and set aside personal and work distractions.
- I care for my safety and the safety of my co-workers.

Awareness

I anticipate potential problems and recognize unusual situations.

- I anticipate that my job task may not go as planned.
- I understand if there are any unusual circumstances or hazards associated with my job task.
- I monitor my work environment carefully, including the activity of my co-workers.
- I prepare to respond quickly and safely, troubleshooting as needed, based on my training and procedures.
- I stop, think, and review procedures, or ask for help, when unsure of what to do, or if the task unexpectedly changes.

Development of Knowledge primarily results from ensuring that high-quality procedures are available for use, when needed, and that job-specific and refresher training on the procedures at appropriate intervals and other aspects of the work task have occurred, as discussed in Chapter 8. Understanding why a procedure has been developed in a specific way, accounting for job requirements, safety hazards, etc., provides more insight into the work task and helps ensure that it will be done correctly. Use of equipment, tools, and personal protective equipment is often an important part of getting work done correctly and safely, so checking the condition and knowing how to use required equipment are critical. Finally, since coworkers often work alongside you to complete many job tasks, making sure that they also have sufficient Knowledge is essential for both their safety and your safety.

Fundamentally, Commitment is taking personal responsibility and being accountable for the work that you are doing, and making sure it is done correctly and safely without shortcuts and other unapproved deviations from job requirements. If there is a need to change a procedure or a better way to do the work task is recognized, the established process for making changes must be used to ensure that the changes are carefully reviewed for potential safety hazards and authorized, so that everyone is completing the work in the same way. Part of Commitment is focusing on the work task and paying attention to appropriate detail in completing the work, rather than being distracted by personal issues or other things going on in the work environment. Caring for your safety and the safety of your coworkers reinforces Commitment to getting the work done safely.

Recognition that a work task does not necessarily occur exactly the same way every time requires effective job planning and increases anticipation of potential problems and appropriate responses. Careful attention to detail and monitoring of the work environment allow early detection and appropriate troubleshooting of any situations that occur, including requests for assistance from other site personnel. Awareness may also reveal that the worker has incomplete Knowledge of the work task, allowing review of the procedure or involvement of other experienced workers for assistance.

4.3 Improving OD

OD issues and problems often are specific to a local site or organization and can vary widely within the same organization or even the same site, based on differing safety culture, leadership, work activities and hazards, geographic location, and other factors. Although common issues may exist broadly within a large organization, improvement activities usually must be evaluated locally to help prioritize possible improvement goals. The steps for improving OD are shown in Figure 4.7 and include [9,13]:

1. **Focus on OD improvement:** Without specific management attention, OD is unlikely to improve in a sustained way. Leadership should develop specific goals about improving OD each year, based on evaluation of site-specific issues. Leadership processes for improving OD are discussed later in this chapter.

2. **Raise awareness on OD:** Make sure that everyone on-site knows what OD is and why it is important. This can be completed by discussing OD with employees in meetings, safety meetings, casually in the workplace, etc. Solicit input from employees on possible causes of OD issues and things that can be done to improve OD. Review and discuss the OD issues associated with incidents.

3. **Evaluate current OD performance:** Sources of data on OD performance are discussed in the next section, including incident data, audit results, performance metrics, and OD self-assessment surveys.

4. **Prioritize top OD improvement issues:** Develop specific goals and plans for improvement based on high-priority issues identified by evaluating current performance. Focus on activities that are targeted to key issues and that can provide tangible results in a reasonable time. Engage site employees in actively being involved in improvement efforts. It is generally better to select a few activities to help ensure that progress is made without diluting the effort across many goals. For each activity, resource needs, milestones, and methods for measuring and recognizing progress should be identified.

5. **Sustain and repeat:** As progress is made in improving OD, new opportunities should be identified based on continued evaluation of site performance. Periodically repeating the earlier steps as needed is desirable in order to measure and sustain progress and to continue improvement.

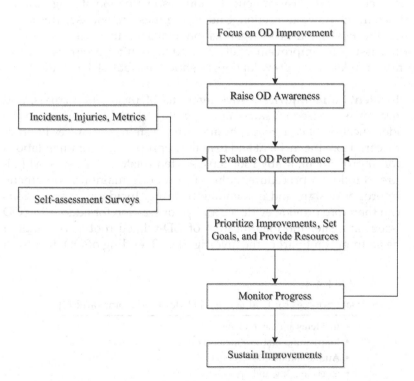

FIGURE 4.7
Steps for Improving OD.

Leaders should keep in mind that, though some activities to improve OD may be complex or require a fair amount of time to enact change, some simpler efforts can also be undertaken. For example, well-designed checklists have proven effective in achieving improved performance in a variety of industries [15]. Checklists remind people of the minimum necessary steps at the time a task is actually being done. They can serve as a timely reminder of the correct procedures for higher risk activities and help verify that actions are completed as intended.

4.4 Evaluating OD

4.4.1 General Approaches

A key part of an OD improvement program is that it must be specific to local conditions at a site, since site operations, activities, and culture can sometimes vary significantly, even in the same organization, company, or business. Any improvement program requires management leadership and commitment, development of a plan, assignment of resources, evaluation of progress, etc., as discussed in the previous section. Underlying these activities is the ability to evaluate the level of OD in an organization or at a site in order to identify and prioritize issues for improvement efforts and to monitor progress. Sources of information related to OD evaluation, as shown in Table 4.4, include [9,13,14]:

- **Incidents and Injuries** – Investigations of incidents, injuries, and near misses based on safety and operational problems focus on the identification of equipment, human, and system root causes in order to learn from the incident and to make appropriate recommendations for improvement. Many root causes often relate to OD, such as failure to follow a procedure, failure to properly maintain equipment, improper design, and inadequate training. Incident investigations can therefore provide evidence and prioritization for addressing OD issues at a site, and identification of OD-related root causes should be an important part of the investigation. Trending of OD data from

TABLE 4.4

Data Sources for Evaluating OD [Modified from 9 and 13]

- Incidents and near misses
- Injuries and first aid/medical cases
- Audits and field inspections
- Safety, quality, and operations metrics
- Assessment surveys

incidents and injuries can also help identify and raise awareness of significant OD issues at a site.

- **Audits** – Traditional safety, health, and environment (SHE); quality; and other audits may observe OD-related issues but may not specifically address them as such in the final report. Essentially, the causes of any gap identified in an audit should be evaluated for possible OD considerations. OD can be specifically evaluated by adding questions to process safety and other audits, by scheduling interviews to talk to site employees about OD, and by reviewing all potential audit findings, resulting in specific OD-related findings for improvement.

- **Metrics** – The best sources of information on current OD performance are metrics that are probably already being collected on various aspects of site operations, including quality, productivity, cost, and many others. These metrics may not normally be looked at to evaluate OD, but can provide great insight without much additional effort. Action items that are overdue, for example, reflect a lack of OD for getting things done on time. Quality problems may have factors related to OD. Low process uptime or utilization may also result in some cases from OD-related problems. The key is to look at metrics for indications of OD issues to help identify opportunities for improving OD. New metrics can also be collected and evaluated, of course, to provide additional insights into OD. Some examples of metrics for evaluating OD are provided in Table 4.5.

- **OD assessment surveys** – OD assessments can be developed using the OD characteristics discussed earlier in this chapter [9]. These can be done using an internet survey tool or through group discussion, depending on the size of the group and the need for analysis of the data. Internet surveys can be convenient to include a large number of employees and to filter data by different groups. If used, free-form entry questions should be included to provide for participant comments on what is currently working well or to suggest other actions that might be needed. Surveys represent employee perspectives and opinions and are therefore not hard data, such as provided by metrics. Often, though, site survey data can be used to help identify OD improvement needs that are not as easily found from other sources. Survey results should be shared with participants and other interested groups.

Of course, OD evaluation must be a continuing process to identify new issues, to measure progress, and to sustain gains already made.

If desired, OD improvement workshops can be developed that involve site employees in identifying and prioritizing potential OD improvement activities [9]. This helps raise employee awareness of OD, engages employees in identifying problems and developing solutions, and helps

TABLE 4.5

Possible Metrics for Evaluating OD [Modified from 9]

- Number of incidents (incidents with OD root cause)
- Number of injuries (injuries with OD root cause)
- Audit findings
- Planned activities completed on schedule
- Overdue findings (incidents, PHRAs, audits)
- Overdue test and inspections, procedure revisions
- Environmental deviations
- Excess waste
- Poor product quality
- Customer complaints
- Higher costs
- Lower uptime or equipment utilization
- Staff turnover rates
---

the organization get started on making improvements. Hard data from incidents and metrics and soft data from self-assessment surveys can be reviewed to provide data for the workshop group to identify and prioritize issues and to develop specific path forwards. Workshops may typically be 4–8 hours, depending on the size of the group and the type of work activities being evaluated, but can be a good way to energize a group about improving OD.

4.4.2 Evaluating Personal OD

Evaluating current performance related specifically to personal OD can be difficult [9,14], so assessment surveys are often used to obtain employee feedback. Some basic questions to consider are shown in Table 4.6. In addition, evaluation of OD as part of incident/injury and near miss investigations can indicate specific issues that can be addressed. For example, if a procedure is not followed by a worker resulting in an injury, careful analysis of why the procedure was not followed as related to Knowledge, Commitment, and Awareness in addition to other potential system failures should provide insight for improving personal OD. Quality and other operational problems may similarly provide an insight into personal OD issues.

The quality and availability of operating and maintenance procedures should be periodically reviewed with workers to ensure they are current and useful. Similarly, the quality and timing of training should be reviewed

TABLE 4.6

Evaluation of Personal OD [Modified from 14]

Knowledge

- Have you clearly set expectations for a procedure-based culture?
- Are procedures used universally or with shift differences?
- Are procedures readily available and comprehensive?
- What is the quality of your procedures? Accuracy? Clarity?
- Are you using effective checklists to supplement procedures?
- Are effective revision practices in place?
- Does everyone understand WHY in addition to WHAT?
- Is high-quality training provided? Trainer qualifications? Formats?
- How do you measure training success?
- Is training frequency appropriate? How do you know?
- How is training on changes provided?

Commitment

- Do you visibly demonstrate personal priority for safety?
- Is your daily practice in alignment with core values?
- Are you providing sufficient and capable resources?
- How do you promote pride in the organization?
- How do you facilitate employee involvement?
- Do you ever talk to the workforce about OD?
- Are you clear about expectations and accountability?
- Does everyone have the opportunity to influence or provide input on work tasks?
- Do you encourage open communication? Follow-up?
- Do you provide appreciation/recognition of good work?
- Are you interested in employee well-being?
- Do employees have opportunities for job growth and development?

Awareness

- Does everyone have a basic understanding of process hazards?
- Do your procedures clearly outline what can go wrong?
- Do procedures provide clear troubleshooting guidance?
- Does training include troubleshooting unusual or unexpected events?
- Is there time to plan jobs and anticipate problems?
- How are people told to monitor their work environment?
- Does your site culture easily support people asking for help?

to evaluate appropriate training frequencies and effectiveness. Some additional issues that can be used to evaluate Knowledge, Commitment, and Awareness via discussion or survey with workers are shown in Table 4.7. See Chapter 8 for additional discussion of procedures, training, and related activities.

TABLE 4.7

Potential Issues Related to Personal OD [Modified from 14]

Knowledge
- Certain tasks are only done occasionally, so it is difficult to remember them
- There is not enough training
- Training quality needs to be improved
- It has been too long since training was done
- There are no procedures or the procedures are not easily available
- Procedures are not clear or they are not current
- Equipment, tools, and PPE are not provided/maintained in good condition

Commitment
- Do not care; just want to get job done quickly
- Do not believe questions for help are encouraged
- Not enough time to plan the job
- Shortcuts are the only way to get job done on time
- Too many distractions in the workplace
- Work is interrupted many times before completion
- Not encouraged to suggest changes to procedures or work practices
- Do not think coworkers really want suggestions or help

Awareness
- Do not have time to think through potential problems
- Do not have time to monitor work areas and coworkers
- Too often distractions (personal or work environment)
- Training/procedures do not clearly cover what can go wrong
- Training/procedures do not provide much troubleshooting help
- Asking for help is not encouraged
- There is no one to ask for help

4.5 Leadership Processes

Efforts to improve OD necessarily require the personal commitment and focus of site leadership. Leaders must [10]:

- Walk the talk at all levels ("Felt Leadership")
- Define expectations and acceptable performance
- Consistently enforce expectations
- Verify implementation status and progress
- Monitor performance data
- Sustain performance.

TABLE 4.8

Leading Improvement of OD [Modified from 16]

- **Focus** – Focus on OD improvement and goals
- **Leadership** – Provide resources and personal leadership
- **Accountability** – Set clear expectations on performance
- **Measurement** – Measure to assess performance and progress
- **Engagement** – Provide and support a trustful work environment to involve employees

---

Leading improvement of OD is a continuous process, where improvement opportunities must be identified and worked on, improvement must be sustained, and then new opportunities must be identified for continued improvement.

The FLAME model [16], as discussed below, describes best practices for leadership efforts in leading improvement of OD [9,13]. Five key areas are included, as shown in Table 4.8:

- **Focus** – A vision has been developed to ensure appropriate focus on OD that provides for a strong, multidisciplinary OD program. The vision and OD program have been effectively communicated to develop awareness and engage site personnel and are supported daily through leadership attention and behaviors.

- **Leadership** – The leader is a role model who actively works to build a strong leadership team committed to continuous improvement. Coaching and other practices are used to structure leadership activities as well as to support the desired safety culture and strong OD performance.

- **Accountability** – Organizational, team, and individual goals include a focus on OD with clear expectations on performance. Goals and commitments are reviewed periodically, feedback to reinforce progress is provided, and recognition is delivered, as appropriate.

- **Measurement** – Metrics, audits, and other measures are defined to continually assess site activities, performance, and progress toward goals. Efforts are made to develop and apply data based on local facility/organizational needs and priorities.

- **Engagement** – Leaders focus on supporting a trustful work environment to engage and support site personnel through consistent leadership principles and behaviors. Programs are personalized in terms of impact to all employees, good communication principles are followed, and interdependent behavior is encouraged and supported. Site personnel know they are important to success and that their contribution is valued.

Other approaches are also available for leading organizational change [17], as discussed in Chapter 2, which can provide additional considerations for leading effective OD programs.

Although OD improvement is the responsibility of all organizational leaders, it can also be helpful to assign an OD leader to provide additional focus and accountability on achieving an effective OD program [16]. In most cases, this would be a part-time assignment, such as the process safety system leaders discussed in Chapter 13. Ideally, this person should meet the following criteria:

- A member of leadership
- Broad responsibilities, not just with a focus on SHE, PS, or OD
- Ability to impact site resources, priorities, and activities.

The intent is that the site OD leader is a line leader who is able to provide significant leadership and influence site activities in support of improved OD. For large organizations or facilities, one or more additional people could be assigned specific day-to-day OD responsibilities to support the OD leader. Some possible activities of the OD leader include:

- Set and support specific site goals to improve OD
- Raise awareness of OD through training and relevant communications
- Ensure OD-related causes are identified as part of incident investigations
- Evaluate and monitor OD performance by analyzing metrics, audits, incidents, quality, uptime, etc., for OD contribution and improvement opportunities
- Conduct and evaluate OD self-awareness surveys at appropriate intervals
- Participate in corporate OD network and team activities.

4.6 Case Study

On April 23, 2004, an explosion and fire, as shown in Figure 4.8, occurred at the Formosa Plastics Corp. in Illiopolis, IL, resulting in five fatalities and three serious injuries [18]. An operator mistakenly opened the bottom valve of a polymerization reactor that was currently in operation, apparently bypassing an active pressure interlock, instead of opening the bottom valve of a nearby reactor that was being cleaned. The result was a large release of

FIGURE 4.8
The Formosa explosion 2004 [18].

hot and pressurized polyvinyl chloride (PVC), which is highly flammable and toxic. The explosion damaged most of the reactor building and a nearby warehouse. As a precaution due to potentially hazardous smoke from the fire, the local community was evacuated near the plant for two days.

A first look at this incident suggests significant operator error in going to the incorrect reactor and bypassing an interlock on an operating reactor. From an OD perspective, it is clear that operator awareness was likely a major contributing factor to this incident, with less evidence that knowledge or commitment were issues. It is not sufficient to stop there. What organizational OD characteristics (and other process safety system failures) also contributed to this incident?

As shown in Figure 4.9, the reactor configuration and appearance of the two reactors (D306 and D310) were very similar. The reactors were labeled but otherwise looked identical, contributing to the operator mistakenly going to the wrong reactor. The bottom valves of the reactors were located on the lower level of the reactor building. From that location, operators did not have local indicators or convenient communication to determine reactor status. Although operators were not authorized to bypass interlocks, the ability to do so was convenient and could be completed without detection.

Previous process hazards and risk analysis studies had suggested revision of bypass procedures and hardware to prevent inadvertent or deliberate manual bypass of the pressure interlock, but these recommendations had not yet been implemented. Previous incidents with similar causes had

FIGURE 4.9
The reactor layout at Formosa [18].

also occurred, but recommendations also had not yet been implemented. These factors together suggest that organizational OD issues also contributed to this incident, and while operator error occurred, better process safety systems and organizational OD may have prevented the incident. The CSB concluded that the company "did not adequately address the potential for human error [18]."

References

1. Collins, Jim. 2001. *Good To Great: Why Some Companies Make the Leap ... And Others Don't*. HarperBusiness, New York.
2. Kletz, Trevor. 2001. *An Engineer's View of Human Error*, 3rd ed. CRC Press, Boca Raton, FL.
3. Center for Chemical Process Safety. 2007. *Guidelines for Risk Based Process Safety*. Wiley-AIChE.
4. Reason, James. 1990. *Human Error*. Cambridge University, New York.
5. Center for Chemical Process Safety. 1994. *Guidelines for Preventing Human Error in Process Safety*. Wiley-AIChE.
6. Dekker, Sydney. 2006. *The Field Guide to Understanding Human Error*. Ashgate Publishing, Aldershot.
7. Center for Chemical Process Safety. 2007. *Human Factors Methods for Improving Performance in the Process Industries*. Wiley-AIChE.
8. Klein, James A. 2005. Operational discipline in the workplace, *Process Safety Progress* 24:228–235.
9. Klein, James A. and Bruce K. Vaughen. 2008. A revised model for operational discipline. *Process Safety Progress* 27:58–65.

10. Center for Chemical Process Safety. 2011. *Conduct of Operations and Operational Discipline*. Wiley-AIChE.
11. Klein, James A., William M. Bradshaw, Lee N. Vanden Heuvel, Donald K. Lorenzo, and Gregory Keeports. 2011. Implementing an effective conduct of operations and operational discipline program. *Journal of Loss Prevention in the Process Industries* 24:98–104.
12. Manuele, Fred. 2011. Reviewing Heinrich: Dislodging two myths from the practice of safety. *Professional Safety*, October, pp. 52–61.
13. Klein, James A. and Bruce K. Vaughen. 2011. Implement an operational discipline program to improve plant process safety. *Chemical Engineering Progress* 107:48–52.
14. Klein, James A. and Eduardo M. Francisco. 2012. Focus on personal operational discipline to get work done right. *Process Safety Progress* 31:100–104.
15. Gawande, Atul. 2011. *The Checklist Manifesto: How To Get Things Right*. Picador.
16. Klein, James A. and Brian D. Rains. 2013. Operational Discipline: Panacea for Performance? *Global Congress on Process Safety*, San Antonio, TX.
17. Kotter, John P. 2012. *Leading Change*. Harvard Business Review Press, Boston, MA.
18. U.S. Chemical Safety Board. 2007. *Vinyl Chloride Monomer Explosion*. Report No. 2004-10-I-IL. www.csb.gov.

Section II

Practical Approaches for Designing Safe Processes

5

Design Safe Processes

Design is not just what it looks like and feels like. Design is how it works.

Steve Jobs

Why We Need to Design Safe Processes

Safe process and equipment design are essential for reducing the process safety risks and for reducing incidents. If process hazards exist, the technical information used for the process and equipment design must be understood so that process safety systems can be designed to manage the process and the equipment integrity. Since safe process design depends on understanding the process hazards (Chapter 6) and how their associated risks are managed (Chapter 7), the development of the process and equipment design (this chapter) must take into account the types of hazards and how their risks will be managed. Hence, the protocol for designing safe processes is iterative. Everyone who designs, fabricates, installs, operates, maintains, and changes the equipment must understand their role when managing process hazards and risks.

**Design safe processes to ensure that appropriate technology
and safeguards are incorporated to control process
risks and achieve safe and reliable operations.**

5.1 Introduction

If the process design or the equipment design is not selected well in the beginning, operational, maintenance, and process safety issues may occur, posing problems once the facility is built. Fortunately, process designers and equipment designers have multiple resources from which to select their engineering controls and designs, including specific codes and standards developed to address the handling of hazardous materials and energies.

These codes and standards establish expectations for the engineering design technologies, with special emphasis on both the processing equipment required to manage hazardous processing conditions, such as high temperatures and pressures, and on the safeguarding equipment required to detect, respond to, and mitigate process deviations and releases, such as interlocks, pressure relief systems, and fire protection systems. Safe process and equipment design must be accurately documented and accessible to everyone working with the hazardous materials and energies.

The eight life cycle phases for processes and equipment are illustrated in Figure 5.1, beginning with the equipment designed to meet the process design. The equipment is fabricated, installed, and commissioned before being used. Then the equipment is operated, maintained, and changed during its useful life. Once the equipment's life is over, it is then decommissioned. The design phase for processes may include bench scale research and development,

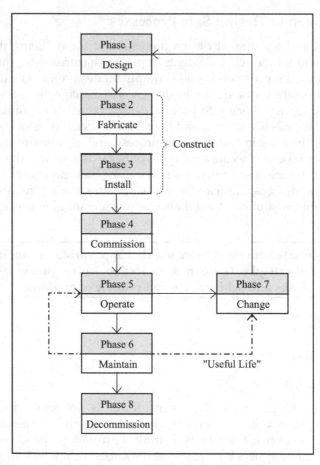

FIGURE 5.1
The eight life cycle phases of a process or equipment.

applying Inherently Safer Process (ISP) designs before scale-up and piloting studies. The construction and commissioning phases include locating the facility, the fabrication and installation of the equipment, and developing operating, maintenance, and emergency response procedures. Once personnel are trained, the process is handed over to operations for routine operation and maintenance during the processes life. Subsequent process or equipment modifications and expansions must be managed safely during the process lifetime before decommissioning and its equipment is removed from service.

Keep in mind that when locating new processes or when locating major expansions of existing process units, the hazards and risks associated with the change must consider where the process will be located before establishing *how* the processes and equipment will be arranged in the new location. Siting and layout of hazardous processes will be impacted by different location-specific risks, such as the location's terrain, weather, and accessibility, and includes how raw materials and products will be transported to and from the facility, and how they will be stored and handled at the facility [1].

5.2 Key Concepts

Choosing the process and equipment design is an iterative method considering different options that are balanced between many factors. In particular, the four risk mitigation strategies used to help identify specific engineering and administrative controls when designing the equipment are based on the "hierachy of controls" introduced in Chapter 3, Table 3.3:

1. inherently safer process design
2. passive control design
3. active control design
4. procedural control design.

The control measures are the steps taken to eliminate or remove a hazard or at least reduce the risk to a low level [2]. The first three controls focus on equipment-specific engineering designs, whereas the fourth focuses on the administrative controls. Inherently safer technologies and design strategies include minimization, substitution, moderation, and simplification. Passive controls include dikes, flame arrestors, and fire proofing. Active controls include pressure relief systems, blow out panels, and deluge systems. And procedural controls include operating, maintenance, and emergency response procedures. All of these controls are used in the protection layers required to reduce process safety risks (Figure 3.6). Personal protective equipment (PPE) is used as the last line of defense between the hazard and personnel.

These risk mitigation strategies apply when selecting the engineering and administrative controls essential for safe operation, as was depicted with the protection layers model in Figure 3.6. Hence, the factors affecting the process and equipment design include inherently safer principles, the processing conditions, such as operating temperatures or pressures, and the materials of construction. This section expands on the key concepts for designing the equipment for safe processes: the engineering design codes, standards, and practices; the technology documentation; and why the process and equipment technology information must be documented and sustained.

5.2.1 Using Codes and Standards

Codes and standards for managing hazardous materials and energies have been developed and established to help process and equipment designers apply safe technologies and the best engineering practices. These engineering practices identify best practices for managing the hazardous materials and energies, such as on the specific chemical process and its chemistries, on specific processing equipment and on best ways to fabricate, install, and maintain such equipment. In the United States, the term "Recognized and Generally Accepted Good Engineering Practices" (RAGAGEP) is used to describe this technology for regulated processes [3]. Essentially, these good engineering practices define minimum acceptable processing equipment design criteria to help ensure that the risks are reduced when handling hazardous materials and energies [2,4]. For comparison, the UK Health and Safety Executive (HSE) recommends codes and standards for equipment designs such that the process operating risks meet its "As Low As Reasonably Practicable" (ALARP) criteria [5–7].

There are many organizations which have developed specific process and equipment standards, such as the American Petroleum Institute (API), the National Fire Protection Association (NFPA), and the British Standards Institute (BSI). These organizations provide the current industrial guidance essential for safely designing, operating, and maintaining the equipment throughout their lifetime. A comprehensive list of many networks and organizations providing technical guidance is provided in the literature [2].

One example of a RAGAGEP standard used to design pressure vessels is the ASME Pressure Vessel Code, Section VIII [8]. There are three divisions of the code:

Division 1 – provides design, fabrication, inspection, testing, and certification requirements for fired or unfired pressure vessels operating at either internal or external pressures exceeding 15 psig. This division specifies the materials of construction, specifies the fabrication methods (i.e., welding), and specifies the examination (test and inspection) requirements.

Division 2 – has more rigorous requirements that vessels for Division 1, allowing for higher design stress values.

Division 3 – provides requirements for pressure vessels operating at pressures generally above 10,000 psi.

Each division includes a product certification package and specific certification marker which is stamped and placed on the coded vessel before it is installed, commissioned and used.

5.2.2 Documenting the Technology

The codes and standards that are used for the equipment design must be documented, kept accurate, and kept accessible [2,9]. This process technology, combined with the process hazards understanding, provides the basis for effectively performing hazards analyses and forms the foundation for developing the safe work practices, safe operating procedures, and safe maintenance procedures. The process hazards and associated risks are used to design effective emergency response programs and drills. Although each organization must determine how best to manage their technology information, it must be controlled, stored, and accessible to everyone working with the hazardous materials and energies. Additional discussion on specific types of documents used for conveying process safety information between groups is provided in Section 5.3.8.

5.2.3 Sustaining the Information

For safe operations, not only must the process design basis and the equipment design basis be documented and readily available, it must be sustained and kept up-to-date. Too many incidents resulting in serious injuries, fatalities, environmental harm, and significant property loss have occurred when the design and operation of critical process equipment were lost over time. When changes are made to the processes or the equipment, the process and equipment design documents must be changed, as well, such that they accurately reflect the current processing conditions and engineering designs. The Bhopal incident briefly summarized in Section 5.4, is a tragic case study of how the robust as-built design and its process technology information degraded over time when the operating and maintenance management disciplines failed to sustain the equipment per its original, safe process and equipment design. If the process technology is not sustained, then even a well-designed process safety system will degrade.

5.3 The Process and Equipment Design Basis

Simply stated, an effective process safety program relies on accurate process technology. The process design basis directly affects how the process safety risk is managed, directly impacting the equipment's design, fabrication, installation, operation, and maintenance. Hence, it is important to choose the process design carefully, especially if there are options for selecting an inherently safer process design, as will be described below. The key concepts noted earlier apply to both the processes and the equipment, focusing on the equipment that must be:

- designed and constructed using accepted codes and standards
- operated within their design limits
- tested and inspected to ensure their integrity during their lifetime.

For safe operations, the hazards and risks must be understood, the proper equipment must be selected, the proper operating limits must be enforced, and the equipment's integrity must be sustained.

5.3.1 Applying Inherently Safer Process Design

Before designing the equipment required to handle the hazardous materials and energies, the processing steps, including the processing chemistry, must be selected. The purpose of Inherently Safer Process (ISP) reviews is to identify opportunities that permanently eliminate or reduce the process hazards. Otherwise, systems will need to be designed that must control the hazards and their risks through additional safeguards [2,10–12]. Additional safeguards, such as pressure relief systems or emergency shutdown systems, add complexity to the process and will add to the operating and maintenance costs over the lifetime of the process. The primary goals of inherently safer process reviews were shown in Table 3.2: minimization, substitution, moderation, and simplification. Inherently safer process design reviews are often conducted from detailed checklists [11]. The greatest benefit from an ISP review occurs in the design or early project stages for new processes, equipment, or operations because safety can be significantly improved without the cost of additional safeguards. Although ISP reviews can be conducted for existing processes to identify potential processing improvement opportunities, the costs associated with such changes may be prohibitive at this point [1,13].

5.3.2 Understanding the Hazards

The material's physical property data, combined with an intrinsic hazard assessment (IHA), form the basis for understanding the toxicity, flammability, explosivity, and chemical reactivity hazards associated with the materials. However, hazards associated with the processing conditions, such as high pressures or vacuum, high or low temperatures, generation or release of stored energy ("potential" energy), and high speed rotating equipment ("kinetic" energy) must be understood, as well. In addition, the hazards associated with nonionizing radiation, low oxygen environments (a special case of toxicity), and acids or bases (special cases of chemical reactivity) may exist and will need to be understood, as well. As will be described in more detail in Chapter 6, the three steps for understanding the hazards through an IHA are (1) identification, (2) assessment, and (3) documentation. With the process hazards information, appropriate engineering and administrative controls can be evaluated, designed, and included in process safety systems used to manage the process safety risks. The evaluation of the risks and how best to manage them are described in greater detail in Chapter 7.

5.3.3 Understanding the Process Risks

Effective understanding and management of the process safety risks depend on the quality of the process and equipment design information (this chapter) and a thorough understanding and assessment of the hazards. The basic risk evaluation method described in Chapter 7, the process hazards and risk analysis (PHRA), uses the process hazards information and assessments to methodically and thoroughly evaluate the risks. In general, PHRAs are used to:

- understand the types and ranges of consequences of potentially hazardous events
- identify the failures or pathways that can lead to these events
- evaluate the process risk of these events
- identify the residual risk, where gaps exist between the tolerable risk and the risk associated with the current design.

These risk evaluations are the driving force for actions requiring the organization to address the gaps by designing and implementing engineering and administrative controls—the protection layers. Inadequate analysis of process risks increases the likelihood that process safety systems and safeguards will be inadequately designed and ineffective, as well, thus increasing the likelihood of serious injuries and incidents. Please refer to Figure 3.6 for an illustration of the eight protection layers which are engineering controls, administrative controls, or a combination of both engineering and administrative controls.

5.3.4 Selecting the Equipment

The equipment selected for safely operating a facility includes equipment for material unloading and loading, for material storage, for material processing (e.g., reactors, columns, separators, effluent disposal, etc.), for utilities (e.g., electricity, steam and inert gases), for process control (e.g., BPCS, instrumentation, SIS, etc.), for consequence mitigation (e.g., relief systems, flame arrestors, fire protection, etc.), and for emergency response. Since the hazards of these materials include their reactivity, toxicity, flammability, combustibility, and explosivity, how they are managed must include their quantities, their processing conditions (e.g., temperatures, pressures, etc.), and what physical state they are in while handling and storing them (e.g., gases, liquids, solids, etc.).

Materials of construction specifications, piping system design, other process unit interconnectivities, and the utility connections are included in the equipment design. Piping system design includes the valve, flange, and thermal insulation specifications, as well as piping support and vibrational analyses. Utility systems include electricity, cooling mediums, instrument air, steam, other heating mediums, fuels, and inert gases. And to ensure safe operations, as has been noted before, the equipment used for the barriers illustrated in Figure 3.6 must be designed, constructed, operated, and maintained for the life of the equipment. Additional details on specific designs for the critical equipment are provided in the literature [2,4,14].

The codes and standards discussed in Section 5.2.1 help designers with both processing, utilities and safeguarding equipment. For example, there are codes and standards for specific pressure vessel safeguarding equipment design. To prevent physically overpressurizing the vessel, pressure vessels must have a pressure relief system designed as its last line of defense. Pressure relief systems are sized depending on the potential overpressure scenarios, consisting of rupture disks, relief valves, or a combination of both. The relief system design includes the inlet and outlet sizes and must conform to the codes used to build the pressure vessel [8]. Industry codes also provide guidance on scenarios that should be considered and on equations for sizing of devices [15]. The potential overpressure scenarios include fire, blocked flow, control valve failure, overheating, power outage, tube rupture, and cooling water failure events. Runaway reactions may cause two-phase flow through the relief system, with specific modeling and design guidance provided by the Design Institute for Emergency Relief Systems (DIERS) [16].

Another example of detailed design, operation, and sustaining guidance for safeguarding equipment is the industry's safety instrumented system (SIS) design standard [17,18]. An SIS is a system of safety instrumented functions (SIFs) composed of sensors, logic solvers, and final elements for the purpose of taking the process to a safe state when predetermined conditions are violated. A schematic of an SIF is shown in Figure 5.2. Each SIF detects and responds to out-of-limit (abnormal) conditions, bringing the process to a functionally safe state. An SIS is independent of the BPCS, with specific safety integrity levels (SILs) defined for each system. The SIF level consists of specific controls which are required to ensure that potentially hazardous conditions are detected and controlled. In addition, the SIS systems meet RAGAGEP ("Recognized and

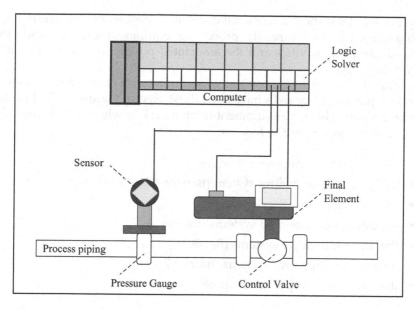

FIGURE 5.2
The parts of a safety instrumented function (SIF).

Generally Accepted Good Engineering Practices") and ALARP level engineering designs ("As Low As Reasonably Practicable").

The codes and standards provide the specifications for the processing and safeguarding equipment used to operate safe processes. The basic difference between processing, utilities, and safeguarding equipment and their associated piping and instrumentation is:

- processing equipment—used in the day-to-day operations to convert the raw materials into intermediate or final products, such as reactors, columns, separators, and pumps
- utilities equipment—used to provide or remove energies to or from the process, such as electricity, steam or hot oil systems, and cooling mediums
- safeguarding equipment—used for the different protection layers, helping reduce the frequency of a loss of containment event or helping to mitigate the consequences of the event once it occurs.

In addition, the design of the piping systems used to convey the materials to and from the processing equipment and other equipment or utilities must consider materials of construction, whether flanges can be used or not, the types of valves, and the associated piping support structures. The instrumentation equipment used to monitor, control, and respond to the processing conditions, such as flow rates, temperatures and pressures must be properly designed, as well. Note that a piping and instrumentation diagram (P&ID) is the foundational document used when performing hazards analyses, as it captures the processing equipment, utilities, and other process interconnectivities and the associated piping and instrumentation in one diagram.

Recall that the different preventive and mitigative protection layers used to reduce process safety risks have been illustrated in Figure 3.6. The protection layers which have an equipment-related design, whether for preventive or mitigative controls, are as follows:

- the processing equipment (Barrier 1)
- the basic process control system (Barrier 3)
- the critical alarms (Barrier 4)
- the safety instrumented systems (Barrier 5)
- the active mitigative systems (Barrier 6)
- the passive mitigative systems (Barrier 7)
- the emergency response equipment (Barrier 8).

At this point, it is worth defining "critical equipment" as used in this book:

Critical equipment: Equipment handling hazardous materials or energies whose failure or malfunction could result in a loss of control or containment of the hazardous material or energy.

The critical equipment should include all the safeguarding equipment identified in hazards and risk analyses that are required to reduce the process safety risks (Chapters 7 and 8). The critical equipment's integrity must be sustained for its entire life cycle (Chapter 9). And all changes to the critical equipment must be effectively managed (Chapter 10). Note that there are equipment critical to the business, such as high-pressure air compressors, which may not have hazardous process safety consequences upon their failure. However, loss of this business critical equipment could cause a significant business interruption. Thus, their integrity should be sustained during their life cycle, as well. A list of typical "critical equipment" used to handle hazardous materials and energies is provided in Table 5.1.

Keep in mind that another type of equipment required when managing the safety in hazardous processes is the Personal Protective Equipment (PPE), the "last line of defense" between the hazard and personnel. PPE designs can be active, such as respirators, or passive, such as flame resistant clothing (FRC). PPE is one aspect of the occupational safety and health efforts at a facility, typically being managed through industrial hygienist experts [4]. PPE is required during normal operations, during maintenance, and during emergencies. Normal operations/maintenance PPE is evaluated

TABLE 5.1

Typical "Critical Equipment" and Associated Instrumentation and Systems Used to Handle or Control Hazardous Materials and Energies

Equipment	Systems
• Pumps (including pumps transferring liquids, compressors, blowers/fans, agitators, and their drivers/motors/turbines/engines) • Pressure vessels • Storage tanks • Eductors • Safeguards (i.e., safeguards identified as "independent protection layers" in PHRAs, Chapter 7) • Dikes, catch tanks, etc.	• Piping systems (including piping components such as valves) • Relief and vent systems and devices (includes overpressure, vacuum, and explosion mitigation equipment) • Emergency shutdown systems • Basic process control systems (including monitoring devices and sensors, alarms, and interlocks) • Safety instrumented systems (SIS) • Fired heaters/furnaces • Flares

Refer to the Barriers identified in Figure 3.6.

using the same hazards analysis protocols described in this book: understand the material's hazards, determine how the materials can adversely affect personnel, select the correct PPE design, and then ensure that the PPE is fabricated, inspected, maintained, and properly donned and used when personnel have the potential for exposure to the hazard. Emergency response PPE includes special gear used when responding to fires or toxic release emergencies.

All PPE has a specific head-to-toe objective when protecting personnel from hazards, focusing on potentially adverse effects to a person's head, ear, eye, face, lungs, body, skin, hand, and foot. Hence, depending on the area's hazards, a combination of required PPE for everyone entering the area may include a hard hat, ear protection, eye protection, a respirator, gloves, FRC, and steel toed-shoes. Often, maintenance activities require additional PPE designed to meet safe work practices such as fall protection, confined space, line breaks, and electrical isolation. And emergency responders have turn-out gear with air supplies to protect them from the thermal radiation and the smoke from fires. There are many different types of PPE which can be used to address the residual risk after all other engineering and administrative controls have been developed and implemented [19]. It is important to recognize that there are engineering codes and standards used to design the PPE, as well [4,20,21].

In summary, there are four basic types of equipment directly associated with a process unit which must be selected using the guidance or requirements provided with specific codes and standards listed in Table 5.2: processing, utilities, safeguarding, and PPE. All of the equipment must be designed, constructed, used, and maintained during its lifetime for effective management of the hazards and risks.

TABLE 5.2

The Different Types of Equipment

	Types of Equipment	Risk Reduction Purpose	
		To help reduce the frequency	To help reduce the consequence
1	Process equipment	√	
2	Utilities equipment	√	√
3	Safeguarding equipment	√	√
4	Personal protective equipment (PPE) (see Note below)		√

Note: PPE is designed to protect a person from "head to toe." In particular, the following equipment may be required depending on the type or types of hazards in the area:

Part of Body Being Protected	Specific Personal Protective Equipment (PPE)
Head	Hard hats
Ears	Ear protection (hearing protection programs, usually covered under separate industrial hygiene efforts)
Eyes	Safety glasses, goggles, eye shields, welding shields
Face	Face shields
Lungs	Inhalation protection (respirator programs, usually covered under separate industrial hygiene efforts)
Arms	Sleeves
Hands	Gloves
Body	Flame resistant clothing (FRC)
Legs	Flame resistant clothing (FRC)
Feet	Safety shoes, boots

5.3.5 Selecting Safe Operating Limits for Process Equipment

Every process has three fundamental types of operating limits, those selected for product quality, those selected for process safety, and those selected for equipment integrity. These limits are based on the process and equipment design basis, with their ranges defined with the following upper and lower limits:

- operating or quality limits
  - such as "standard operating conditions" (SOCs) or "standard operating parameters" (SOPs)

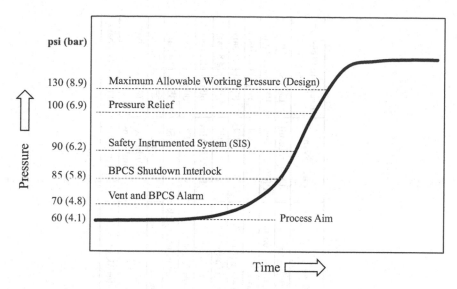

FIGURE 5.3
The different types of operating limits.

- safety limits
  - such as "safe operating limits" (SOLs), "critical operating limits" (COLs), or "critical operating parameters" (COPs)
- critical or equipment design limits
  - such as "the never to exceed" operating parameters.

The BPCS is used to monitor and adjust the processing conditions, such as temperatures and pressures, during normal operations (refer to Barrier 3 in Figure 3.6). When these SOC or SOL limits are maintained, the process is operating under normal conditions for making quality intermediates or products. As processing conditions deviate from the normal operating set points, the BPCS will have designated alarm set points for "high" or "low" SOL requiring either a manual or automatic response. Interlock set points are typically used for high–high or low–low COLs or COPs. However, if conditions reach the upper or lower *equipment* design limits, SIS systems (Barrier 5) are often selected as an engineered, independent layer of protection. In addition, for conditions with potential vessel overpressurization, other engineering controls, such as relief systems, are implemented to help reduce the risk.

Continuing the discussion in Chapter 3 to help manage a runaway reaction (see Figure 3.3), example pressure set points could be selected for the different operating limits, as is illustrated in Figure 5.3 and summarized in Table 5.3. Quality products cannot be made under unsafe operating conditions when these operating, safety, and critical limits are properly selected.

TABLE 5.3

An Example of the Pressure Set Points Illustrated in Figure 5.3

Limit	Limit Name	Example Response	Limit "Name"	Example Pressure Setpoints
Upper equipment design	*Never to be exceeded* limit	Maximum Allowable Working Pressure	Design Limit	130 psi (8.9 bar)
		Pressure relief (e.g., pressure relieve valve)	PRV Set Point	100 psi (6.9 bar)
Upper safety	High High Safety Limit	SIS action with alarm/interlock	HHSL	90 psi (6.2 bar)
	High Safety Limit	BPCS shutdown action with alarm/interlock	HSL	85 psi (5.9 bar)
Upper operating or quality	High Operating/Quality Limit	BPCS control action with alarm/interlock	HQL	70 psi (4.8 bar)
			Process Aim	60 psi (4.1 bar)
Lower operating or quality	Lower Operating/Quality Limit	BPCS control action with alarm/interlock	LQL	50 psi (3.4 bar)
Lower safety	Low Safety Limit	BPCS shutdown action with alarm/interlock	LSL	35 psi (2.4 bar)
	Low Low Safety Limit	SIS action with alarm/interlock	LLSL	25 psi (1.7 bar)
			"Ambient" Conditions	≈14.7 psi (≈1 bar)
Lower equipment design	*Never to be exceeded* limit	Pressure relief (e.g., vacuum breaker)	Design Limit	<14.7 psi (<1 bar)

5.3.6 Establishing Safe Equipment Layout

How the equipment is arranged—its layout within the process unit—is important when flammable or combustible materials are being processed. Tightly packed equipment increases equipment confinement and congestion, adversely affecting ease-of-access by operations, maintenance, and emergency responders. With greater separation distances between equipment, the consequences of loss of containment incidents are reduced with the effectiveness of the larger distances being strongly influenced by the types of hazards and how far their impact could be. For example, greater distances help reduce the impact of fires to surrounding areas by reducing the exposure to and intensity of the thermal radiation. Or for explosions, greater distances between equipment help reduce potential equipment congestion density and thus help reduce the magnitude of the blast waves (especially vapor cloud explosions [VCE]). Greater distance between equipment and occupied buildings allows for increased blast wave decay, thus reducing potential consequences to equipment, buildings, and their occupants.

For toxic release effects, greater distances help reduce the impact on personnel due to increased diffusion and dilution of the toxic gas or vapor concentrations as they disperse further from the source of the leak. Since consequences resulting from toxic clouds may be incurred further than from thermal radiation or blast waves from a particular release source, the lag time between the release time and potential exposure time to personnel within the facility or the surrounding community is increased with greater distances between the release location and those potentially affected. This lag time provides additional time for warning people downwind whether they should shelter-in-place or quickly evacuate the soon-to-be affected area. Additional discussion for reducing risk by strategically arranging equipment handling hazardous materials is provided in the literature [1].

5.3.7 Establishing Equipment Integrity

The equipment integrity for the life of the equipment is established through understanding, monitoring, and managing the distinct phases in an equipment's life cycle. These phases were illustrated in Figure 5.1:

1) Design, using equipment-related codes and standards based on the process design
2) Fabricate, based on the equipment design
3) Install, based on the equipment and process design
4) Commission, based on construction verification (a prestart-up review)
5) Operate, based on safe operating limits
6) Maintain, based on tests and inspections required for sustaining equipment integrity

7) Change, due to, in part, proposed process or equipment improvements

8) Decommission, due to process and equipment improvements or the results of the tests and inspections.

It is essential that everyone understands the equipment design when fabricating, installing, operating, maintaining, and changing the equipment. Often within major projects, the fabricate and install phases are combined into a single "construct" phase, bridging the design phase to the operating and maintaining phases with one procurement/construction-focused phase.

The following pressure vessel illustration helps depict how the different phases of the equipment's life cycle relate to one another and help manage the equipment during its lifetime. When high processing pressures are required for a reaction, there are special codes and standards which must be used to *design* the pressure vessels [8]. The equipment is *fabricated* and *installed* based on the equipment design specifications. The equipment is *commissioned* before it is *operated* within its safe operating limits. During its useful operating life, the equipment's integrity must be *maintained* by adhering to the recommended tests and inspections. *Changes* must be effectively managed and must always refer to the original design documentation. And when the equipment's useful life is over, it is then *decommissioned* and removed from service.

5.3.8 Documenting the Process and Equipment Design Specifications

The documents used to record the process and equipment information must be available to everyone who works with any phase of the equipment's life cycle. The technical documents provide the information for developing and effectively implementing all of the process safety systems, such as for:

- assessing the process hazards
- evaluating the process risks
- designing, operating, maintaining, changing, and monitoring the processes and equipment
- training personnel
- responding to incidents.

Weaknesses in the accuracy of the technology information will adversely affect the effectiveness of the systems designed to manage the risks.

The sources for developing the different diagrams include fundamental research reports, safety data sheets (SDSs), material and energy balances, equipment design specifications, control system design, process unit and equipment layout, building layout, design, construction and occupancy, area accessibility and escape routes. This information includes the process chemistries, processing conditions, and process hazards. The different diagrams used for evaluating process safety risks include block or process flow diagrams (BFDs, PFDs), piping and instrumentation diagrams (P&IDs), and Cause and Effect (C&E) diagrams. Some companies provide specific guidelines and standards for diagram formatting consistency between facilities. A summary of some of the different types of process and equipment diagrams and their uses is shown in Table 5.4.

TABLE 5.4

Some Diagrams Used for Documenting Process and Equipment Information

	Diagram	Source	Use
1	Chemical interaction, compatibility, or reactivity matrix	Process design basis Fundamental research reports Safety data sheets	Process and equipment design Procedures and safe work practices Process hazards analyses Training
2	Block flow diagram (BFD)	Process design basis	Process hazards analyses Training
3	Process flow diagram (PFD)	Process design basis Material and energy balances	Process hazards analyses Training
4	Piping and instrumentation diagram (P&ID)	Process design basis Material and energy balances Equipment design specifications (includes piping/transmitters/controls)	Process hazards analyses Training
5	Electrical loop/line diagram	Process design basis Equipment design specifications Control system design	Process hazards analyses Training
6	Interlock logic cause and effect diagram (C&E)	Process design basis Equipment design specifications Control system design	Process hazards analyses Training
7	Facility plot /building plan/ structural design	Process unit and equipment layout Building layout and occupancy Building design and construction Accessibility and escape routes	Facility siting studies Quantitative risk assessments Process hazards analyses Emergency response and planning Training
8	Hazardous area electrical classification drawing	Process design conditions, flammable material dispersion contours and process unit and equipment layout	Process and equipment design Facility siting studies Quantitative risk assessments Process hazards analyses Training

Controls must be in place to manage changes to both electronic documentation and printed copies to ensure that everyone has the latest process technology information. Unfortunately, we have a history of incidents which resulted in fatalities, injuries, environmental harm and significant facility damage when changes were made to the process or equipment without addressing proper management of change procedures [22]. Case studies of incidents which occurred due to lack of process technology information and understanding are presented in Section 5.4 below.

5.4 Case Studies

This section presents case studies which occurred due to poor understanding of the process design, with this lack of knowledge resulting in inadequately developed and implemented process safety systems. The process technology information, documenting the risks of the hazardous materials and energies, is used for selecting:

- inherently safer process and/or equipment designs
- safe operating limits
- safeguards and protection layers
- training materials
- preventative maintenance programs
- emergency response drills.

Effective process safety systems use this information to share with others the process hazards and risks, to manage and handover information on equipment and process changes, to monitor systemic or operational gaps, and to help identify potential technological and systemic weaknesses. The first case study focuses on the lack of understanding of the fundamental chemistry and process scale-up, whereas the second case study illustrates where communications between different groups broke down, resulting in poor communication of the hazards and risks between those making the decisions. The third case study presents an all-too-familiar story where change was simply ineffectively managed.

Case Study 5-1: T2 Industries, Inc., 2009

An unrecognized runaway reaction led to the catastrophic incident at T2 Industries, resulting in fatalities and significant property damage [23]. The exothermic reaction was not understood, leading to inadequate change management when the pilot scale reactor was scaled up to a production size. The poorly designed heat transfer and relief system for the larger unit failed, resulting in the over-pressurization and explosion of the pressure vessel which could not relieve the pressure quickly enough. The employees managing the emergency, unaware of the hazards, were standing near the reactor as the relief system vented and died when the reactor exploded.

The investigation revealed that the engineers did not understand the process chemistry—the potential for a runaway reaction—and improperly designed the heat transfer system, the reactor's relief system, and the tests and inspections when changing the scale of the process from the pilot plant to production runs. Hence, the training for and operation of the process did not address the hazards. In addition, they did not monitor and investigate the warning signals from earlier production runs and had not developed an effective emergency response plan. The process safety systems were ineffective due to the fundamental lack of understanding of the process chemistry and how to manage its risks. This case study is summarized in Table 5.5.

TABLE 5.5

Two Case Studies on Incidents Resulting from Poor Understanding of the Process and Equipment Design

Process Safety System Chapter	How the Process Technology Information is Used	Case Study 5-1 T2 Industries, Inc. [23]	Case Study 5-2 Deepwater Horizon [24]
5 Design safe processes	Understand and apply inherently safer process and/or equipment designs; Understand and apply safe operating limits; Design equipment and operate processes and equipment with up-to-date technologies	Did not understand runaway reaction with improper design of the heat transfer system upon scale-up	Did not use inherently safer process designs; Did not design with up-to-date technology
6 Identify and assess process hazards	Understand the hazardous materials and energies; Understand the reactivity, potential side reactions, or contaminant effects	Unaware of runaway reaction potential	Unaware of the effects of water pressure on the mitigation plan for a well blowout

(Continued)

TABLE 5.5 (*Continued*)

Two Case Studies on Incidents Resulting from Poor Understanding of the Process and Equipment Design

	Process Safety System Chapter	How the Process Technology Information is Used	*Case Study 5-1* T2 Industries, Inc. [23]	*Case Study 5-2* Deepwater Horizon [24]
7	Evaluate and manage process risks	Understand which hazards to control; Understand and evaluate risks correctly; Understand which safeguards and protection layers are needed	Did not address potential runaway reaction overpressurization scenario and design proper relief system	Did not address engineering and administrative controls
8	Operate safe processes	Have process hazards and risk information that can be shared across groups; Ability to train employees and contractors effectively (awareness and skill-based)	Did not have runaway reaction information to share with and properly train employees	Did not have the "big picture" of the combined impact of the different group's decisions
9	Maintain process integrity and reliability	Ability to design effective preventative maintenance (PM) programs	Did not have back up system or adequate maintenance to ensure a working heat transfer system	Did not test blowout design
10	Change processes safely	Ability to design for and make safe equipment and process changes; Ability to handover changes safely between groups	Did not review hazards when changes made on scale-up from pilot plant to production	Did not adequately share changes made between different groups
11	Manage incident response and investigation	Ability to schedule and design proper emergency response drills; Understand hazards and know how to respond and recover safely	Did not design safe response to the unknown runaway reaction	Did not have an effective emergency response plan
12	Monitor process safety program effectiveness	Ability to measure and determine systemic or operational gaps; Ability to address potential technological and systemic weaknesses	Did not monitor and recognize runaway reactions occurring when changes were made for increased production runs	Did not properly address failed tests before proceeding with operations

Case Study 5-2: Deepwater Horizon

The explosion, fatalities, and sinking of the Deepwater Horizon rig at the Macondo well in the Gulf of Mexico occurred in 2010 [24]. Mechanical failures, faulty engineering designs, and poor communication between the groups managing the rig and its operation set the stage for the incident. When flammable gases broke through the platform, ignited, and exploded, 11 people died. After burning for a few days, the platform sunk and oil spewed into the Gulf for almost 90 days one mile (1.6 km) deep underwater from the uncontrolled well. The investigation revealed that up-to-date engineering and administrative controls, including inherently safer process designs, were not used, and that failures in the tests were ignored. In addition, the emergency response plan did not address the technical issues of plugging the uncontrolled well at such depths. The process safety systems were ineffective due to the poor selection of the process technologies, and how their risks were addressed. This case study is summarized in Table 5.5, as well.

Case Study 5-3: Bhopal

The worst chemical industry disaster occurred in Bhopal, India, on December 3, 1984. A methyl isocyanate (MIC) leak spread beyond the facility's fence line and caused the death by poisoning more than 2,000 people and injuring more than 200,000 people [25–28]. Reports and analyses on what happened at Bhopal have emphasized the loss of the equipment integrity after its initial design and installation. The management systems for safely operating, maintaining, and changing the process and equipment failed to sustain the initial, robust design. For Bhopal in particular, the following operational conditions and maintenance design specifications were not sustained:

- the set points for high-temperature alarm indicators were changed and set too high
- the temperature indicators were not maintained, and hence, were unreliable
- the process control's refrigeration loop designed to control reaction exotherms had been removed
- the storage tank's SOLs were exceeded when the process design level maximum had been increased from 50% to 85% tank capacity

- the emergency scrubber had been taken offline and was not started up quickly enough once the release occurred
- the flare had not been repaired and hence was inoperative at the time of the release.

Poor operational discipline with the process safety systems required to manage the operating, maintenance, and equipment changes had set the stage for the disaster [29]. The robust, as built design, and its process technology information degraded over time when the operating and maintenance management disciplines failed to sustain the equipment per its process and equipment design [30]. In conclusion, if the process technology is not sustained, then even a well-designed process safety system will degrade.

5.5 Measures of Success

☑ Design safe processes based on inherently safer principles with a thorough understanding of the process hazards and risks.

☑ Design safe process and safeguard equipment using accepted engineering practices and standards combined with a thorough understanding of the process hazards and risks.

☑ Keep readily available, comprehensive, and accurate process technology/process safety information, focusing on the process and equipment design documentation, such as:

   o The process chemistry, including side reactions
   o The material and energy balances
   o The drawings (e.g., process flow diagrams, piping and instrumentation diagrams, area electrical classification drawings, etc.)
   o Lists of equipment critical to safely operate the processes (e.g., pressure relief systems, safety instrumented systems, and other protection layers).

☑ Use the process and equipment design documentation to support the other process safety system activities, including when evaluating for process risks, developing operating and maintenance procedures, making process or equipment design changes, developing emergency response plans, and training personnel across all disciplines at the facility.

References

1. Center for Chemical Process Safety. 2017 (in press). *Guidelines for Siting and Layout of Facilities*. Wiley-AIChE.
2. Center for Chemical Process Safety. 2012. *Guidelines for Engineering Design for Process Safety*, 2nd ed. Wiley-AIChE.
3. U.S. OSHA. 1992. 29 CFR 1910.119: *Process Safety Management of Highly Hazardous Chemicals*. www.osha.gov.
4. Crowl, Daniel A. and Joseph F. Louvar. 2011. *Chemical Process Safety, Fundamentals with Applications*, 3rd ed. Prentice Hall, Upper Saddle River, NJ.
5. Baybutt, Paul. 2014. The ALARP principle in process safety. *Process Safety Progress* 33:36–40.
6. UK HSE. 2016. *Policy and Guidance on Reducing Risks As Low As Reasonably Practicable in Design*. www.hse.gov.uk.
7. UK HSE. 2016. *Principles for Cost Benefit Analysis (CBA) in Support of ALARP Decisions*. www.hse.gov.uk.
8. American Society of Mechanical Engineers. 2015. *Boiler and Pressure Vessel Code, Section VIII, Pressure Vessels, 2015 Edition*. www.asme.org.
9. Center for Chemical Process Safety. 1995. *Guidelines for Process Safety Documentation*. Wiley-AIChE.
10. Hendershot, Dennis. 2010. A summary of inherently safer technology. *Process Safety Progress* 29(4):389–392.
11. Center for Chemical Process Safety. 2009. *Inherently Safer Chemical Processes: A Life Cycle Approach*, 2nd ed. Wiley-AIChE.
12. Kletz, Trevor A. and Paul Amyotte. 2010. *Process Plants: A Handbook for Inherently Safer Design*, 2nd ed. CRC Press, Taylor and Francis, Boca Raton, FL.
13. Hendershot, Dennis. 2011. Inherently safer design—not only about reducing consequences! *Process Safety Progress* 30(4):351–355.
14. Green, Don and Robert H. Perry. 2008. *Perry's Chemical Engineers' Handbook*, 8th ed. McGraw-Hill, New York.
15. American Petroleum Institute. 2014. *API Standard 521: Pressure-Relieving and Depressuring Systems*, 6th ed. (reference to ISO 23251:2006E). www.api.org.
16. Design Institute for Emergency Relief Systems (DIERS). 2016. American Institute of Chemical Engineers. www.aiche.org.
17. International Electrotechnical Commission. 2016. *IEC 61511-1:2016 RLV, Redline Version. 2016. Functional Safety—Safety Instrumented Systems for the Process Industry Sector—Part 1: Framework, Definitions, System, Hardware and Application Programming Requirements*. webstore.iec.ch.
18. Vaughen, Bruce K., Jeffry O. Mudd, and Bryan E. Pierce. 2011. Using the ISA 84/HAZOP/LOPA procedure to design a safety instrumented system for a fumed silica burner. *Process Safety Progress* 30:132–137.
19. Ziskin, Michael H. 2013. Select appropriate chemical protective clothing. *Chemical Engineering Progress* 109(9):26–33.
20. U.S. OSHA. No date. 29 CFR 1910: *Subpart I, Personal Protective Equipment*; 1910.132: *General Requirements*; 1910.133: *Respiratory Protection*; 1910.135: *Head Protection*; 1910.136: *Foot Protection*; 1910.137: *Electrical Protective Equipment*; and 1910.138: *Hand Protection*. www.osha.gov.

21. UK HSE. 2013. *Personal protective equipment (PPE) at work, A brief guide.* INDG174 (rev2). www.hse.gov.uk.
22. Center for Chemical Process Safety. 2008. *Guidelines for the Management of Change for Process Safety.* Wiley-AIChE.
23. U.S. Chemical Safety Board. 2009. *T2 Industries, Inc., Runaway Reaction.* Report No. 2008-3-I-FL. www.csb.gov.
24. BP. 2010. *Deepwater Horizon Accident Investigation Report.* www.bp.com/en/global/corporate/gulf-of-mexico-restoration/deepwater-horizon-accident-and-response.html.
25. Center for Chemical Process Safety. 2008. *Incidents That Define Process Safety.* Wiley-AIChE.
26. Kletz, Trevor A. 2009. *What Went Wrong? Case Histories of Process Plant Disasters and How They Could Have Been Avoided*, 5th ed. Butterworth-Heinemann/IChemE.
27. Institution of Chemical Engineers (IChemE). 2014. *Remembering Bhopal—30 years on.* IChemE Loss Prevention Bulletin, Issue 240.
28. Willey, Ronald J. 2014. What are your safety layers and how do they compare to the safety layers at Bhopal before the accident? *Chemical Engineering Progress* 111(12):22–27.
29. Vaughen, Bruce K. 2015. Three decades after Bhopal: What we have learned about effectively managing process safety risks. *Process Safety Progress* 34(4):345–354.
30. Bloch, Kenneth 2016. *Rethinking Bhopal: A Definitive Guide to Investigating, Preventing, and learning From Industrial Disasters*, 1st ed. Elseivier Press/IChemE.

6

Identify and Assess Process Hazards

The beginning is the most important part of the work.

Plato

Why We Need to Identify and Assess Process Hazards

Recognizing and assessing process hazards at a facility are the foundation of an effective process safety program. If process hazards are not identified and understood, how can appropriate process safety safeguards and systems be implemented and maintained to manage the risks associated with those hazards? Although this may seem obvious, there are many examples of capable people working with significant process hazards that did not adequately identify, assess, understand, and manage them. The result is often catastrophic injuries and incidents, such as the examples discussed in Chapter 1. By properly identifying and assessing process hazards, appropriate safeguards are included in the process design, and process safety systems are implemented to ensure ongoing effective risk management.

> **Identify and assess process hazards to ensure that process risks can be properly evaluated and managed.**

6.1 Introduction

Proper design and implementation of process safety systems for managing process risk rely on the thorough identification and evaluation of the process hazards and associated potentially hazardous events. Comprehensive assessment of process hazards in a facility, regardless of its size or purpose, is therefore an important starting point in the design of the desired process and in the implementation of a process safety program to help ensure safe and reliable operations. A thorough evaluation of all materials present in a process, and the process conditions under which they are used, is essential. Effective applications of the process safety systems and methodologies discussed in this book are dependent on this basic process design technology and process safety information (PSI), which are developed as part of hazard assessment. It is not possible to conduct risk management reviews or effectively plan for possible incident response, as discussed in Chapters 7 and 11, for example, if process technology and safety documentation are incomplete, hard to find, or missing.

Process hazards are often present in petrochemical processes as well as in many other industries unless they have been eliminated or minimized through proper design, including an application of the inherently safer process technology, as discussed in Chapters 5 and 7. An intrinsic hazard assessment (IHA), or equivalent, is completed to ensure that all process hazards have been properly identified, assessed, and documented to provide a detailed understanding of the basic data for managing process hazards that are present [1]. The IHA is part of the process technology and safety documentation for a process and should not be confused with process hazards and risk analysis (PHRA), which is discussed in Chapter 7. The IHA should be completed first, and the results are then used as input for the PHRA. The PHRA evaluates the types of hazardous process events that are possible based on the process hazards identified in the IHA, methodically identifies failures that can lead to these events, and ensures that appropriate protection layers are provided to adequately manage process risk. An application of inherently safer technologies to eliminate or reduce process hazards therefore can greatly reduce the need for the methodologies described in this chapter and Chapter 7.

If a comprehensive IHA or equivalent is not done or is not done well, the PHRA may miss critical information, and the result may be poor design and ineffective implementation of appropriate process safety safeguards and systems, resulting in potentially catastrophic incidents. For example, the U.S. Chemical Safety Board has concluded in several incident investigations that basic process hazard identification and assessment were among the root causes of serious incidents, as shown in Table 6.1.

The goal of an IHA is to identify the types and levels of process hazards within a specified process boundary. The primary types of process hazards discussed in this book include toxicity, flammability, and others, as shown in Table 1.2. Some examples of chemical hazards of interest in hazard assessment, based on the Globally Harmonized System of Classification and Labelling of Chemicals (GHS) [7], are shown in Table 6.2. The key parameters for determining the level of process hazard vary for each type of process hazard and are discussed in further detail in this chapter.

It is important to highlight that the detail necessary for completing a comprehensive IHA is generally not found solely in a safety data sheet (SDS), material safety data sheet (MSDS), International Chemical Safety

TABLE 6.1

Examples of CSB Incidents with Hazard Assessment Findings

CSB Incident Investigation	Year	Finding
D. D. Williamson [2]	2003	… did not have adequate hazard analysis systems to identify feed tank hazards, nor did it effectively use contractors and consultants to evaluate and respond to associated risks.
Hayes Lemmerz [3]	2003	The hazards of aluminum dust were neither identified nor addressed.
BLSR Operating [4]	2003	… did not identify the potential flammability hazard, properly class and describe the material, or inform employees and contractors of the hazard.
T2 Laboratories [5]	2007	… did not recognize the runaway reaction hazard associated with the product it was producing.
Texas Tech [6]	2010	Academic institutions should ensure that practices and procedures are in place to verify that research-specific hazards are evaluated and mitigated.

TABLE 6.2

Common GHS Chemical Hazards Related to Process Safety [Modified from 7]

Health and Environmental Hazards	Physical Hazards
• Acute toxicity • Skin corrosion/irritation • Serious eye damage/eye irritation • Respiratory or skin sensitization • Target organ systemic toxicity • Aspiration toxicity	• Explosives • Flammable gases, liquids, and solids • Oxidizing gases, liquids, and solids • Self-reactive and self-heating substances • Pyrophoric liquids and solids • Organic peroxides

Card (ICSC), or a similar information source, which are often considered a primary source of hazard information. For example, the U.S. EPA and U.S. OSHA have said that "MSDSs can provide adequate chemical hazards information, but not necessarily process hazards information [8]," and the U.S. Chemical Safety Board has also discussed MSDS limitations [9–10]. The specific material properties provided in these documents do not provide sufficient information for assessing the process hazards, where process conditions and material interactions are often very important, especially for chemical reactivity. Although multiple sources of information should always be used to ensure proper assessment of process hazards, as discussed in this chapter, a basic understanding of information on SDSs and chemical labels is desirable [7,11].

6.2 Key Concepts

6.2.1 Physical Property Data

Basic physical property data should be available and compiled, as appropriate, to support process design, to determine operating conditions, and to conduct IHA and PHRA. This includes raw materials, intermediates, products, waste streams, and emissions. Not all data are needed for every material, but common properties that may be useful are provided in Table 6.3. In many cases, physical property data can be obtained from:

- Online databases [12–15]
- Physical property and process simulation software [16]
- Vendors and suppliers
- Books and journals
- Experimental measurement.

Property data for mixtures can be difficult to find but is often evaluated using group contribution calculations, software, or laboratory tests. The Webwiser tool [14] provides an excellent summary of physical property and emergency response information.

6.2.2 Completing the Intrinsic Hazards Assessment

All the chemicals, materials, chemistry (planned and inadvertent), and normal and abnormal operating conditions (such as temperature, pressure, or concentration) within a process must be evaluated to ensure adequate documentation of the hazards assessment. The basic steps for systematically completing an IHA are shown in Figure 6.1. They include the following [1]:

TABLE 6.3

Some Useful Physical Properties

Physical Properties	Temperature Dependent Properties
• Physical state/appearance	• Density
• Molecular weight	• Vapor pressure
• Freezing, melting, boiling point	• Heat capacity
• Particle size	• Heat of vaporization
• Critical temperature, pressure	• Viscosity
• Electrical conductivity	• Surface tension
• Dielectric constant	• Thermal conductivity
• Corrosivity	• Solubility

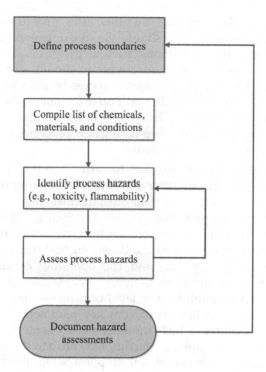

FIGURE 6.1
Intrinsic hazard assessment (IHA) [Modified from 1].

1. **Define the process boundaries:** By setting boundaries, the scope of the hazard assessment can be managed. For large sites, it may make sense to define the boundary as a subsection of the overall process or facility; for smaller sites, it may be the entire facility. Factors include the scope of operations, the number and type of materials, processing conditions, and location on the site. A warehouse or tank farm that contains a variety of materials stored as liquids under ambient conditions will have significantly different levels of process hazards than a manufacturing building that has some of the same materials in pipes and vessels in mixtures at high temperature and pressure. The reactor section of a process may be very different than the separation or finishing sections. An R&D facility or university similarly will be very different than a large-scale chemical processing plant, with hazard assessments likely completed for individual laboratories or projects.

2. **Compile a list of boundary constituents and conditions:** Within each process boundary, a complete list of all the raw materials, intermediates, products, utilities (such as natural gas or nitrogen), and materials of construction (such as used for packing, piping, gaskets,

seals, or hoses) is compiled. Amounts, rates, phases (vapor/liquid/solid), compositions, etc., should be documented. Normal and abnormal operating conditions, such as pressure and temperature and other types of process information, should also be listed, as appropriate. Connections between process boundaries should also be considered to evaluate the potential for materials or operating conditions from one area either purposefully or inadvertently affecting another area, possibly introducing new process hazards or interactions that must also be assessed.

3. **Identify and assess the process hazards:** The types and levels of process hazards vary in different process boundaries depending on the types and quantities of materials, intrinsic physical properties, and the operating conditions compiled in Step 2. All materials and interactions should be evaluated for toxicity, flammability, reactivity, and other process hazards, based on both pure and mixture properties, as appropriate, and the operating conditions. Some materials may present more than one process hazard, such as both toxicity and flammability. The level of process hazard (as defined by the facility) is determined from key parameters, as discussed later in this chapter, which should be documented as part of the assessment. For example, the facility may wish to document a list of all flammable materials with a flash point below 100°F (37.8°C) as a high hazard.

4. **Document the hazard assessment:** The results of the hazard assessment should be documented in detail and shared with affected personnel, as appropriate. Documentation should include:

   – Process boundary limits
   – Material lists and inventories
   – Process conditions
   – Types and levels of process hazards
   – Potential consequences (such as human health effects).

   Hazard assessment documentation should be stored so that it can be accessed easily and should be communicated to a variety of personnel, as needed, including R&D, design, engineering, operations, and maintenance.

If this information is not compiled formally as an IHA, it is important that the equivalent process safety information be provided for use in PHRAs and other process safety system activities. Identification of hazards in university or chemical research laboratories should be documented using a similar approach, but additional considerations may also be important [17].

6.3 Toxicity Hazards

Toxicity is a measure of the extent that a hazardous material can harm human health [18]. When assessing toxicity hazards, both the intrinsic and potential health effects should be considered:

- intrinsic health effects result from the impact of exposure to the material
- potential health effects result from the potential reaction or phase change of the material that can occur, for example, when an atmospheric release occurs or the material is heated.

For example, chlorosulfonic acid (HSO_3Cl) readily decomposes in the presence of water to release hydrochloric acid (HCl) and sulfuric acid (H_2SO_4). An inadvertent release into the atmosphere could lead to reaction with water due to humidity, rain, or other contact with water. A hazard assessment should therefore include the possibility of exposure to all three materials.

The IHA of toxicity hazards requires that the following information be evaluated:

- **Health impacts** – document the entry pathways of how the toxic substance may enter or impact the body and the resulting health effects that can occur, such as exposure to and inhalation of ammonia with possible respiratory irritation, lung damage, or fatality
- **Toxic levels of concern** (LOCs) – provide the maximum levels of exposure that can lead to various toxicity effects for a given time of exposure, or duration, such as the level where no significant health consequence usually results for up to a one hour exposure.

Note that many materials that are not normally considered toxic, such as flammable materials like acetone or toluene, may also have toxicity effects that need to be assessed. Depending on regulatory or facility requirements, the impact of toxic materials on animal health and the environment should also be assessed, but these cases are outside the scope of this book.

Exposure to toxic chemicals was responsible for the many deaths and injuries that occurred in Bhopal, India, in 1984—often considered the worst incident in the history of the chemical industry, as discussed in Chapters 1 and 11. Due to reaction with water, a large amount of methyl isocyanate (MIC) was accidentally released from a storage tank at night, resulting in thousands of deaths and injuries in the surrounding community [19]. MIC exposure at low ppm levels can cause severe eye, skin, and lung effects, potentially resulting in death:

> The unspeakable was happening. Driven by the wind, the wave of gas was catching up with the flood of humanity trying to escape. Out of their minds with terror, people with shredded clothes and torn veils ran in all

directions, trying to find a pocket of breathable air. Some, whose lungs were bursting, rolled on the ground in awful convulsions. Everywhere the dead with their greenish skins lay side by side with the dying, still wracked with spasms and with yellowish fluid coming out of their mouths [20].

Not all releases of toxic materials lead to fatalities or irreversible health effects, of course, but many industrial and academic toxicity incidents have occurred, resulting in significant numbers of injuries or fatalities. These incidents reinforce the need to properly assess toxicity hazards to help ensure that appropriate safeguards to prevent or mitigate exposure are provided.

6.3.1 Assessing Toxicity Hazards

Assessment of toxicity can include both short- and long-term health effects [18]:

- **Acute toxicity** – harmful effects from a single or short-term exposure, such as an unexpected release or spill in a manufacturing facility or laboratory
- **Chronic toxicity** – harmful effects that occur from long-term continuous or repeated exposure, such as from smoking or persistent low levels in a work environment.

Hazard assessment for process safety usually focuses primarily on acute toxicity exposures that may result from sudden, unintended releases or exposures resulting from failure of process safeguards. The exposure may be one time only for a short duration, but could lead to fatality or severe, irreversible health effects, such as at Bhopal. Another example is an unexpected release of and exposure to chlorine from a cylinder that can lead to pulmonary and respiratory health effects, possibly ultimately resulting in fatality. In contrast, chronic exposures to a variety of chemicals may occur at low concentrations in the workplace for long durations, which can lead to adverse health effects, such as cancer or other diseases. For example, repeated inhalation of silica dust at low levels over time can lead to silicosis, resulting in impaired lung function. Chronic toxicity effects, since they do not result from a single exposure, such as a process safety incident, are normally evaluated and managed by occupational health or industrial hygiene professionals.

The health effects of toxic exposures depend both on the entry pathway into the body and the resulting impact on target systems once in the body. Entry can occur due to:

- inhalation
- skin or eye contact
- ingestion.

TABLE 6.4

GHS Acute Toxicity Effects on Humans [7]

Acute Toxicity	Cat 1	Cat 2	Cat 3	Cat 4
Oral (mg/kg)	≤5	>5 ≤50	>50 ≤300	>300 ≤2000
Dermal (mg/kg)	≤50	>50 ≤200	>200 ≤1000	>1000 ≤2000
Gases (ppm)	≤100	>100 ≤500	>500 ≤2500	>2500 ≤5000
Vapors (mg/l)	≤0.5	>0.5 ≤2.0	>2.0 ≤10	>10 ≤20
Dusts (mg/l)	≤0.05	>0.05 ≤0.5	>0.5 ≤1.0	>1.0 ≤5

Cat 5:

- Anticipated oral LD50 between 2000 and 5000 mg/kg
- Indication of significant effect in humans[a]
- Any mortality at class 4[a]
- Significant clinical signs at class 4[a]
- Indications from other studies[a]

[a] If assignment to more hazardous class is not warranted.

For example, inhalation of acetone, which is often considered a relatively non-toxic, flammable solvent, can irritate eyes, nose, and throat, as well as impact the respiratory and central nervous systems at levels well below its flammability limits (see Section 6.4). The IHA should document the possible health effects of exposure to the material, including common entry pathways and the resulting target systems and impacts.

A key consideration for a chemical release is whether the potential exposure may impact the ability of people to evacuate during the emergency, due to, for example, impacts on the respiratory, circulatory, or central nervous systems. Availability of appropriate personal protective equipment to protect against releases and for emergency response must also be evaluated and provided, as part of the process safety program. Table 6.4 provides a summary of acute toxicity categories (category 1–5, with 1 equaling the highest hazard) for oral, dermal, and inhalation effects on humans, based on the Globally Harmonized System of Classification and Labelling of Chemicals (GHS) [7]. Additional GHS tables that assign toxicity hazard categories are available for skin corrosion, eye effects, and other toxicity concerns [7].

Chemical hazard labels provide a simple and readily available method for quickly documenting or assessing the level of hazard associated with different substances. NFPA 704 [21] ranks health, flammability, and instability in a simple colored diamond with ratings from 0 to 4 (highest hazard). Some examples of NFPA health rating include H-1 for acetone, H-2 for ethanol, H-3

TABLE 6.5

Chemical Label Ratings [Modified from 7 and 21]

NFPA/HMIS Ratings	GHS Ratings
0. Minimal hazard	1. Severe hazard (highest)
1. Slight hazard	2. Serious hazard
2. Moderate hazard	3. Moderate hazard
3. Serious hazard	4. Slight hazard
4. Severe hazard (highest)	5. Minimal hazard

for ammonia, and H-4 for chlorine. The HMIS system for chemical labels, developed by the American Coatings Association, is similar to NFPA 704 and provides health, flammability, physical hazard, and personal protection information for different substances [22]. HMIS ratings also range from 0 to 4 (highest hazard). The GHS classification [7] provides category ratings from 1 (highest hazard) to 5, which importantly *reverses* the severity classifications provided by the NFPA and HMIS ratings. This is shown in Table 6.5. Specific definitions used to define the rating levels for the different hazard types can typically be obtained from SDSs and other sources [7,14,21,22]. A wide range of additional chemical hazard labels for specific hazard types are available [7].

Exposure limits, such as permissible exposure limits (PELs), are available [15] for many chemicals and are used for both occupational and process safety as well as emergency response. In addition, Acute Exposure Guideline Levels (AEGLs) [23] or Emergency Response Planning Guidelines (ERPGs) [24] provide toxic levels of concern (LOCs), defined as threshold concentrations (e.g., ppm), for exposure to chemicals for defined durations (e.g., 60 minutes), which are useful for modeling and responding to short-term acute exposure releases. The LOCs (AEGL/ERPG) define the maximum concentrations for exposure to hazardous chemicals that lead to different injury levels for general populations:

- LOC-1 – minor irritation or discomfort that does not impair the ability to evacuate during the release and is not permanent
- LOC-2 – irreversible or long-lasting health effects that may also impair the ability to evacuate during the release
- LOC-3 – life-threatening health effects that may result in fatality.

ERPGs are provided for up to a one hour exposure, and AEGLs are provided for exposures from 10 minutes to eight hours, although in this book, one hour exposures only will be used. Table 6.6 provides a list of PELs and AEGLs for various materials. For example, exposure to anhydrous ammonia for up to one hour at a maximum concentration of 30 ppm (AEGL-1) would result only in reversible, nonpermanent health effects such as respiratory or eye irritation for most people. Similarly, exposure to 160 ppm (AEGL-2) for up to one hour

TABLE 6.6

Toxicity Levels of Concern [Modified from 14, 15, and 23]

Chemical	NFPA Health Rating	U.S. OSHA PEL (ppm)	AEGL-1 60 minutes (ppm)	AEGL-2 60 minutes (ppm)	AEGL-3 60 minutes (ppm)
Acetone	H-1	1000	200	3200	5700
Acrylonitrile	H-4	2	NA	1.7	28
Ammonia	H-3	50	30	160	1100
Chlorine	H-4	1	0.5	2	20
Ethyl mercaptan	H-2	10	1	120	360
Hydrogen sulfide	H-4	20	0.51	27	50
Methanol	H-1	200	530	2100	7200
Sulfur dioxide	H-3	5	0.2	0.75	30
Styrene	H-2	100	20	130	1100
Toluene	H-2	200	67	560	3700
Vinyl Chloride	H-2	1	250	1200	4800

could result in irreversible health effects, such as skin or eye burns and respiratory damage. Exposure to 1,100 ppm (AEGL-3) for up to one hour could lead to death. Programs are available for modeling the distances to these LOCs for differing release and meteorological conditions [15], which can be used for PHRAs and emergency response planning.

Use of ERPGs when available is generally preferred since the American Industrial Hygiene Association (AIHA) has an active program for development of new ERPGs. Additional parameters may be useful, especially when AEGLs and ERPGs are not available, including:

- Immediately dangerous to life and health (IDLH) and short term exposure limit (STEL) [15]
- Temporary emergency exposure limits (TEELs) [25]
- Dangerous toxic load (DTL), specified level of toxicity (SLOT), or significant likelihood of death (SLOD) [26].

These and other parameters should be used with caution with appropriate review by qualified toxicologists, because they are less commonly used and their basis may not be as complete as AEGLs and ERPGs. For mixtures, a weighted average of the LOC-x values can be used:

$$\frac{1}{(\text{LOC-x})_{\text{mix}}} = \sum_{i=1}^{n} \frac{y_i}{(\text{LOC-x})_i}$$

where y_i is the mole fraction of chemical i.

The safety hazard index (SHI) is also sometimes used as a measure of both substance volatility and toxicity [1]. The SHI is calculated as follows:

$$SHI = \frac{\left(\text{Vapor Pressure }(\text{atm})\times 10^{6}\right)}{\left(\text{LOC-3}(\text{ppm})\right)}$$

SHI can be useful in identifying substances that have greater potential for impacting larger areas when released due to high volatility and high toxicity. The SHI for ammonia, for example, is 13,153, and for chlorine is 385,197, partly reflecting the much lower AEGL-3 of chlorine (20 versus 1,100 for ammonia).

The warning properties of substances should also be considered. One reason for the high loss of life at Bhopal was the lack of warning properties at low enough levels that could have helped people evacuate from the area. The most common warning property simply is that many releases will be visible, warning people of the presence and location of a release. In other cases, a release may not be visible, but people will be able to smell the substance. The odor threshold of many materials is well below the AEGL-2 limit, including for example, ammonia, which has an odor threshold of 5–53 ppm and an AEGL-2 of 160 ppm. The odor, therefore, is a warning property that makes people aware of a release and helps them to evacuate safely. In contrast, carbon monoxide is an odorless gas with an AEGL-2 of 83 ppm. For this reason, detectors are often required in houses to warn of carbon monoxide due to faulty heating systems in the winter. The odor threshold for the methyl isocyanate released at Bhopal is 2.1 ppm, and the AEGL-2/3 are 0.067/0.20 ppm. Useful summaries of toxicity limits, health effects, warning properties, and other information are available [13–15], and technical summary documents are provided for both AEGLs [23] and ERPGs [24].

6.3.2 Case Study

On August 4, 2002, a large release of chlorine, as shown in Figure 1.2a, occurred at the DPC Enterprises near Festus, Missouri [27]. A flexible transfer hose failed during tank car unloading, releasing approximately 48,000 pounds of chlorine and resulting in emergency evacuation of the plant and surrounding community. Three workers at the company had minor exposures to chlorine during cleanup operations, and 63 people from the community requested medical evaluation for respiratory issues. The material of construction used in the flexible hose braiding was found to be stainless steel rather than Hastelloy, due to shipping and quality assurance failures. Chlorine monitors, alarms, and automatic shutoff valves were provided to stop the flow of chlorine in case of a release, but several of the automatic valves failed to completely close, allowing the release to continue. Chlorine has an AEGL-2 of 2.0 ppm and an AEGL-3 of 20 ppm. Exposure to chlorine causes burning of the eyes, nose, and mouth as well as potentially severe respiratory effects. Chlorine would be documented as a high-level toxicity hazard.

A more devastating release of chlorine from a tank car occurred due to a train derailment near Graniteville, SC in 2006 [28]. Due to an improperly lined switch, a train derailed and collided with another train. Three derailed cars that contained chlorine ruptured, releasing approximately 60 tons of chlorine. The train engineer and eight other people were killed. A total of 5,400 people from the surrounding area were evacuated, and more than 500 people went to the hospital. Total damages were estimated at $6.9 million.

Some additional incidents involving release of high-level toxicity materials include:

- Seveso, Italy, 1976 – The rupture disk on a reactor burst, releasing a toxic cloud containing dioxin and several other hazardous materials. The release impacted a large, densely populated area, which was quickly evacuated. Dioxin (TEEL-2 = 0.0023 ppm) causes eye and skin irritation, choracne, and potential reproductive impacts. About 200 cases of chloracne were treated, but thousands of animals died as a result of the release [19]. This incident led to the 1982 Seveso Directive in Europe to help prevent similar major accidents.

- Jilin, China, 2005 – An explosion occurred in a benzene chemical plant due to a blockage in a nitration tower, followed by several additional explosions. Approximately 100 tons of benzene (AEGL-2 = 800 ppm), nitrobenzene (TEEL-2 = 20 ppm), and other toxic materials were released into the nearby river, causing a 50-mile long toxic area in the river. Since benzene is a suspected occupational carcinogen, use of river water for human consumption was temporarily discontinued, potentially impacting millions of people [29].

- Gumi, South Korea, 2012 – A sudden release of hydrofluoric acid from a tank car killed five workers and led to an 8-ton release that caused significant damage to area crops and livestock [30]. Hydrogen fluoride has an AEGL-2 of 24 ppm and AEGL-3 of 44 ppm. Exposure causes eye and skin burns and severe respiratory effects.

- LaPorte, Texas, 2014 – An accidental release of methyl mercaptan while troubleshooting process problems, resulted in four fatalities [31]. Methyl mercaptan is a colorless gas with a strong, objectionable odor that impacts the respiratory and central nervous systems. The AEGL-2 for a 60-minute exposure is 23 ppm.

6.4 Flammability Hazards

Assessment of flammability hazards is based on the relative ability of a substance or material to burn with a flame under specific conditions, potentially leading to a variety of fire or explosion events. Flammability results from fire, which is the rapid oxidation of combustible materials, releasing light, heat, and combustion products [32]. Flammability hazards are based on the fire triangle, as shown in Figure 6.2. For a fire to occur, the following must be present in appropriate concentrations:

- **fuel**, in a flammable range
- **ignition source**, of sufficient strength
- **oxidant**, most often simply oxygen in the air.

If one element of the fire triangle is missing, then a flame, and therefore a flammability event, cannot occur.

In assessing flammability hazards, the primary focus is on the properties of the fuel, since the availability of an ignition source and air is often assumed and will be evaluated separately in the PHRA (see Chapter 7). The IHA of flammability hazards requires that the following information be evaluated:

- **flammability characteristics**, related to potential fuels, including ignition and oxygen requirements, per the fire triangle
- **injury and damage thresholds**, which can cause burn injuries and explosion damage.

Assessment of fire hazards is discussed in Section 6.4.1, and assessment of explosion hazards is discussed in Section 6.4.2. In addition, toxicity of both the

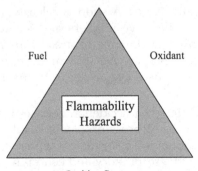

FIGURE 6.2
Conditions necessary for a fire (fire triangle).

fuel and potential combustion products must be assessed for toxicity hazards, as discussed in the previous section.

A major flammability incident occurred on the Piper Alpha Oil Platform in the North Sea in 1988 [19]. A large flammable hydrocarbon release occurred from the relief valve on a standby condensate pump, which was activated when the main pump stopped. The hydrocarbon vapors ignited, blowing down a firewall, which allowed the fire to spread rapidly, leading to additional explosions. The oil rig was eventually totally destroyed at a loss of over $1 billion. A total of 167 crew members were killed in the inferno that encompassed the oil rig:

> Flames were igniting his hair and burning layers of skin off his face, and every time he tried to scream, it was like swallowing fire... And then there were the bodies. Some men had passed out after 15–20 minutes of breathing in smoke, and parts of the staircase were blanketed with their bodies... The din of the escaping gas, roaring all around, and the sharp crackle of fire was deafening [33].

The potential for significant injury, property loss, and environmental harm is always present when flammable materials are present or inadvertently released, as evidenced by the many industrial incidents that have occurred. Proper assessment of flammability hazards helps ensure that appropriate safeguards to prevent or mitigate fires and explosions are provided.

6.4.1 Assessing Flammability Hazards

Several types of fire events can occur, including [34,35]:

- **flash fire** – rapid combustion of a flammable gas or vapor in air in which the flame propagates slowly enough that no significant overpressure is generated.
- **fireball** – similar to a flash fire, but in a fuel-air cloud that is typically of longer duration than a flash fire.
- **pool fire** – liquid pool where the vapors above the liquid burn, often for an extended time.
- **jet fire** – typical during a high-pressure release of a flammable material that forms a jet or vapor plume that burns.

The possible events must be evaluated in the PHRA, as discussed in Chapter 7, for the specific materials and conditions involved to ensure that appropriate safeguards are provided. The flammability of a liquid stored at ambient conditions, for example, will be very different than when the liquid is processed at a higher temperature.

Chemical hazard labels based on NFPA 704 [21] were discussed earlier for providing a simple and readily available method for quickly assessing the level of hazard associated with different substances. The flammability ratings

TABLE 6.7

Flammability Levels of Concern [Modified from 14]

Chemical	NFPA Flammability Rating	Flash Point (closed cup)	LFL (% vol)	UFL (% vol)	Autoignition Temperature
Acetone	F-3	−20°C	2.5%	12.8%	651°C
Acrylonitrile	F-3	−1°C	3%	17%	481°C
Ammonia	F-1	NA	16%	25%	651°C
Chlorine	F-0	NA	NA	NA	NA
Ethyl mercaptan	F-4	−45°C	2.8%	18%	299°C
Hydrogen sulfide	F-4	NA	4%	44%	260°C
Methanol	F-3	12°C	6%	36%	464°C
Sulfur dioxide	F-0	NA	NA	NA	NA
Styrene	F-3	34°C	0.9%	6.8%	490°C
Toluene	F-3	4°C	1.1%	7.1%	480°C
Vinyl Chloride	F-4	−78°C[a]	3.6%	33%	472°C

[a] Open cup.

of several materials are shown in Table 6.7, with higher ratings indicating a higher level of flammability hazard. For example, the flammability rating for ammonia is F-1, for aniline is F-2, for acetone is F-3, and for ethane is F-4. As previously discussed, these ratings are typically available on SDSs and from other sources [14]. GHS designations, once again reversed from NFPA, are available [7].

Key flammability parameters include [1,32,34–37]:

- **Flash point** (F_p) – the minimum temperature at which a flammable substance gives off sufficient vapor to form an ignitable mixture with air near the surface of the liquid or within the vessel used. Just above this temperature is the less-commonly used fire point, where the flame can be sustained, once ignited. Flash point is tested in both closed-cup and open-cup devices, although closed-cup tests are recommended since they usually provide a lower value than open-cup tests and therefore generally provide a more conservative value [37]. Numerous methods are also available for estimating flash point that can be convenient for determining the value for mixtures. Liquids below a flash point of 100°F (37.8°C) are considered Class I flammable liquids, liquids with flash points between 100°F and 140°F (37.8–60°C) are considered Class II combustible liquids, and liquids with flash points above 140°F (60°C) are considered Class III combustible liquids [32], although the category designations for GHS [7] are slightly different.

- **Lower and upper flammability limits** (LFL, UFL) – the minimum and maximum concentration of vapor or gas in air below or above

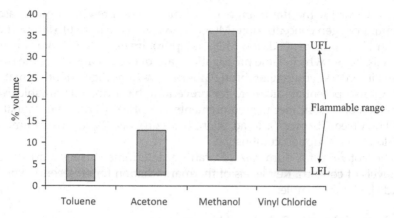

FIGURE 6.3
Flammability range.

which propagation of flame does not occur in the presence of an ignition source. The LFL and UFL define the flammability range, where the ignition of the vapor in air is possible, as shown in Figure 6.3. If a large flammable release occurs, it is generally assumed that at least part of the cloud will be in the flammable range allowing ignition if (1) an ignition source is present and (2) the temperature is higher than the flash point. The flammability range is usually reported for normal ambient conditions but will vary as temperature and pressure are changed.

Values of the flash point and flammability limits for several materials are provided in Table 6.7.

Key ignition parameters include [32,37]:

- **Minimum ignition energy** (MIE) – the minimum spark energy (mJ) required to ignite a flammable atmosphere. MIE depends on the concentration of the flammable material in air and is usually the lowest near the stoichiometric concentration. MIE values of hydrocarbons are quite low around 0.2 mJ, which is lower than static electricity that can be generated by walking on a carpet under certain conditions. Conductive and semiconductive liquids and other materials can also accumulate sufficient electrostatic charge to permit ignition.

- **Autoignition temperature** (AIT) – the lowest temperature at which vapors ignite spontaneously from the heat of the environment. AIT is typically well above ambient temperature for many materials, but is an important consideration when high processing temperatures are possible or, for example, when materials are being heated in drying ovens.

Oxygen in air for the fire triangle is assumed to be present in most cases. A minimum oxygen concentration (MOC), however, is required to support combustion [32,37]. In general, the MOC is approximately 8–10 % vol for many materials, below which flame propagation cannot occur. Dilution of the oxygen content in process equipment through inerting with nitrogen or other materials is therefore a potential safeguard for preventing hazardous flammable events. Evaluation of specific inerting requirements and practices must be evaluated in the PHRA (see Chapter 7). In addition, oxidizing agents, such as ammonium perchlorate, can support combustion.

The potential for burn hazards can be determined using thermal radiation levels of concern. Key levels of thermal radiation for 60 second exposures include [38], for example:

- **Potential pain**—2 kW/m² (2 kj/s-m²)
- **Potential second degree burn**—5 kW/m² (5 kj/s-m²)
- **Potential fatality**—10 kW/m² (10 kj/s-m²).

Similar to toxicity, where the dose is dependent on the toxicity level of concern and the time of exposure, serious burns can occur for both shorter and longer exposures at different thermal radiation levels. The ability of people to evacuate from near a fire to limit the potential duration of exposure is therefore an important consideration in evaluating possible harmful effects.

6.4.2 Assessing Explosion Hazards

A detailed evaluation of explosion hazards is beyond the scope of this book, but explosion hazards must be identified and assessed as part of the IHA. Explosivity describes the degree to which a combustible material in air is explosive. Explosions involve the acceleration of flame speeds, leading to the rapid evolution of a pressure wave that has the potential to do damage and injury, including (1) thermal radiation hazards, discussed in the previous section, and (2) debris effects, where equipment fragments can be propelled over great distances. The main types of explosions with respect to flammability hazards include [32,34,35]:

- **Deflagration** – energetic combustion that propagates through a flammable vapor or material at subsonic flame speeds, driven by turbulent mixing and rapid transfer of heat, creating a damaging pressure impulse
- **Detonation** – highly energetic combustion that propagates at supersonic flame speeds due to a strong shock wave.

Deflagrations are the most common explosion event involving flammable materials, although certain materials may transition to detonations in the right conditions. The transition from flash fire to deflagration to possible detonation is dependent on material properties, but also these additional factors that increase turbulent flow and flame speed:

- **Confinement** – such as spaces surrounded by walls, internal flow in pipes, or other process equipment or barriers that restrict flow in one or more directions
- **Congestion** – such as densely-packed process equipment and piping that both restricts flow and increases turbulence.

A rule of thumb for deflagrations is that the maximum pressure generated can be 8–10x the initial pressure. For detonations, the maximum pressure can be 30–50x the initial pressure [37]. Typically, if flammability hazards have been identified as part of the IHA, then explosion hazards may also be possible.

For an explosion to occur, the following conditions must be present, based on the fire triangle:

- the fuel concentration has to be within flammable range
- the oxidant concentration has to be sufficient to support combustion
- an ignition source has to be present, and
- congestion and confinement are typically required, depending on the specific fuel characteristics.

The maximum deflagration pressure and the rate of pressure rise are important parameters that can be measured for a variety of fuels. Once a fuel ignites, the rate of flame acceleration determines the maximum overpressure generated in a confined and congested environment. The fundamental or laminar burning velocity is a measure of the rate of flame propagation that can occur and indicates how reactive a material might be in terms of generating high deflagration pressures and increased rates of pressure rise [34].

Major types of explosion events include [34,35]:

- **Vapor cloud explosion** (VCE) – large flammable releases that occur above the flash point of materials in the cloud, which when ignited, result in damaging deflagrations depending on the size and location of the cloud, reactivity of the fuel, and degree of confinement and congestion.
- **Boiling liquid expanding vapor explosion** (BLEVE) – rapid release and vaporization of volatile, flammable materials usually due to external heating of a closed container (such as a drum or vessel) leading to catastrophic container rupture and vapor ignition.
- **Vessel burst** – catastrophic failure of a vessel caused by a rapid increase in pressure, possibly resulting from uncontrolled heating, chemical reaction, failure of pressure instrumentation and devices, or internal deflagration within the equipment.

The damage that results from explosions can be estimated from the resulting magnitude of overpressure that occurs. Key levels include [34]:

- Damage to glass windows and houses: 0.15–1 psi (1.0–6.9 kPa)
- Human eardrum damage: 2.5–12.2 psi (17.2–84.1 kPa)
- Significant building damage depending on construction: 2–10 psi (13.81–68.9 kPa)
- Fatality from direct blast effects: 14.5–30 psi (100.0–206.8 kPa).

Debris resulting from the explosion can also cause injuries and damaging effects.

6.4.3 Case Study

On January 13, 2003, a vapor cloud deflagration and fire occurred at BLSR operating near Houston, Texas [4]. Two vacuum trucks were unloading basic sediment and water waste that had been collected from natural gas well sites. The waste was usually separated from flammable condensate prior to collection and shipment and was considered a nonflammable liquid with a flash point higher than 141°F (60°C). Accidental collection of flammable condensate as part of this process, however, can lead to the waste stream potentially containing higher than expected levels of flammable materials. During unloading of the vacuum trucks, hydrocarbon vapors were released and ignited, leading to a vapor cloud deflagration that killed one worker, seriously burned another worker, completely destroyed the two vacuum trucks, and damaged the facility. The U.S. Chemical Safety Board investigation concluded that the potential flammability hazard was not recognized, and therefore systems were not in place to ensure safe handling of the waste stream.

Some additional major incidents involving release of flammable materials include:

- Flixborough, UK, 1974 – A massive release of cyclohexane at high temperature and pressure formed a large vapor cloud that exploded killing 28 employees and injuring 89 people [9].
- Pasadena, TX, 1979 – A large flammable release from a polyethylene reactor led to a large vapor cloud that ignited killing 23 workers and injuring over 130 people [9].
- Texas City, TX, 2005 – A hydrocarbon vapor cloud explosion resulted in 15 fatalities, 180 injuries, and over an estimated $3 billion in damages [39].
- Hemel Hempsted, UK – The Buncefield fire and explosion occurred at an oil-products storage terminal resulting from overflow of petrol from a large storage tank, which ultimately engulfed more than 20 storage tanks injuring 43 people [40].

- Paraguaná, Venezuela, 2012 – A large flammable vapor cloud exploded, resulting in 47 fatalities and 135 injuries [41].
- Reynosa, Mexico, 2012 – A flammable vapor cloud ignited, resulting in 30 fatalities and over 45 injuries [42].
- West, TX, 2013 – A fire and explosion occurred at a fertilizer blending and distribution facility, where a detonation of ammonium nitrate killed 15 people and injured 260 [43].

6.5 Combustible Dust Hazards

Small particle size powders and fine dusts can present an explosion and fire hazard if dispersed in air in sufficient concentration [44]. Most natural and synthetic organic materials, such as sugar, corn, and polyethylene, as well as some metals and nonmetallic inorganic materials are combustible. Combustible dusts may occur due to:

- Manufactured products – such as powdered sugar, corn starch, or polymeric materials, including fines that are part of more coarse powder materials
- Processing wastes – such as fine dusts created during handling, processing, or finishing operations, such as sanding or grinding.

Very small amounts of combustible dusts that are allowed to accumulate, due to poor housekeeping or other causes, can be hazardous. The U.S. Chemical Safety Board (CSB), for example, determined that large dust accumulations under 0.25 inches (6.4 mm) deep resulted in a catastrophic explosion at the West Pharmaceutical Plant in 2003 that killed six employees [45]. A CSB report in 2006 identified 281 dust explosions from 1980 to 2005 that resulted in 119 fatalities, 718 injuries, and major property damage, which in several incidents exceeded $100 million [9]. Industries involved included food, lumber, chemical, metal, plastics, and others.

The fire triangle, as discussed in the previous section, is relevant for dust explosions and fires as well:

- **fuel**, small-particle size combustible dusts, dispersed in air in an explosible range
- **ignition source**, of sufficient strength
- **oxidant**, most often simply oxygen in the air.

If one element of the safety triangle is missing, then a dust explosion cannot occur. As noted earlier, dusts must be dispersed or suspended in air and must

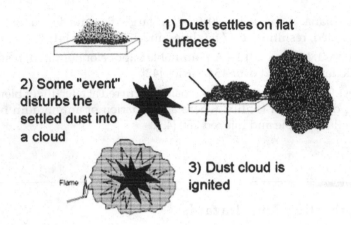

1) Dust settles on flat surfaces

2) Some "event" disturbs the settled dust into a cloud

3) Dust cloud is ignited

Flame

FIGURE 6.4
Secondary dust explosion [10].

be in an explosible concentration range, similar to the flammable range for flammable vapors, for combustion to occur. If dust has accumulated on process equipment, for example, it must first be dispersed in air, possibly due to poor clean-up practices or a separate, primary explosion as shown in Figure 6.4. The particle size must also be relatively small as discussed in the next section. To understand this requirement, compare the ease of lighting sawdust (30–600 µm particle size) versus wood kindling or larger logs in a fireplace.

Once dispersed in air, at least part of the dust cloud will likely be in the explosible concentration range, and an explosion can occur if an ignition source is present. As discussed in the previous section, confinement and congestion are usually required for an explosion. As a result, the dust explosion penta-gon [9], as shown in Figure 6.5, is sometimes used to illustrate the require-ments for dust explosions. The primary focus for preventing dust explosions is

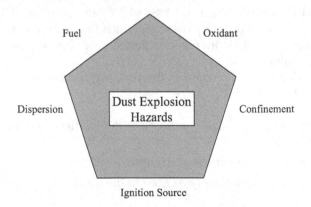

FIGURE 6.5
Dust explosion pentagon [10].

minimizing dust accumulation in the workplace via equipment design, such as vessel integrity, ventilation, and dust collection systems, and good housekeeping practices.

In assessing dust explosion hazards, the primary focus is on the properties of the combustible dust. The availability of an ignition source and air is often assumed and will be evaluated separately in the PHRA (see Chapter 7). In addition, the PHRA will also evaluate the possibility of dust dispersion and confinement, which are also required for an explosion. The IHA of combustible dust hazards requires that the following information be evaluated:

- **dust properties**, related to potential fuels, including ignition and oxygen requirements
- **damage thresholds**, which can cause burn injuries and explosion damage.

A major dust explosion incident occurred at the Imperial Sugar Company in Port Wentworth, GA in 2008 [46]. A series of dust explosions due to significant accumulation of combustible sugar dusts resulted in massive destruction at the facility. A primary explosion inside an enclosed steel conveyor belt dispersed the accumulated sugar dust into the air, triggering additional severe secondary sugar dust explosions. A total of fourteen workers were killed and 36 were injured:

> She heard a faraway boom and another louder one... Then the room blew up. Flames leaped at her. The blast tossed her head over heels. She landed more than a dozen feet away. She crawled through rubble. Debris pinned another worker to the floor. Some of the floor was gone. She saw exposed pipes and wiring. When she reached what she thought was a stairwell, it was gone too [46].

The potential for significant injury, property loss, and environmental harm is always present when combustible powders and dusts are allowed to accumulate, as evidenced by the many industrial incidents that have occurred. Proper assessment of combustible dust hazards helps ensure that appropriate safeguards to prevent dust explosions are provided.

6.5.1 Assessing Combustible Dust Hazards

Not all powders and dusts are combustible. Inorganic materials, including stable oxides, such as calcium carbonate, silicon dioxide, and titanium dioxide, salt, baking soda, and many other dusts are not combustible and do not pose an explosion hazard. Organic powders, metals, plastics, and other materials, however, are potentially combustible and must be properly identified and assessed as potential process hazards. The National Fire Protection

Association (NFPA) has defined combustible dust as "a combustible particulate solid that presents a fire or deflagration hazard when suspended in air or some other oxidizing medium over a range of concentrations, regardless of particle size or shape [47–51]." Requirements for a Dust Hazards Assessment (DHA) have recently been documented in NFPA 652 [50].

Key combustible dust parameters include [44,47–50]:

- **Particle size** – Smaller particle powders and dusts have greater total surface area that can support rapid combustion. NFPA has defined dust as "any finely divided solid, 420 μm or 0.017 in, or less in diameter [48–50]." Particle sizes below this level should be assumed to be potentially combustible, and in some cases, larger particle sizes may also be combustible. Coarser powders may also contain a fraction of very fine dusts that may present a hazard. Testing should be done to determine explosivity conclusively if there is an uncertainty. For comparison, the period in this sentence is approximately 600 μm, and the particle size of table salt is approximately 100–150 μm. Particle size is an important factor that can significantly impact the values of other parameters.

- **Minimum explosible concentration** (MEC) – When dust particles are dispersed in air, the dust cloud concentration (g/m^3) must be above the MEC to support sustained combustion, similar to the flammability range discussed in the previous section. Typical MEC values range from 10 to 100 g/m^3, which in most cases, is a high enough concentration to significantly impair visibility although all dense clouds should be assumed to be hazardous. Industrial hygiene professionals generally focus on dust exposure health hazards at concentrations well below the MEC.

- **Dust deflagration index** (K_{st}) – K_{st} is a measure of the relative explosion severity of dusts, with higher values indicating greater severity, based on the maximum burning rate (bar-m/s) of a dust cloud of ideal concentration under turbulent conditions. K_{st} hazard levels include ST 1 (weak explosion), ST 2 (strong explosion), and ST 3 (very strong explosion) [46].

- **Limiting oxygen concentration** (LOC)* – LOC is the minimum oxygen level (vol %) required for combustion of a dust cloud at any concentration, similar to the minimum oxygen concentration (MOC), which was discussed in the previous section. The typical LOC range of 8%–15% can be used to determine inerting requirements when combustible dusts are being handled.

- **Minimum ignition energy** (MIE) – MIE is the minimum spark energy (mJ) required to ignite a dust cloud and support combustion,

---

* Not to be confused with LOC used for level of concern earlier in this chapter.

which is typically in the range of 10–100 mJ. These levels are much higher than the spark energy required to ignite a flammable vapor cloud (<1 mJ). As particle size decreases for a powder, the MIE also decreases.

The GESTIS-DUST-EX database is available online to provide data on over 4,500 dust samples [52]. Injury and damage thresholds for dust explosions are the same as levels of concern that were discussed in Section 6.4.2.

6.5.2 Case Study

On January 29, 2003, a major dust explosion and fire occurred at West Pharmaceutical Services in Kinston, NC [45], where rubber drug delivery components were manufactured. Visible polyethylene dust accumulation at the facility was limited due to cleaning practices at the facility. The investigation determined, though, that combustible dust materials accumulated above suspended ceiling panels at a depth of 0.25 inches or possibly more after being carried by operating area ventilation. Although the investigation did not conclusively identify a primary explosion that dispersed this dust, it is believed that the dust above the ceiling panels fueled the catastrophic dust explosions that severely damaged the facility. Six workers were killed and 38 were injured.

Some additional major incidents involving dust explosions include:

- Westwego, LA, 1977 – Grain dust explosion with 36 fatalities [44]
- Harbin, China, 1987 – Linen flax dust explosion with 58 fatalities [44]
- Huntington, IN, 2003 – Aluminum dust explosion with 1 fatality (see Figure 1.2d) [3]
- Gallatin, TN, 2011 – Iron dust explosion with 2 fatalities [53].

6.6 Chemical Reactivity Hazards

Chemical reactivity hazards are more complicated than the other process hazards discussed in this chapter since chemical reactivity is not always an inherent property of a substance. Chemical reactivity can occur due to the instability or self-reaction of pure materials, often due to increasing temperature, and also due to interactions with other substances. Although the first of these is relatively easy to characterize, the second is much more difficult due to the nearly infinite number of different materials and conditions. Therefore, substances that by themselves may be considered stable can react with any number of other substances or materials under differing conditions due to purposeful or inadvertent interaction or mixing.

Water, for example, is a stable substance that can react with innumerable other substances depending on the chemical interaction and variables such as the temperature. The Bhopal incident, discussed earlier in the chapter, was the result of water inadvertently being mixed with methyl isocyanate (MIC) leading to an uncontrolled reaction and the subsequent loss of containment of toxic MIC into the surrounding community. In some cases, materials that appear stable at one temperature may react violently at a higher temperature. Many incidents have occurred where chemical reactivity hazards were not identified in what was mistakenly considered a mixing process of stable materials. Because chemical reactivity hazards for most materials result from a wide range of possible interactions, SDSs are particularly problematic for identifying and assessing chemical reactivity hazards and should not be relied on.

A U.S. Chemical Safety Board (CSB) report on chemical reactivity over 20 years identified 167 major incidents, resulting in over 10 fatalities, 1,000 injuries, and $1 billion in property damage [54]. A key finding of the CSB report was that only 25% of these incidents occurred in chemical reactors, with other incidents occurring in process equipment, storage, and other locations. Chemical reactivity hazards occur whenever potentially incompatible substances and materials interact and must be carefully identified as part of the IHA. Since many chemical reactivity incidents also result in production of or loss of containment of toxic and/or flammable substances, domino effects of fires, explosions, and toxic exposures from these incidents are frequent.

In assessing chemical reactivity hazards, the potential for controlling intended reactions and for preventing unintended and uncontrolled reactions is a primary focus [55–58]. Uncontrolled or runaway reactions in a reactor system are characterized by accelerating self-heating, generally resulting in rapid and sharp increases in vessel temperature and pressure, as shown in Figure 3.3. In severe cases, the reactor pressure rating can be exceeded, leading to catastrophic reactor failure and the possible release of toxic and/or flammable materials. Many conditions contribute to the occurrence of runaway reactions, including:

- improper formulation [59,60]
- inadequate heating/cooling system design or operation
- material addition errors, including inadvertent mixing or contamination
- accumulation of reactants due to processing problems
- mixing or agitation errors
- process control errors.

Fortunately, chemical reactivity hazards can be minimized through safe formulation practices, reaction modeling, reactor design, and operating standards, which must be evaluated in the PHRA (see Chapter 7). The possibility of uncontrolled reactions occurring in other process equipment and in storage must also be carefully evaluated.

IHA of chemical reactivity hazards requires that the following information be evaluated:

- **Reaction chemistry** – Any intended reaction chemistry should be understood and documented, including reaction steps, the possibility for side and runaway reactions, standard operating conditions, consequences of deviating from these conditions, initiators or catalysts, and mass and energy balances.

- **Thermal and chemical stability** – The stability of self-reactive and mixed materials to temperature and decomposition must be understood to determine appropriate storage, handling, and processing conditions and practices. Some reactions may be very slow at low temperatures but increase very rapidly as the temperature rises. In many cases, self-heating due to reaction over time can lead to serious process events [61].

- **Chemical compatibility interactions** – Development of a chemical compatibility interaction matrix is essential for understanding what chemical and material reactions within a process boundary may be possible, whether due to intended chemistry or inadvertent mixing.

- **Thermodynamic and kinetic data** – The thermodynamic heat generation potential of reactions must be understood to determine the degree to which they are exothermic (generate heat) or endothermic (require heat). Kinetics must be evaluated to understand reaction rates and sensitivity to changing operating conditions.

- **Reaction products and byproducts** – Reaction products, intermediates, and byproducts must be identified and assessed to evaluate the potential for formation of unstable, reactive substances, toxicity or flammability hazards, and gases that can increase system pressure.

Ultimately, you must know the potential for and severity of uncontrolled reactions, what can cause them, how to recognize them, and how to respond and control them. Key parameters for assessing chemical reactivity hazards are discussed in Section 6.6.1. Development of chemical compatibility interaction matrices is discussed in Section 6.6.2. These discussions are necessarily introductory and can be supplemented with many available references on evaluation of chemical reactivity hazards [55–57].

A major runaway reaction and explosion occurred at the Morton International, Inc. in Paterson, NJ in 1998 [62]. While making the 33rd batch of a yellow dye based on reaction of ortho-nitrochloro-benzene and 2-ethylhexyl amine, the reaction accelerated beyond the cooling capacity of the reactor leading to a high-temperature deviation in processing. A secondary, high-temperature decomposition occurred that blew the hatch off the reactor and released its contents. The flammable vapors ignited, injuring nine workers:

> The reaction in the kettle was vibrating the second-floor steel decking and there was a very loud rumbling, which other workers on the site thought was the sound of a train passing... the explosion blew them to the mid-level landing and flash fires spread throughout the building... They suffered second- and third-degree burns, requiring their hospitalization in intensive care for five days... "The pressure was so much on me, I couldn't move. It had pinned me against the wall. It seemed like forever [62]."

The potential for significant injury, property loss, and environmental harm is always present when potentially reactive substances are purposefully or inadvertently mixed, as evidenced by the many industrial incidents that have occurred. Proper assessment of chemical reactivity hazards helps ensure that appropriate safeguards are provided to allow safe reaction chemistry and to prevent or mitigate unintended and runaway reactions with potential toxicity and flammability effects.

6.6.1 Assessing Chemical Reactivity Hazards

Many chemicals can be screened for reactivity using standard tables, such as lists of pyrophoric materials, chemical structures susceptible to peroxide formation, water-reactive chemicals, oxidizers, and polymerizing compounds [55]. In addition, functional groups can also be screened for potential reactivity for materials that are not on these lists. The U.S. EPA has also prepared a generic chemical compatibility chart that can be used to identify possible reactions between different chemical classes [63]. A detailed evaluation tool is discussed in more detail in Section 6.6.2.

Chemical hazard labels based on NFPA 704 [21] were discussed earlier for providing a simple and readily available method for quickly assessing the level of hazard associated with different substances. The reactivity (instability) ratings are provided with the health and flammability ratings, with higher ratings indicating a higher level of instability or reactivity hazard with respect to the pure material. For example, the instability rating for hydrogen cyanide is R-1, for butadiene is R-2, for ammonium nitrate is R-3, and for peracetic acid is R-4. As discussed, these ratings are typically available on SDSs and from other sources [14]. A chemical hazard label may also contain letters such as O for oxidizer and W for water reactive, which are also useful for understanding potential chemical reactivity hazards. The GHS [7] labels provide categories for classifying self-reactive substances, as shown in Table 6.8, pyrophorics, and organic peroxides.

TABLE 6.8

GHS Self-Reactive Substance Types [7]

Type	Criteria
A	Can detonate or deflagrate rapidly, as packaged.
B	Possess explosive properties and which, as packaged, neither detonates nor deflagrates rapidly, but is liable to undergo a thermal explosion in that package.
C	Possess explosive properties when the substance or mixture as packaged cannot detonate or deflagrate rapidly or undergo a thermal explosion.
D	• Detonates partially, does not deflagrate rapidly and shows no violent effect when heated under confinement; or • Does not detonate at all, deflagrates slowly and shows no violent effect when heated under confinement, or • Does not detonate or deflagrate at all and shows a medium effect when heated under confinement.
E	Neither detonates nor deflagrates at all and shows low or no effect when heated under confinement.
F	Neither detonates in the cavitated bubble state nor deflagrates at all and shows only a low or no effect when heated under confinement as well as low or no explosive power.
G	Neither detonates in the cavitated state nor deflagrates at all and shows no effect when heated under confinement nor any explosive power, provided that it is thermally stable (self-accelerating decomposition temperature is 60°–75°C for a 50 kg package), and, for liquid mixtures, a diluent having a boiling point not less than 150°C is used for desensitization.

The heat (or enthalpy) of reaction (ΔH_r) is an important measure for evaluating chemical reactivity hazards. The heat of reaction is the total heat generated (exothermic) or required (endothermic) by the reaction, where the heat of reaction for exothermic reactions is negative. Although endothermic reactions are generally of less concern than exothermic reactions, they must still be evaluated carefully, and in particular, the formation of unstable, toxic, flammable, or gaseous products can be hazardous. The heat of reaction is calculated as:

$$\Delta H_r = \sum (m_i \Delta H_{f,i})_{\text{products}} - \sum (n_j \Delta H_{f,j})_{\text{reactants}}$$

where the heats of formation of the products and reactants can be obtained for many materials online [13]. Heats of reaction can also be estimated from functional groups for some complex reactions or typically are measured experimentally using calorimetry [64]. Although all reactions should be evaluated in a PHRA (see Chapter 7), in general, heat of reactions exceeding −100 cal/g are considered high, requiring careful evaluation. The kinetics of the reaction should also be evaluated, although this is outside the scope of this book. For example, the heat of reaction for oxidative rusting of iron is −1,200 cal/g, which is much higher than the heat of reaction of −420 cal/g for nitration of benzene, although the rate of rusting is comparatively very slow.

Some additional key chemical reactivity parameters include [55–57]:

- **Reaction type** – The type of reaction, such as oxidation, nitration, hydrogenation, polymerization, condensation, esterification, and many others, provides high level, basic information on reaction mechanisms, thermodynamics, and kinetics. Nitration, for example, is typically a highly exothermic reaction that proceeds at high rates, and esterification, in contrast, is typically less exothermic and proceeds at slower rates. Although all reactions should be evaluated in more detail, the overall reaction type can quickly provide insight into basic chemical reactivity hazards.

- **Onset temperature** (T_{onset}) – the lowest temperature where decomposition or reaction increases the temperature due to higher heat of reaction than heat loss. Above the onset temperature, the rate of reaction may accelerate due to the increasing temperature until a maximum self-heat rate $(dT/dt)_{max}$ is obtained. Onset temperatures near ambient or possible process conditions indicate that reaction is possible and must be carefully evaluated.

- **Self-accelerating decomposition temperature** (SADT) – the lowest temperature at which self-accelerating thermal decomposition may occur for a peroxide or other self-reactive substance in its largest size shipping container. Above the SADT, thermal decomposition of the substance occurs raising the temperature and leading to increasingly faster decomposition rates. The SADT therefore determines the safe handling and storage requirements for the substance, such as ambient or refrigerated storage.

- **Adiabatic temperature rise** (ΔT_r) – the maximum temperature increase resulting from reaction under adiabatic conditions. Because no heat is lost to the environment, all heat generated by the reaction is used to increase the temperature.

- **Maximum attainable temperature** (T_{max}) – the maximum temperature resulting from reaction under adiabatic conditions $(T_{max} = T_{onset} + \Delta T_r)$.

In most cases, especially for reactions that have not been well characterized, calorimetry should be used to experimentally determine these key parameters and to understand reaction kinetics [64]. Kinetic models are also often developed and can be useful for hazard assessment. Typical forms for documenting chemical reactivity data and hazards are available [55].

6.6.2 Developing a Chemical Compatibility Matrix

A chemical compatibility interaction matrix is developed for all the substances and materials that have been identified in a process boundary, as well as possible substances that may inadvertently be present, to identify potential reactions, either intended or resulting from inadvertent mixing of materials. The Chemical Reactivity Worksheet (CRW) [65,66] is a convenient and free tool that easily supports developing interaction matrices and provides underlying documentation to allow analysis of reaction predictions. The CRW includes:

- pure component intrinsic hazards, including NFPA 704 ratings, flammability, and reactivity for thousands of substances
- a worksheet to predict specific binary interactions for up to 100 component mixtures with supporting documentation that provides for additional research
- the ability to use functional groups for predictions and to add additional substances to the database.

CRW predictions do not account for ternary and greater interactions, so that results for complex mixtures should normally be reviewed by a chemist or other knowledgeable resource, modelled, or investigated experimentally. CRW has been included in the CAMEO emergency software suite [16] and is available as a free download [65].

Figure 6.6 shows the CRW chemical compatibility worksheet for the chemicals listed in Tables 6.6 and 6.7. The worksheet shows the binary interactions listed as:

- Y—compatible
- N—incompatible
- C—caution
- SR—self-reactive material.

Each binary interaction box can be selected and a summary of the predicted hazards is shown at the bottom of the screen. For the selected chlorine–acetone pair, for example, the worksheet shows the hazard summary in the lower left corner. Additional information on potential gas products, reaction documentation, and mixture comments is also provided. Selecting the chemical rather than the binary interaction box provides hazard information for the pure chemical. User data on predicted interactions, test results, other information can be added as available. The worksheet can be saved to an Excel spreadsheet and summary reports are available. Figure 6.7 shows the compatibility matrix for the Bhopal incident discussed earlier. The predicted hazards are that an exothermic reaction can occur at ambient temperature; the polymerization reaction may become intense and may cause pressurization; and reactions products may be corrosive, flammable, and toxic.

The CRW should be supplemented with information on substance–material interactions where possible. For example, the interaction of some chemicals with various materials of construction or process materials (e.g., packing, gaskets) may be of interest to help prevent reactions and ensure

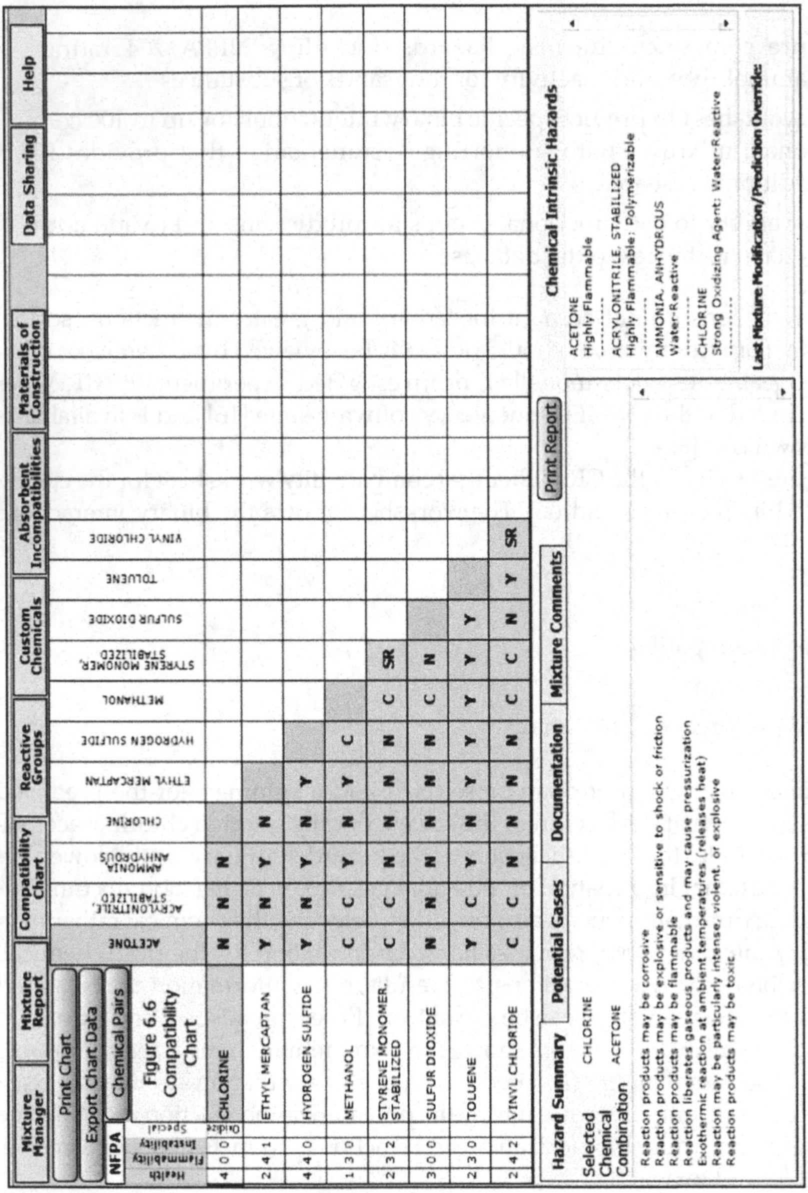

FIGURE 6.6
Chemical reactivity worksheet compatibility chart [Modified from 65].

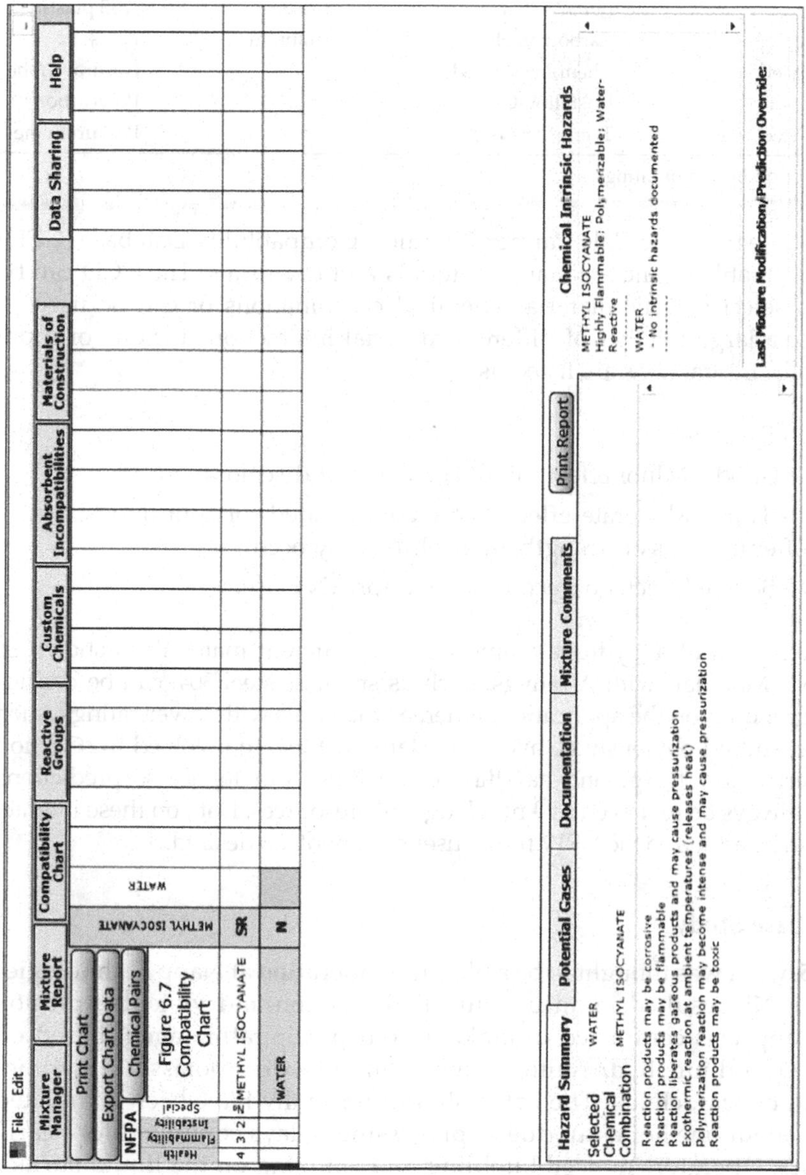

FIGURE 6.7
Compatibility chart for Bhopal incident [Modified from 65].

TABLE 6.9

Material Compatibility Ratings for Anhydrous Ammonia [Modified from 67]

Excellent (A)	Good (B)	Fair (C)	Severe (D)
EPDM	Buna N (Nitrile)	Silicone	ABS plastic
Neoprene	Carbon Steel	Titanium	Brass
Polypropylene	ChemRaz (FFKM)		Natural rubber
PTFE	Hastelloy-C®		Polycarbonate
Stainless Steel 304	Tygon® (E-3603)		Polyurethane

See text for definition of ratings.

process integrity. The Cole-Parmer Chemical Compatibility Database (CCD) [67] is available online for many materials and chemicals. The CCD can be used to select specific material/chemical combinations or can be used to evaluate a large number of different material interactions for one or more chemicals. Interactions are listed as:

- A – Excellent
- B – Good – Minor effect, slight corrosion or discoloration
- C – Fair – Moderate effect, not recommended for continuous use. Softening, loss of strength, or swelling may occur
- D – Severe Effect, not recommended for ANY use.

The interaction of anhydrous ammonia with different materials is shown in Table 6.9. Materials with A ratings, such as stainless steel 304, can be considered depending on the application, whereas materials with lower ratings, such as brass, should obviously be avoided. Many factors are involved in selection of materials for storage and handling of hazardous materials, so predictions should always be reviewed by knowledgeable resources. Data on these interactions can be added to the CRW file as user comments, if desired.

6.6.3 Case Study

In 1995, what was thought to be a blending operation at Napp Technologies in Lodi, NJ, resulted in an uncontrolled reaction and violent explosion [68]. Napp was contracted to make a gold precipitating agent based on blending sodium hydrosulfite, aluminum powder, potassium carbonate, and benzaldehyde. The batch should normally have been completed within about one hour, but due to processing delays continued for nearly 24 hours. Operators noticed bubbling and smoking on the liquid surface in the blender and the release of a foul-smelling gas. Sodium hydrosulfite and aluminum powder are water-reactive chemicals that may have come into contact with water leaking from the blender cooling system. Sodium hydrosulfite gives off sulfur dioxide and hydrogen sulfide in an exothermic reaction with water, which becomes self-sustaining. Aluminum powder reacts exothermically with water to produce hydrogen gas. After about 24 hours, loud hissing noises were heard and an explosion occurred that destroyed a majority of the facility. Five employees were killed. Calorimetric tests completed as part of the incident investigation confirmed that small amounts of water could induce a powerful runaway reaction at relatively low temperature. The adiabatic temperature rise with aluminum powder present was 486°C.

Some additional major chemical reactivity incidents include:

- Hanover Township, PA 1999 – A process vessel containing hydroxylamine exploded when the first commercial batch was being distilled to purify the hydroxylamine. Five people were killed and 14 were injured in the explosion [69].

- Toulouse, France, 2001 – The decomposition and explosion of large amounts of ammonium nitrate killed 30 people, injured over 2000, and caused over $2 billion in damage [19].

- Pascagoula, MS, 2002 – Steam leaked into a distillation column, unexpectedly heating mononitrotoluene and causing a decomposition that exploded the column and injured three employees [70].

- Morgantown, NC, 2006 – Formulation changes led to an exothermic reaction that exceeded the reactor cooling capability, causing a runaway acrylic polymerization and subsequent vapor cloud explosion that killed 1 worker and injured 14 [71].

- Jacksonville, FL, 2007 – Loss of cooling on a reactor system resulted in a runaway reaction, causing 4 fatalities and 13 injuries [5].

6.7 Other Process Hazards

Not all process hazards fit into the major categories discussed in the chapter: toxicity, flammability, combustible dusts, and chemical reactivity. Other types of process hazards that may be identified and need to be assessed include:

- High pressure or vacuum
- High or low temperature
- Generation or release of stored energy
- High speed rotating equipment
- Nonionizing radiation
- Low oxygen environments (special case of toxicity)

- Acid/base (special case of chemical reactivity)
- Other process hazards identified within a process boundary.

Identification of process hazards is, of course, critical to ensuring that they have been evaluated appropriately [72].

Assessment of each process hazard must necessarily be specific to the hazard. In general, the assessment process includes (see Figure 6.1):

- **identification** – the first step is always recognition of the process hazard
- **assessment** – specific to the process hazard, based on literature search, previous studies, calculation, experimentation, and involvement of knowledgeable resources
- **documentation** – the assessment data and findings must be documented as part of the IHA for the process and for use in the PHRA (see Chapter 7).

On June 10, 2008, workers at the Goodyear Tire and Rubber Company in Houston, TX, closed an isolation valve between an ammonia heat exchanger (shell side) and a pressure relief valve to replace a burst rupture disk under a relief valve, as shown in Figure 6.8 [73]. When the rupture

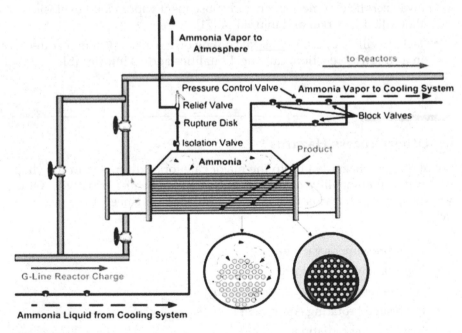

FIGURE 6.8
Heat exchanger rupture incident [73].

disk was replaced, the isolation valve was not reopened. The next day, a block valve was closed to isolate the pressure control valve from the heat exchanger, and a steam line was connected to the heat exchanger (process side) to clean the piping. The steam not only flowed through the process side heat exchanger tubes, but also heated the liquid ammonia in the heat exchanger shell side. As the ammonia was heated, the pressure increased and could not be relieved due to the closed isolation and block valves. The pressure eventually violently ruptured the heat exchanger shell with debris killing one worker and ammonia being released, exposing several additional workers.

6.8 Measures of Success

☑ Design and implement an effective process safety system for identifying and assessing process hazards.

☑ Keep readily available, comprehensive, and accurate material and process hazards documentation as part of the process technology/process safety information.

☑ Complete a thorough intrinsic hazard assessment (IHA), or equivalent, for process hazards that are present, including toxicity, flammability, reactivity, and others.

☑ Use the material and process hazards documentation to support other process safety system activities, such as designing the process and equipment, evaluating for process risks, developing operating and maintenance procedures, making process or equipment design changes, developing emergency response plans, and training personnel across all disciplines at the facility.

References

1. Dharmavaram, Seshu and James A. Klein. 2012. An introduction to assessing process hazards. *Process Safety Progress* 31:266–270.
2. U.S. Chemical Safety Board. 2004. *D. D. Williamson and & Co., Inc., Catastrophic Vessel Failure*. Report No. 2003-11-I-KY. www.csb.gov.
3. U.S. Chemical Safety Board. 2005. *Hayes Lemmerz International, Inc., Aluminum Dust Explosion*. Report No. 2004-01-I-IN. www.csb.gov.
4. U.S. Chemical Safety Board. 2003. *BLSR Operating, Ltd., Vapor Cloud Deflagration and Fire*. Report No. 2003-06-I-TX. www.csb.gov.

5. U.S. Chemical Safety Board. 2009. *T2 Laboratories, Inc., Runaway Reaction*. Report No. 2008-03-I-FL. www.csb.gov.
6. U.S. Chemical Safety Board. 2010. *Texas Tech University Lab Explosion Case Study*. Report No. 2010-05-I-TX. www.csb.gov.
7. U.S. OSHA. No date. *A Guide to the Globally Harmonized System of Classification and Labeling of Chemicals (GHS)*. www.osha.gov/dsg/hazcom/ghsguideoct05.pdf.
8. U.S. EPA and U.S. OSHA. 1997. *EPA/OSHA Joint Chemical Accident Investigation Report, Napp Technologies*. EPA 550-R-97-002. www.epa.gov/oem/docs/chem/napp.pdf.
9. U.S. Chemical Safety Board. 2007. *Barton Solvents: Static Spark Ignites Explosion Inside Flammable Liquid Storage Tank*. Report No. 2007-06-I-KS. www.csb.gov.
10. U.S. Chemical Safety Board. 2006. *Combustible Dust Hazard Study*. Report No. 2006-H-1. www.csb.gov.
11. Willey, Ronald J. 2012. Decoding safety data sheets. *Chemical Engineering Progress* 108(6):28–31.
12. DIPPR. Design Institute for Physical Property Data. dippr.byu.edu.
13. NIST Chemical WebBook. National Institute of Standards and Technology. webbook.nist.gov/chemistry.
14. Webwiser. National Institutes for Health. U.S. National Library of Medicine. webwiser.nlm.nih.gov.
15. National Institute for Occupational Safety and Health. *NIOSH Pocket Guide to Chemical Hazards*. www.cdc.gov/niosh/npg/.
16. U.S. Environmental Protection Agency. CAMEO (Computer-Aided Management of Emergency Operations). www.epa.gov/cameo.
17. Leggett, David J. 2012. Identifying hazards in the chemical research laboratory. *Process Safety Progress* 31:393–397.
18. Plog, Barbara A. and Patricia J. Quinlan. 2001. *Fundamentals of Industrial Hygiene*, 5th ed. National Safety Council.
19. Center for Chemical Process Safety. 2008. *Incidents That Define Process Safety*. Wiley-AIChE.
20. Lapierre, Dominique and Javier Moro. 2002. *Five Past Midnight in Bhopal: The Epic Story of the World's Deadliest Industrial Disaster*. Grand Central Publishing, New York.
21. NFPA. 2012. *NFPA 704: Standard System for the Identification of the Hazards of Materials for Emergency Response*. National Fire Protection Agency. www.nfpa.org.
22. HMIS. 2016. *Hazardous Materials Identification System*. American Coatings Association. www.paint.org/programs/hmis.html.
23. U.S. Environmental Protection Agency. *Acute Exposure Guideline Levels for Airborne Chemicals (AEGLs)*. www.epa.gov/oppt/aegl/.
24. American Industrial Hygienists Association. *Emergency Response Planning Guidelines™ (ERPGs)*. www.aiha.org/get-involved/AIHAGuidelineFoundation/EmergencyResponsePlanningGuidelines/Pages/default.aspx.
25. www.atlintl.com/DOE/teels/teel.html.
26. UK HSE. www.hse.gov.uk/chemicals/haztox.htm.
27. U.S. Chemical Safety Board. 2003. *Chlorine Release, DPC Enterprises*. 2002-04-I-MO. www.csb.gov.
28. www.ntsb.gov/investigation/summary/RAR0504.html.
29. en.wikipedia.org/wiki/Jilin_chemical_plant_explosions_2005.

30. www.nature.com/news/alert-over-south-korea-toxic-leaks-1.12369.
31. U.S. Chemical Safety Board. 2016. *DuPont LaPorte Facility Toxic Chemical Release.* Interim Report and video. www.csb.gov.
32. NFPA. 2008. *Fire Protection Handbook,* Volumes 1 and 2. 20th ed.
33. www.dailymail.co.uk/news/article-1031994/The-day-sea-caught-20-years-Piper-Alpha-explosion-survivors-finally-able-tell-story.html.
34. Center for Chemical Process Safety. 2010. *Guidelines for Vapor Cloud Explosion, Pressure Vessel Burst, BLEVE and Flash Fire Hazard.* Wiley-AIChE.
35. Center for Chemical Process Safety. 1999. *Guidelines for Consequence Analysis of Chemical Releases.* Wiley-AIChE.
36. Crowl, Daniel A. 2012. Minimize the risks of flammable materials. *Chemical Engineering Progress* 108(4).28–33.
37. Bodurtha, Frank T. 1980. *Industrial Explosion Prevention and Protection.* McGraw-Hill, New York.
38. ALOHA. response.restoration.noaa.gov/oil-and-chemical-spills/chemical-spills/resources/thermal-radiation-levels-concern.html.
39. U.S. Chemical Safety Board. 2007. *Refinery Explosion and Fire.* Report No. 2005-04-I-TX. www.csb.gov.
40. Buncefield Major Incident Investigation Board. 2008. *The Buncefield Incident 11 December 2005: The final report of the Major Incident Investigation Board,* Volume 1. HSE Books, Sudbury.
41. Venuzuala. 2013. *PDVSA, Evento Clase A Refineria de Amuay.* Gobiemo Boivariano de Venuzuela, Ministerio dei Poder Popular de Petroleo y Minera. www.pdvsa.com/interface.sp/database/fichero/publicacion/8264/1632.PDF.
42. usatoday30.usatoday.com/news/world/story/2012/09/20/29-dead-46-injured-in-mexico-pipeline-fire/57809722/1.
43. U.S. Chemical Safety Board. 2016. *West Fertilizer Company Fire and Explosion.* Report No. 2013-02-I-TX. www.csb.gov.
44. Eckhoff, Rolf K. 2003. *Dust Explosions in the Process Industries, Third Edition: Identification, Assessment and Control of Dust Hazards.* Gulf Professional Publishing.
45. U.S. Chemical Safety Board. 2004. *Dust Explosion.* Report No. 2003-07-I-NC. www.csb.gov.
46. U.S. Chemical Safety Board. 2009. *Sugar Dust Explosion and Fire.* Report No. 2008-05-I-GA. www.csb.gov.
47. Frank, Walt and Guy R. Colonna. 2015. On-going developments in addressing combustible dust hazards. *Process Safety Progress* 34:24–30.
48. Colonna, G. R., W. L. Frank, and S. A. Rodgers. 2012. *NFPA Guide to Combustible Dusts.* National Fire Protection Association (NFPA), Quincy, MA.
49. NFPA. 2013. *NFPA 654: Standard for the Prevention of Dust Explosions from the Manufacturing, Processing, and Handling of Combustible Particulate Solids.* National Fire Protection Association (NFPA), Quincy, MA.
50. NFPA. 2016. *NFPA 652: Standard on the Fundamentals of Combustible Dust.* National Fire Protection Association (NFPA), Quincy, MA.
51. Murphy, Michelle R. 2016. Making sense of combustible-dust hazard analysis. *Chemical Engineering Progress* 112(4):28–32.
52. GESTIS-DUST-EX database. IFA: Institute for Occupational Safety and Health of the German Social Accident Insurance. staubex.ifa.dguv.de/explosuche.aspx?lang=e.

53. U.S. Chemical Safety Board. 2011. *Metal Dust Flash Fires and Hydrogen Explosion*. Report No. 2011-4-I-TN. www.csb.gov.
54. U.S. Chemical Safety Board. 2002. *Improving Reactive Hazard Management*. Report No. 2001-01-H. www.csb.gov.
55. Johnson, Robert W., Steven W. Rudy, and Stephen D. Unwin. 2003. *Essential Practices for Managing Chemical Reactivity Hazards*. AIChE, New York.
56. Barton, John and Richard Rogers. 1997. *Chemical Reaction Hazards*, 2nd ed. Gulf Publishing.
57. Center for Chemical Process Safety. 1995. *Guidelines for Chemical Reactivity Evaluation and Application to Process Design*. Wiley-AIChE.
58. Kletz, Trevor. 2009. *What Went Wrong? Case Histories of Process Plant Disasters and How They Could Have Been Avoided*, 5th ed. Gulf Professional Publishing.
59. Balchan, S., J. A. Klein, and F. G. Klein. 1995. Process Safety of Polymer Resin Manufacturing: A 20-Year Perspective. *Loss Prevention and Safety Promotion in the Process Industries*, Antwerp, Belgium.
60. Klein, J. A. and A. S. Balchan. 1996. Safe Formulation and Manufacture of Acrylic Resins. *International Conference and Workshop on Process Safety Management and Inherently Safer Processes*, Orlando, FL.
61. Klein, James A. and Gordon R. Mros. 2006. Characterization and safe handling of reactive initiator solutions. *Process Safety Progress* 25:303–310.
62. U.S. Chemical Safety Board. 1998. *Chemical Manufacturing Incident*. Report No. 1998-06-I-NJ. www.csb.gov.
63. EPA Chemical Compatibility Chart. No date. ehs.yale.edu/forms-tools/epa-chemical-compatibility-chart.
64. Crowl, Daniel A. and Jason M. Keith. 2013. Characterize reactive chemicals with calorimetry. *Chemical Engineering Progress* 109(7):26–33.
65. Center for Chemical Process Safety. No date. *Chemical Reactivity Worksheet*. www.aiche.org/ccps.
66. Gorman, Dave, Jim Farr, Rob Bellair, Wade Freeman, Dave Frurip, Al Hielscher, Harold Johnstone et al. 2014. Enhanced NOAA chemical reactivity worksheet for determining chemical compatibility. *Process Safety Progress* 33:4–18.
67. Cole-Parmer Chemical Compatibility Database. 2016. www.coleparmer.com/Chemical-Resistance.
68. U.S. EPA and U.S. OSHA. 1997. *EPA/OSHA Joint Chemical Accident Investigation Report*. Napp Technologies, Lodi, NJ.
69. U.S. Chemical Safety Board. 2002. *The Explosion at Concept Sciences: Hazards of Hydroxlamine Case Study*. Report No. 1999-13-C-PA. www.csb.gov.
70. U.S. Chemical Safety Board. 2003. *Explosion and Fire*. Report No. 2003-01-I-MS. www.csb.gov.
71. U.S. Chemical Safety Board. 2007. *Runaway Chemical Reaction and Vapor Cloud Explosion*. Report No. 2006-04-I-NC. www.csb.gov.
72. Center for Chemical Process Safety. 2010. *A Practical Approach to Hazard Identification for Operations and Maintenance Workers*. Wiley-AIChE.
73. U.S. Chemical Safety Board. 2011. *Heat Exchanger Rupture and Ammonia Release in Houston, Texas*. Report No. 2008-06-I-TX. www.csb.gov.

7

Evaluate and Manage Process Risks

Risk comes from not knowing what you are doing.

Warren Buffett

Why We Need to Evaluate and Manage Process Risks

If process hazards are present, they must be managed through appropriate safeguards designed to prevent serious injuries and incidents. It is essential to methodically and thoroughly:

- ensure that all process hazards have been identified and assessed
- understand the types and ranges of consequences of potentially hazardous events
- identify the failures or pathways that can lead to these events
- evaluate the process risk of these events, based on event frequency and consequences
- provide multiple protection layers to effectively manage these risks.

Insufficient analysis of process risks increases the likelihood that process safety systems and safeguards will be inadequately designed and ineffective, thus increasing the likelihood of serious injuries and incidents. Proper identification and assessment of process hazards, as discussed in Chapter 6, and the evaluation and appropriate reduction of process risk, as discussed in this chapter, help ensure that effective process safety systems and safeguards are in place to achieve safe and reliable operations.

Evaluate and manage process risks to ensure safe process design and that process safety systems are properly designed, implemented, and maintained.

7.1 Introduction

Proper design and implementation of process safety systems for managing process risk necessarily rely on the thorough evaluation of potentially hazardous event scenarios that may be associated with the process. This requires a methodical approach for identifying potential events, ranging from small to worst case, and then evaluating both their likelihood (event frequency) and their severity (event consequences). Processes that use toxic materials, for example, may have a range of possible hazardous events depending on the materials involved, processing steps, operating conditions, type and source of possible release or exposure, existing safeguards, physical location relative to people and other process equipment, weather conditions, and many other factors. A range of hazardous events must be identified for all of the process hazards in the process. Each event needs to be evaluated for potential safety and health effects, process risk, and appropriate protection layers to reduce risk and help ensure the event does not occur. Thorough evaluation of process risk is, of course, dependent on the types and levels of process hazards that have been documented, as discussed in Chapter 6, and the intended process and equipment design, operating steps, and processing conditions that have been documented, as discussed in Chapter 5.

Although basic risk evaluation studies have often been called process hazard analysis (PHA) [1], identification and assessment of process hazards should normally be conducted and documented as part of the process technology and safety information, as discussed in Chapter 6. More recently, PHA has been described as hazard identification and risk analysis (HIRA) [2] to emphasize that PHA is really a risk evaluation and management methodology. In this book, we choose the middle ground and describe basic risk evaluation as process hazards and risk analysis (PHRA), which is similar to the familiar PHA terminology but better emphasizes the risk evaluation and management purposes of the study.

The PHRA is designed to methodically identify, evaluate, and manage process risks. Comprehensive and current process safety information, intrinsic hazard assessments, process and equipment design, and intended operating steps and processing conditions must be available to support detailed analysis of process risks. Identification of potentially hazardous events, and the equipment or operating failures and pathways that lead to these events, allows the PHRA team to determine process risk based on estimated frequencies and consequences. Current or planned safeguards are evaluated for effectiveness and reliability, and additional protection layers are provided, if appropriate, to achieve desired risk criteria that may be set by the individual organization's risk tolerance guidelines or, in some cases, by regulations in some countries. The PHRA also establishes many of the process safety system requirements, such as maintenance of safeguard functionality, that must be continuously managed to sustain process safety program performance, as discussed in Part III.

7.2 Key Concepts

7.2.1 What Is Process Hazards and Risk Analysis (PHRA)?

The PHRA is a structured, comprehensive methodology to ensure thorough identification, evaluation, and management of process hazards and risks. If done well, appropriate protection layers are provided to reduce process safety risk and to help ensure safe and reliable operation of potentially hazardous processes. If not done well, process incidents can result from toxic releases, fires and explosions, out-of-control chemical reactions, and other process events that can lead to fatalities, environmental harm, and major damage, destruction, and disruption. Some examples of major incidents related to ineffective PHRAs are shown in Table 7.1.

The PHRA consists of several complementary studies as shown in Figure 7.1. Specific requirements may vary somewhat based on country regulations or the organization's guidelines, but usually include:

- **Hazards identification (HI)** – review of the intrinsic hazard assessment (IHA) or equivalent documentation, as described in Chapter 6, to identify and assess the process hazards based on the scope of the PHRA. The PHRA is intended to evaluate the risks associated only with process hazards rather than general hazards and other occupational health and safety issues.

- **Consequence analysis (CA)** – identification and understanding of possible hazardous events associated with the process hazards that have been identified, such as toxic releases, fires, and explosions. Events may range from small releases to worst case, catastrophic events, depending on a variety of factors resulting from loss of engineering and administrative controls. In some cases, detailed quantitative consequence modeling may be needed to adequately understand the potential consequences of significant events.

- **Hazardous event evaluation (HEE)** – evaluation of process equipment and operational failures, based on HAZOPs and other methodologies (see Section 7.5), to understand how potentially hazardous events can occur, to evaluate safety and health effects, and to ensure that adequate engineering and administrative controls are provided to manage risk.

- **Human factors (HF)** – evaluation of human error, human factors, and the work environment to identify how human activity can lead or contribute to hazardous events.

- **Facility siting (FS)** – evaluation of process location and building construction to understand the potential event impact on surrounding

TABLE 7.1

Examples of CSB Incidents with PHRA Findings

CSB Incident Investigation	Year	Finding
Morton [3]	1998	Neither the preliminary hazard assessment conducted by Morton in Paterson during the design phase in 1990 nor the process hazard analysis conducted in 1995 addressed the reactive hazards of the Yellow 96 process.
Sonat [4]	1998	Sonat management did not use a formal engineering design review process or require effective hazard analyses in the course of designing and building the facility.
Concept Sciences [5]	1999	Basic process safety and chemical engineering practices—such as process design reviews, hazard analyses, corrective actions, and reviews by appropriate technical experts–were not adequately implemented.
D. D. Williamson [6]	2003	D. D. Williamson did not have adequate hazard analysis systems to identify feed tank hazards, nor did it effectively use contractors and consultants to evaluate and respond to associated risks.
Sterigenics [7]	2004	The process hazard analysis program at the Ontario facility did not fully identify and evaluate the hazard associated with an explosive concentration of ethylene oxide (EO) reaching the oxidizer.
T2 Laboratories [8]	2007	One of the T2's design consultants identified the need to perform a hazard and operability study (HAZOP, a type of PHA) during scale-up. A comprehensive HAZOP likely would have identified the need for testing to determine the thermodynamic and kinetic nature of the reaction, as well as the limitations of the cooling and pressure relief systems. CSB found no evidence that T2 ever performed the HAZOP.
Bayer CropScience [9]	2008	The PHA team did not validate the assumptions in the PHA including accuracy of the standard operating procedure (SOP), conformance to the SOP, and control of process safeguards.
Veolia [10]	2009	No record existed of a process hazards analysis (PHA) to evaluate the siting of the lab/operations building so close to the operating units.

areas, with special emphasis on control room design. Facility siting may sometimes be referenced as facility/stationary source siting to include the requirements of the U.S. EPA Risk Management Plan, as appropriate.

- **Inherently safer process (ISP)** – evaluation of opportunities for permanently eliminating or reducing process hazards.
- **Layers of protection (LOPs)** – evaluation of engineering and administrative controls, such as safety instrumented systems, pressure

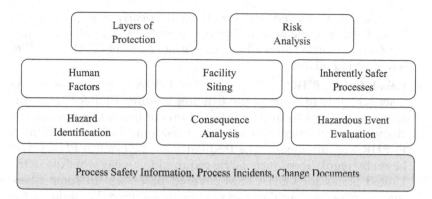

```
┌─────────────────┐      ┌─────────────────┐
│   Layers of     │      │      Risk       │
│   Protection    │      │    Analysis     │
└─────────────────┘      └─────────────────┘

┌─────────────────┐  ┌─────────────────┐  ┌─────────────────┐
│     Human       │  │    Facility     │  │ Inherently Safer│
│    Factors      │  │     Siting      │  │    Processes    │
└─────────────────┘  └─────────────────┘  └─────────────────┘

┌─────────────────┐  ┌─────────────────┐  ┌─────────────────┐
│     Hazard      │  │   Consequence   │  │ Hazardous Event │
│  Identification │  │    Analysis     │  │   Evaluation    │
└─────────────────┘  └─────────────────┘  └─────────────────┘

┌────────────────────────────────────────────────────────────┐
│  Process Safety Information, Process Incidents, Change Documents │
└────────────────────────────────────────────────────────────┘
```

FIGURE 7.1
Sections of a PHRA.

relief, and other safeguards, intended to help prevent or mitigate process events.

- **Risk analysis (RA)** – evaluation of the frequency and consequences of possible hazardous events, as appropriate, dependent on the specific equipment and operational failures that cause the event and actual or planned protection layers.

These studies are discussed in more detail later in this chapter. Companies may sometimes develop guidance, where permitted by regulation, that allows for some smaller scale, R&D, or other projects with low process risk to be conducted with a simplified PHRA scope. These determinations should be made carefully to always ensure that process hazards have been thoroughly evaluated and are managed appropriately.

7.2.2 When Is a PHRA Done?

PHRAs are conducted for different stages of a facility's life cycle in several major categories:

- **New facilities/processes** – Several PHRAs are usually completed during the design and project phases of new facilities or processes to ensure that appropriate engineering and administrative controls are included in the design and that sufficient funds have been allocated for the project. Depending on the project scope and process hazards, a screening or design PHRA should be conducted early in the design process to ensure major process risks and protection layers have been evaluated and included in the project scope and to identify any "unknowns" that require further follow-up. Application of Inherently Safer Process principles at this stage is most appropriate because the associated costs of changes in early design are much lower than once the new facility has been built. A final or baseline PHRA is conducted after all process design elements, such as P&IDs, have been finalized.

- **Revalidation PHRA** – Revalidation PHRAs are conducted on a set schedule to confirm the existing baseline PHRA is still current and valid or to update it based on equipment or operational changes, process incidents, or other considerations such as changes to PHRA requirements. The frequency of the cyclical PHRA may be set by regulations, such as the required five-year schedule in the United States for OSHA PSM-covered processes. In some cases, the organization's guidance may provide additional criteria for the review frequency based on the types of process hazards, level of risk, complexity of the process, number and severity of process incidents, and degree of change at the facility. This may result in revalidation reviews occurring more frequently than required by regulation in order to ensure that process risks are being managed appropriately. Key considerations for determining if a complete or partial redo of the baseline PHRA is required or if revalidation and update are acceptable are provided in Figure 7.2. Many companies routinely conduct a redo PHRA every several revalidations to help ensure the PHRA is current. Revalidation PHRAs are discussed in Section 7.11.

- **Mothball, dismantle, or startup PHRA** – When processes are being shut down or removed, PHRAs should be conducted to help ensure that construction activity can be completed safely. These studies consider safe removal of hazardous materials, dismantling of process equipment, impact on surrounding facilities, and other factors. New PHRA studies should also be completed if these facilities are later started up again.

- **Management of change (MOC) or Project PHRA** – Changes to facilities, process equipment, and operations are continually made to support safe and reliable operations, increased capacity, new product introductions, new technology, etc. These changes must be reviewed in detail, as discussed in Chapter 10, to ensure that the safety of the process is not compromised. In many cases, especially for major changes in complex, high hazard processes, additional PHRA studies will be required to evaluate the safety of the change in detail.

Depending on the scope of the PHRA, the study may require only a few hours, a few days, or many weeks for completion. Large-scope PHRAs require extra attention for scheduling, team member composition and availability, and other factors to ensure an acceptable PHRA is completed.

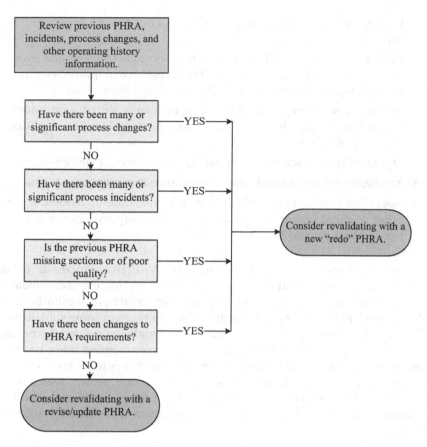

FIGURE 7.2
PHRA Revalidation.

7.2.3 Who Does a PHRA?

PHRAs must be conducted by multi-disciplinary teams who have expertise in engineering and process operations, including:

- specific knowledge and experience with the process being evaluated
- training and experience for successfully completing PHRAs with appropriate methodologies.

This typically means that the team includes members who have specific technical and engineering background in the design and operation of the process, hands-on operating and maintenance experience, and in-depth training on how PHRAs should be conducted. Team members therefore generally include:

- **PHRA leader**, usually an engineer or supervisor with experience in the process
- **PHRA resource or facilitator**, who has previous training and experience in conducting PHRAs; the PHRA resource in many cases may be an independent contractor with specialized experience
- **Technical resource**, typically an engineer or scientist who understands the design, chemistry, and other technical and engineering aspects of the process
- **Operator/technician**, who has hands-on operating experience
- **Mechanic**, who has hands-on maintenance experience
- **Specialists**, who may include control, instrument, modeling, or other relevant technical, operating, or reliability experience that is needed part time to support the PHRA.

Team members need to be fully engaged in the PHRA process to ensure that detailed review and discussion occur to fully evaluate and complete the various PHRA methodologies. Ideally, all team members should be available for every PHRA meeting, though in lengthy reviews some substitutions with other qualified representatives may occasionally be needed. Some team members may also fill more than one of the recommended roles, based on their experience and background. Typically, the PHRA team will consist of approximately four to ten members, but large-scope project teams working with new or complex technology may have many additional members. In some cases, regulations specify the minimum team membership and experience.

7.2.4 How Is a PHRA Done?

The typical steps for completing a PHRA are shown in Figure 7.3, including:

- **Establish scope, charter, and schedule** – The scope and boundaries of the PHRA should be finalized, a charter should be prepared that includes the objectives, resources, and timing for the PHRA, and a meeting schedule should be established. Although not generally required by regulation, the charter establishes work requirements for conducting and completing the PHRA and can be helpful if problems, such as team member availability, occur.
- **Select and train team** – The team composition should be documented as part of the charter, and the team should be trained on how the PHRA will be conducted. Training may be needed periodically to introduce new studies, such as Layer of Protection Analysis, or for new team members to understand methodologies being used.

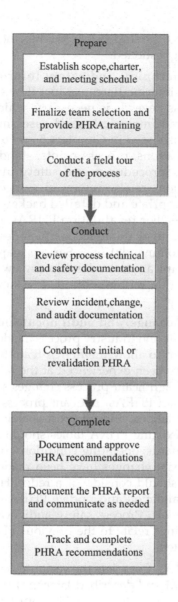

FIGURE 7.3
PHRA Steps.

- **Tour the process (recommended)** – For existing facilities, the team can tour the process area to observe equipment and operations directly and to verify process design information. For new or proposed changes to facilities where equipment has not been constructed or installed, the team may spend additional time reviewing process diagrams or models. Team members should conduct additional field tours, as needed, to address issues which may arise during the course of the study, such as when the accuracy of process drawings may be questioned.

- **Review process technical and safety information** – Review of the raw material, intermediate, product flow streams, their properties, and process hazards is necessary, which are based on the process design and intrinsic hazard assessment (or equivalent) documentation discussed in Chapters 5 and 6. Equipment design basis and specifications, process prints, standard operating conditions, operating procedures, and safety instrumented system (SIS) documentation are also reviewed, as needed, to ensure that the team has appropriate and detailed background on the process and its safeguards. For revalidation PHRAs, the previous PHRA report is reviewed to evaluate its completeness and to determine the status of recommendations. Serious gaps in this information should prompt a site leadership review to determine how required information will be obtained to support completion of the PHRA.

- **Review incident, change, and audit documentation** – Summaries of relevant company and industry process incidents and near misses should be reviewed to allow the PHRA team to evaluate operating problems that have been experienced at the facility or similar processes elsewhere. Significant process changes should also be evaluated. In revalidation PHRAs, relevant process incidents, previous PHRA recommendations, and all process changes are reviewed to determine if the existing PHRA tables should be updated for new equipment, safeguards, operating conditions, or incident scenarios. If enough significant changes have been made since the previous PHRA, the team should consider if a redo PHRA should be completed or if a revalidation and update PHRA is acceptable, as discussed in Section 7.11. Process safety audits can also be reviewed to see if any findings relate to the previous PHRA or how PHRAs should be conducted.

- **Conduct the PHRA** – Complete the various sections of the PHRA as shown in Figure 7.1 and described later in this chapter. Since additional information or analysis may be required between scheduled meetings, an iterative process may be needed to work through some sections.

- **Document and approve recommendations** – Specific recommendations for managing the risks associated with the process will usually be developed, based on PHRA team consensus on the need for further risk reduction. Recommendations should clearly discuss what actions are needed, why they are needed, and what hazardous events could occur. General recommendations to study or evaluate something should be avoided, unless clear actions are associated with them once the study has been completed. Recommendations are documented and then reviewed and approved by site leadership. In some cases, interim recommendations may be needed to address short-term gaps, recognizing that capital projects or process shutdowns may be required before some recommendations can be completed. Some recommendations may be rejected by site leadership, in which case, the justification must be documented. Permitted justifications are usually limited by regulatory guidance to cases where (1) the recommendation is not needed to improve safety, (2) an alternate equivalent approach can be used, (3) there was a factual error made by the PHRA team, or (4) the recommendation is not feasible.

- **Document and communicate the report** – A complete report of all sections of the PHRA must be documented for future use and to meet relevant regulations. If a revalidation PHRA is being conducted, the previous PHRA report may only need to be selectively updated based on the team's review of process incidents, process changes, and previous PHRA recommendations, as discussed in Section 7.11. Results of the PHRA should be communicated to other employees and contractors, as appropriate, since most will not have participated in the PHRA meetings. A full review of all aspects of the PHRA may not be necessary, but completion of the PHRA report should be shared along with relevant aspects of the review and major recommendations. Regulations may specify requirements for documenting, communicating, and storing PHRA reports, which are generally kept for the life of the process. A PHA quality and completeness checklist is available [11].

- **Track and complete recommendations** – All PHRA recommendations must be assigned with responsibilities and timing for completion. They should then be tracked through appropriate action item tracking systems to ensure that they are completed correctly and as scheduled. PHRA recommendations should never be closed simply because a study has been completed or a work order or MOC has been issued. Recommendation closure must indicate that an action meeting the intent of the recommendation has actually been completed, not that it is simply planned for or will occur in the future. In some cases, corporate policy or regulations may specify the maximum time allowed to complete recommendations, based on associated risks or other factors.

7.3 Process Hazards Identification

The process hazards are identified and assessed via an intrinsic hazard assessment (IHA) or equivalent [12], as described in Chapter 6, and included as part of the process safety information prior to the PHRA. IHA documentation includes process hazards related to:

- intrinsic material properties, such as toxicity and flammability
- material interactions, such as chemical reactivity
- process conditions, such as pressure and temperature
- process energies, such as high-speed rotating equipment.

The PHRA team reviews the process hazards, confirms that all process hazards have been identified, and documents a summary of the process hazards within the scope of the PHRA. Toxic materials can be listed, for example, based on the toxicity levels and human health effects. Similarly, flammable materials can be listed by flash point and flammability range. The chemical interaction matrix should be reviewed to identify and summarize potential chemical reactions. If the final design or process safety information are incomplete or not available, the PHRA must normally be stopped until the missing information is provided to the team.

The PHRA team may also review previous PHRA reports if available, process incidents and near misses, management of change documentation, and other operating information to confirm that all process hazards have been identified, as needed. During the initial meeting and field tour, the team should also consider whether additional process hazards are present. Safety and other hazards that are not process hazards should be documented separately, if observed, as they are outside the scope of the PHRA.

7.4 Consequence Analysis

Consequence analysis [13] should be conducted to identify and evaluate the range of potentially hazardous events that can result from loss of engineering and administrative controls, which are discussed in more detail in Section 7.9. The failure of all or some of these controls can lead to or contribute to potentially hazardous events with varying consequences that must be understood by the PHRA team. Consequence analysis includes small events with relatively minor consequences that occur more frequently to worst case events that occur rarely but may have catastrophic consequences. The PHRA team must understand the range of possible events and the protection layers to safeguard against significant event consequences for effective management of process risks.

Depending on the scope and requirements of the PHRA, the consequence analysis may vary from a relatively quick, but thorough, review of possible hazardous events to a lengthy, detailed evaluation involving complex modeling that has previously been developed by experts as part of detailed facility siting or similar reports. The consequence analysis should be reviewed or developed early in the PHRA schedule. This helps the PHRA team understand the types and severities of hazardous events that are possible and helps the team ensure thorough and high-quality review of associated process risks. Additional information or resource needs can also be identified and obtained, if needed, to support team understanding of complex events. The consequence analysis may be periodically reviewed during the PHRA to refresh the team on possible events and to allow revision if needed based on other parts of the PHRA.

If a previous or separate consequence analysis study is not available, the PHRA team should identify the range of potentially hazardous events that can occur and document for each event the following:

- **Scenario**, such as the operating conditions, types and amounts of materials involved, duration, and other relevant information. For example, a toxic release resulting from a pipe failure would include the toxic materials that are present, the operating temperature and pressure, flow rates, the size of the pipe or of the hole in the pipe, and the duration of the release. Scenarios should be developed for smaller, higher frequency events to worst case, lower frequency events to bracket the possible hazardous events that can occur. In some cases, it is also desirable to vary the success or failure of different engineering and administrative controls identified later in the PHRA to better understand the range of events that are possible.

- **Consequences**, such as toxic exposure, fire, explosion, and vessel burst. In some cases, one scenario may need to be listed more than once and evaluated for different consequences. For example, when both toxic and flammable materials are part of a release, the scenario would be listed as both a toxic event and a fire/explosion event for further evaluation. The consequences should consider the potential for human health effects (e.g., toxic exposures or burn injuries), significant property damage, and environmental damage.

- **Area**, such as the geographic area potentially exposed to the consequences of the event scenario. The area impacted is estimated based on appropriate levels of concern, including appropriate safety margins, for the type of event. For example, the distance to the ERPG levels as discussed in Chapter 6 for toxic releases determines the area most likely impacted by exposure injuries, and the distance to the lower and upper flammability range (LFL and UFL) for flammable materials determines the area most likely impacted by fire and explosion events. The area impacted may be estimated qualitatively in some cases, and in other cases, may require detailed consequence modeling [14–15] and weather information to make quantitative predictions. Potential impacts may be both on-site and off-site, and in particular, more quantitative predictions of the area impacted may be required to evaluate scenarios where off-site populations may be affected. Modeling can also be used to determine the maximum events, such as the size (e.g., release rate and duration) of a toxic release scenario, that limit the impacts to on-site locations and may be useful in evaluating the effectiveness of possible safeguards.

- **Effects**, such as more-detailed evaluation of the safety and health effects of exposure of personnel to toxic materials or burn injuries from fires and explosions, as well as subsequent property and environmental damage. For example, the effects of exposure to toxic materials should include discussion of exposure pathways, target organs, impact on emergency evacuation, and, in some cases, the number and severity of potential injuries. Similarly, thermal exposure limits can be used to evaluate potential burn injuries from process flares or fire events.

In many PHRA studies, more qualitative evaluation of event consequences is acceptable, such as listing the consequences as small, medium, or large, depending on regulatory and facility requirements. In this approach, the following or similar definitions may be used:

- small – consequences are limited near the immediate process area
- medium – consequences are limited to the general facility
- large – consequences can occur off-site, possibly impacting public areas.

The following sequence, as shown in Figure 7.4, is helpful for conducting the detailed CA studies, although, in some cases, separate reports may also be available for the team to review, as needed:

Develop scenarios – review the process hazards identified in the hazard identification section of the PHRA and systematically list potential hazardous events that may result from the failure of administrative and engineering controls. For example, start with toxicity process hazards and list the range of possible events for all toxic materials and mixtures. Events may vary based on different safeguard failures, potential human health effects from exposure, location in the process, operating conditions, and other specific

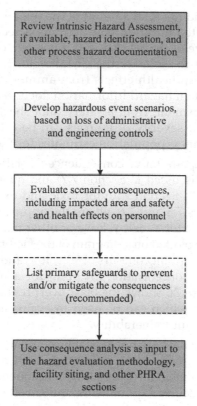

FIGURE 7.4
The steps for a consequence analysis.

scenario parameters, such as pipe or hole size, release rates, and typical weather conditions, if appropriate, that describe a range of possible events from small impact to worst case.

For example, an ammonia leak could occur from a small hole in a pipe or could result from the catastrophic failure of a piping component or storage tank. Effects from exposure to ammonia can range from eye and respiratory irritation to possible fatality. Once all toxicity events have been listed, continue to flammability and other process hazards until all potential events have been identified.

Evaluate consequences – For each event scenario, the rate/amount and duration of material releases or other event consequences should be estimated, appropriate levels of concern should be used to determine the impacted area, and the safety and health consequences from toxic exposure or other effects should be described. For example, the rate of release of anhydrous ammonia can be determined from the operating conditions and size of hole at the release point, and the total amount of the release can be calculated from the estimated duration. The area impacted can be evaluated based on the size of the release, and if needed, consequence models can be used to estimate the distance to the ERPG values of ammonia. Human health effects from ammonia exposure can be described based on hazard information documented in the intrinsic hazard assessment, from a safety data sheet, or other source. The number and severity of injuries can be estimated if desired from the occupancy/population estimates for the impacted area.

Examples of quantitative consequence modeling used in facility siting studies (discussed in Section 7.7) and quantitative risk analysis (discussed in Section 7.10) for toxic and explosion events are shown in Figures 7.5 and 7.6, where the appropriate levels of concern (e.g., ERPG-2 for a toxic release or other parameters discussed in Chapter 6) are plotted on a diagram of the facility. Figure 7.5 shows the calculated endpoint contours of 90%, 10%, and 1% of the released material's lethal concentration (in ppm) for a release duration of 5 minutes. Figure 7.6 shows the blast overpressure contours of 3.0, 0.9, and 0.6 psig which can be used to evaluate building damage levels and occupant vulnerability.

If quantitative data are not available, qualitative estimates, based on team operating experience, process incidents, simple calculations, etc., are often suitable for estimating consequences, depending on regulatory and facility requirements, and are often described simply as small (local impact), medium (plant-wide impact), and large (offsite impact), as previously discussed. Frequently, this information may be provided as part of facility siting evaluations provided by modeling experts.

Identify safeguards (recommended) – The primary safeguards provided to either prevent or mitigate each event scenario should also be listed, as discussed in Section 7.9. This provides a high level view of the administrative and engineering controls currently in place and is useful to the PHRA team in the Hazardous Event Evaluation section of the PHRA. The Bow Tie diagram [33] and other barrier methods discussed in Chapter 3 can be used to help visualize the different barriers that need to be in place to help ensure that significant hazardous event scenarios do not occur.

The results of consequence analysis reports should also be provided to resources working on facility siting and other parts of the process safety program, such as incident response planning, as discussed in Chapter 11.

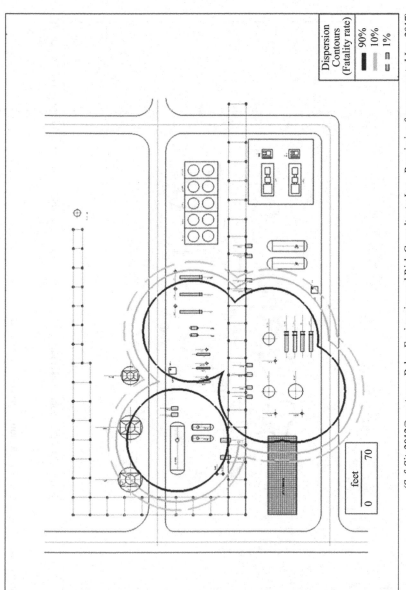

(SafeSite2012© contours, Baker Engineering and Risk Consultants, Inc. —Permission for use granted Jan. 2017)

FIGURE 7.5
An example of consequence modeling—Toxic dispersion contours.

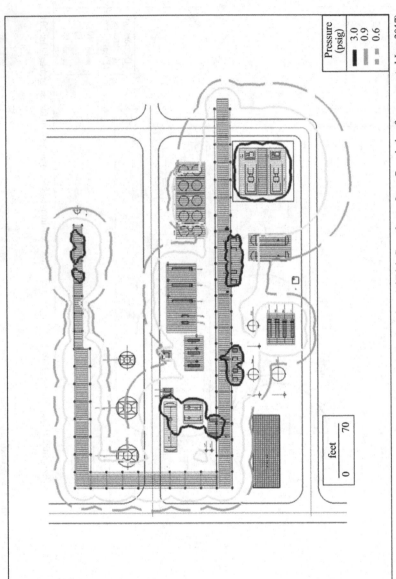

(SafeSite2012© contours, Baker Engineering and Risk Consultants, Inc. –Permission for use granted Jan. 2017)

FIGURE 7.6
An example of consequence modeling—explosion overpressure contours.

7.5 Hazardous Event Evaluation

Hazardous Event Evaluation (HEE) requires application of a detailed methodology [1] to:

- identify specific equipment and operational failures, such as pump failure or human error in following procedures, and to determine whether they can lead to potentially hazardous events, such as the events identified in the consequence analysis
- evaluate the availability and reliability of engineering and administrative controls for preventing the failure or for mitigating the potential event consequences
- assess the process risk associated with the potential event consequences in terms of the frequency rate of the failure and the effectiveness of the safeguards
- determine the need for upgrading existing safeguards or for providing new safeguards to improve process safety gaps, based on PHRA team evaluation and process risk criteria, such as a qualitative risk matrix, if provided.

The HEE methodology is therefore intended to identify all the pathways via equipment and operational failures that can lead to potentially hazardous events and to ensure that sufficient protection layers are provided, based on PHRA team evaluation and applicable risk criteria. The result is a comprehensive and very detailed list of possible engineering and administrative control failures, their consequences, and associated safeguards. If desired, the Bow Tie diagrams [33] or similar methods, as discussed in Chapter 3, can be used to visualize significant event scenarios and safeguards. Layer of protection analysis (LOPA) is often used to quantify the effectiveness of safeguards in achieving desired risk criteria (see Section 7.10).

In many cases, a specific equipment or operational failure may not lead to a hazardous process event, resulting in only product quality problems or other operational but not safety difficulties. Many failures will however lead to hazardous events requiring more detailed analysis of the safeguards and process risk. The frequency of equipment or operational failures can be estimated from published data or from operational experience [16–17]. The combination of the failure frequency rate and safeguard availability and effectiveness is used to evaluate the ultimate event consequences and frequency, commonly using qualitative risk matrices and/or LOPA. Additional discussion of risk analysis is provided in Section 7.10.

Many methodologies are available to support conducting a comprehensive HEE, as shown in Table 7.2, and in some PHRAs, it may be beneficial to use a combination of methods. The main methods are HAZOP and What-if/

TABLE 7.2

Hazard Event Evaluation Methodologies [Modified from 1]

Method	Application
What-if	Simple technique involving brainstorming of what-if and other questions to understand the consequences of equipment and operations failures.
Checklist	Keyword and/or detailed checklists to either be completed as written or to help guide a brainstorming session to develop questions about possible equipment and operations failures.
What-if/Checklist	A combination method where brainstorming of questions is completed followed by use of keyword and/or detailed checklists to develop additional questions to help ensure comprehensive consideration of possible equipment and operations failures.
Hazard and operability (HAZOP)	Very commonly used structured method where deviations from specified process design and operating conditions are evaluated to identify possible resulting hazardous events. HAZOP uses standard guidewords and process variables to develop the list of deviations which are then evaluated for possible consequences.
Failure model and effects analysis (FMEA)	A systematic method for evaluating possible equipment-related failure modes and the effects or consequences that can result from the failures.
Fault tree analysis (FTA)	Primarily an event frequency methodology, FTA is used to develop a graphical diagram based on a top event, such as an explosion, and a logic tree of the various failures that can lead to the event. Quantitative estimates of the top event frequency can be made based on the failure frequencies in the logic diagram.

Checklist, as discussed in the following sections. In all cases, trained leaders must be available to ensure proper application of the methodologies. HAZOP, for example, would not be suitable for design reviews when process design and drawings have not been finalized. Typical columns for documenting PHRAs in a worksheet, for example, are shown in Table 7.3. Although worksheets may be suitable for simple PHRA evaluations, commercial software [18] is often used to provide standardized approaches and analysis tools to facilitate conducting and documenting PHRA results and recommendations.

7.5.1 HAZOP

HAZOP, which is short for HAZards and OPerability study, is the most commonly used HEE methodology [1,18]. A HAZOP is a highly structured methodology which rigorously addresses potential deviations from the intended process design and operating conditions that can lead to

TABLE 7.3

Example Hazardous Event Evaluation Worksheet

No.	Question/ Deviation	Answer/Deviation Causes	Consequences	Severity (S)	Frequency (F)	Risk Ranking	Safeguards	Recommendations / Action Items
1.1	High flow	Valve fails open	Overfill tank with toxicity hazard in local area	C3	F2	R2	High flow alarm; High level alarm	Consider providing a redundant high level transmitter with alarm
1.2	Low/no flow	Valve fails closed	See high pressure				Low flow alarm	
1.3	Reverse flow	No cause						
1.4	Misdirected flow	No cause						
1.5	High temperature	Heat controller fails high	See high pressure					
1.6	Low temperature		No consequence					
1.7	High pressure	Low/no flow; heat controller fails high	Release with toxicity hazard in local area	C3	F1	R1	High pressure alarm; High pressure interlock; Pressure relief valve	
1.8	Low pressure		No consequence					

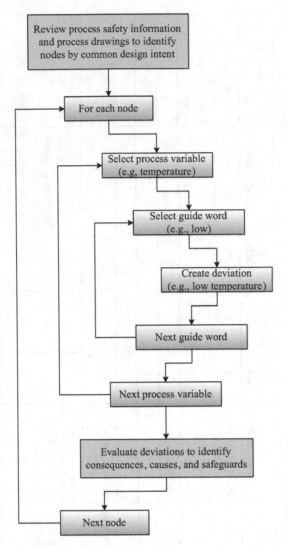

FIGURE 7.7
The HAZOP Methodology.

potentially hazardous events. A well-defined process design is required to support the detail needed to conduct HAZOPs, so it is usually not used for early project design reviews. The steps for completing a HAZOP are shown in Figure 7.7 and include:

- **Identify nodes** – Nodes are segments or subsections of the process, typically specific equipment or piping lines that have a clear design intent or function. Nodes are identified from the piping and instrumentation diagrams (P&IDs), which must be available for the entire

TABLE 7.4

HAZOP Methodology [Modified from 1]

Example Guide Words	Example Process Variables
No	Flow
Less	Pressure
More	Temperature
Part of	Level
As well as	Time
Reverse	Composition
Other than	Reaction
Sooner/later than	Speed
	Agitation
	pH

process, and should be selected carefully based on equipment function, compositions, flow rates, and/or operating conditions. Nodes include major pieces of process equipment, such as reactors, columns, feed tanks, and storage vessels, and piping lines between equipment. Nodes generally include common equipment, such as valves and instruments in a section of pipe that have common design intent.

- **Process variables** – For each node, relevant process variables should be described that define the intended design and operating conditions. Table 7.4 provides a list of common process variables, though not all will be needed for each node. Each process variable will be examined in turn in the HAZOP.

- **Guide words** – For each process variable in the node, a set of guide words is used to develop a list of possible deviations to be evaluated by the PHRA team. Table 7.4 provides a list of common HAZOP guide words, which again may not all be needed for each process variable or node. Each guide word will be used in turn with the group of specified process variables for the node to create the list of deviations to be evaluated.

- **Evaluate deviations** – The combination of the process variable and the guide word defines the process deviation (e.g., high flow) which must be considered by the PHRA team to understand the significance and potential for a hazardous event scenario. Available safeguards must also be identified and evaluated for impact on event frequency, consequences, and risk (see Section 7.10). Examples of how the process variables and guide words are combined to determine deviations for team review are shown in Table 7.5.

HAZOPs requires a well-trained leader and team to make sure the methodology is applied correctly, and depending on the scope of the review, plenty

TABLE 7.5

Example HAZOP Deviations [Modified from 1]

Guide Word / Process Variable	Flow	Temperature	Pressure	Level	Composition or State	Reaction	Time	Sequence
More	High Flow	High Temperature	High Pressure	High Level	Additional Phase	High Rate	Too Long	Step Too Late
Less	Low Flow	Low Temperature	Low Pressure	Low Level	Loss of Phase	Low Rate	Too Short	Step Too Early
None	No Flow		Vacuum Pressure	No Level		No Reaction	Not Started	Step Left Out
Misdirected	Misdirected Flow							
Reverse	Back Flow				Change of State	Reverse Reaction		Step Backwards
Part Of	Missing Ingredient				Wrong Concentration	Incomplete Reaction		Part of Step Left Out
As Well As	Contaminant		High/Low Pressure Interface	Liq/Liq Interface	Contaminants	Side Reaction		Extra Action in Step
Other Than	Wrong Ingredient			Loss of Containment	Wrong Material	Wrong Reaction	Wrong Time	Wrong Action Taken

Other deviations: Loss of containment, agitation/mixing, speed, inerting, sampling, maintenance, pH, and others, as needed.

of time should be scheduled for completion. This method can be tedious with detailed review of deviation after deviation, so the PHRA meeting schedule should provide breaks or changes in activity to keep team energy high, if needed.

The level of detail required by the HAZOP methodology helps identify the many different equipment and operating failure pathways that can lead to potentially hazardous events. Since HAZOPs are focused on deviations from normal or standard, steady state conditions, the PHRA team should also consider other process states such as startup, shutdown, maintenance, and emergency operations, as appropriate. In some cases, additional check-lists, including topics such as loss of containment or external factors such as utility loss or extreme weather conditions, can be helpful by providing perspective outside of the normal design to help ensure that the HAZOP is comprehensive.

7.5.2 What-If/Checklist

The What-if/Checklist methodology [1,18] is a relatively easy-to-apply HEE method, which combines the what-if and checklist approaches. Since it is sim-ple, it is a good method for small-scope reviews, especially if the process/equipment being reviewed is not complex. The What-if/Checklist approach is also good for early stages of projects where a final design has not been developed. The steps for completing a What-if/Checklist review are shown in Figure 7.8 and include:

- **Brainstorming questions** – the what-if methodology consists of directed brainstorming of questions about the process which may have potential for hazardous events. The method relies on creativ-ity and team member experience in developing and listing relevant questions without initially evaluating them. Questions often consist of "what if," but other types of questions, based on who, what, where, when, why, and how can also be used. P&IDs can be used to help focus the scope of questioning, but general questions based on over-all knowledge of the process should also be developed. Depending on the complexity of the process, use of smaller segments based on function, composition, operating conditions, etc., can be used to develop questions in a more focused way.
- **Checklist questions** – After question brainstorming, checklists are used to help ensure the comprehensive development of relevant questions. Checklists are reviewed to see if additional questions are developed by the team to add to the question list. There is no need to answer every question on the checklist, unless specified by the organization's policy. Checklists can consist of key words, detailed questions, and/or specialized lists on specific subjects, such as loss

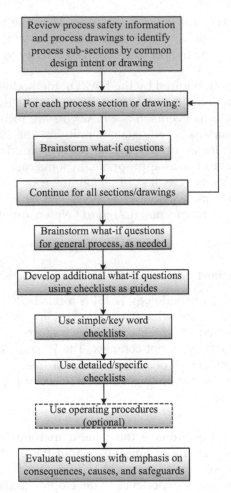

FIGURE 7.8
The what-if/checklist methodology.

of containment, specific process equipment, or process technologies (e.g., acrylic polymerization systems). Operating procedures, incidents, and other technical and safety documentation can also be used as checklists to stimulate additional questions, if desired. Topics for commonly-available checklists are shown in Table 7.6.

- **Question evaluation** – Once all questions have been developed, they must be answered to determine if hazardous events are possible, to identify potential consequences, and to evaluate available safeguards. Questions can be reviewed by the PHRA team together, or they can be assigned to different team members to develop draft responses. The team must then meet later to review the draft answers to accept, reject, or revise them. Depending on the scope of the PHRA, the

TABLE 7.6

Typical What-If/Checklist Topics
[Modified from 1]

Material properties and handling
Chemical reactivity
Process equipment and piping
Control, instrumentation, and alarms
Operations and sampling
Protective and emergency equipment
Storage
Facilities and layout
Utilities and heating
Maintenance, test, and inspection activities
Waste streams and disposal
Codes and standards
Construction
Dismantling, mothballing, and removal

---

number of questions can be quite large, so it is often practical to have team members work separately or in small groups to develop the initial answers and then to schedule subsequent time for team review.

A key advantage of the What-if/Checklist method is that questions can be asked about any subject that comes to mind and is not limited by the specific design of the process. Team training and an effective leader are needed, however, to keep the team focused to some degree and to ensure that questions are comprehensive, so that as many potential equipment and operating failures can be identified as possible. It can be helpful, for example, for the What-if/Checklist leader to be knowledgeable about HAZOP process variables and guide words to help stimulate a wide range of questions from the team and to provide some additional structure to the review. Different structured What-if/Checklist approaches have been developed that combine some of the structure associated with the HAZOP method and the creativity associated with the What-if/Checklist method [19–20].

7.5.3 Other Methodologies

A number of other HEE methodologies are available, which can be used individually or more commonly in combination with the HAZOP or What-if/Checklist methods. These include:

- **Failure mode and effect analysis (FMEA)** – The FMEA methodology is often used in equipment reliability and other evaluations to explore the causes and consequences of specific equipment failures.

FMEAs are used to evaluate the impact of equipment failures on the process and whether the failures can lead to potentially hazardous events. Each piece of equipment is examined for all possible failure modes, such as pump failures of on (when it should be off), off (when it should be on), leaks, etc. The effects of each failure mode in terms of immediate and system consequences are determined, and safeguards can be evaluated for effectiveness. FMEAs can be helpful for complex systems to identify the many failure modes, to determine different failures that can lead to the same event, and to distinguish common mode failures that can lead to multiple consequences.

- **Fault tree analysis (FTA)** – FTA is primarily an event frequency methodology, but it is sometimes listed as a HEE methodology. FTA builds a graphical diagram based on a top event, such as a runaway reaction or loss of containment, and builds a logic tree of the various failures that can lead to the event. By using known or estimated probabilities for the failures, the frequency of the top event can be determined quantitatively. FTA requires expert leaders and is generally used to help quantify specific event frequencies, when needed. As a result, it is often used to support risk analysis when the cost of quantification is justified. In particular, FTAs are used when documenting safety instrumented system reliability.

- **Layer of protection analysis (LOPA)** – See Section 7.10.

The details on these and other methods are outside the scope of this book, but are discussed elsewhere [1,21].

7.6 Human Factors

Human factors [18,22] must be addressed throughout the PHRA. The PHRA, in addition to an equipment and technical focus, is intended to reduce the likelihood of human error and also to anticipate it to ensure that appropriate safeguards are in place to prevent incidents. Issues such as human performance, workplace and equipment design, instrumentation and control system design, human response, and other topics shown in Table 7.7 must be evaluated. In the PHRA, human factors are usually considered in the:

- **HEE methodology** – human actions are considered by (1) identifying potential human errors as causes of deviations from design intent and (2) evaluating if operators will have adequate warning,

TABLE 7.7

Human Factors Evaluation Topics [Modified from 11 and 22]

Operating procedures, job tasks, and safe work practices
Training and performance
Instrumentation and alarm design and response
Control system design and layout
Human/machine interface, including feedback display and controls
Facility, process, and equipment design, layout, and labeling
Fitness for duty (alcohol, drugs, stress, and fatigue)
Ergonomics (accessibility, lifting, and display)
Safety culture and practices
Maintenance procedures and activities
Emergency procedures and response
Workplace distractions
Workplace environment, including housekeeping
Workloads, schedules, and staffing
Communications
Labeling and signs
Operational discipline programs

time, information, equipment/controls, and training/procedures to respond to and prevent or mitigate potentially hazardous events.

- **Human factors checklist** – specialized human factors checklists are available [11] or can be developed to provide specific questions for the PHRA team to consider based on the topics in Table 7.7. The checklist is usually completed separately, but in some cases may also be used to develop additional questions for a What-if/Checklist review, depending on the organization's guidance.

- **Human factors field tour** – a separate field tour is sometimes conducted with sole focus on human factors. The PHRA team must be able to visualize the workplace and equipment design, equipment accessibility, emergency response locations, etc. For example, if emergency spill equipment is located such that operators would need to possibly walk through a spill to get to the equipment, then new or multiple storage locations may be needed. Similarly, if emergency valves are located such that some people would not be able to quickly reach them, they may need to be made more accessible. In some cases, human factors reviews that include field tours may be conducted with human factors specialists separately from the main PHRA study.

7.7 Facility Siting

Facility siting evaluates the potential impacts of hazardous events at all locations on the site, with emphasis on the design and location of control rooms and occupied buildings to help protect site personnel and critical safety equipment [18,23–25,35]. A facility siting review, usually based on a checklist and/ or review of a specialized study completed separately, is conducted as part of each PHRA. At larger sites, a site-wide FS review is often completed based on more quantitative analysis that consolidates the results from all site PHRAs to develop a comprehensive view of potential hazardous events and their impacts on personnel and buildings throughout the facility. The PHRA consequence analysis primarily evaluates the types and range of hazardous events for the scope of the PHRA being conducted. The site facility siting review collects all site PHRA consequence analysis and facility siting results to further evaluate the potential event impacts for the entire facility. If a facility is covered by both the U.S. OSHA PSM and EPA Risk Management Plan regulations, the study may be documented as a combined facility/stationary source siting review.

The intent is to determine if control room and occupied building design and location place site personnel at higher risk of serious injury. For example, if a control room is located in a high hazard area, is it designed to protect occupants by withstanding a potential explosion? Similarly, if an office building is within the potential hazard zone of a toxic release, is it properly designed to protect building occupants or are other appropriate shelter-in-place or safe haven locations available? All potentially occupied buildings must be evaluated, including permanent buildings, temporary buildings, portable buildings, and trailers. In the BP Texas City fire and explosion in 2005 [26], 15 workers were killed and 170 were injured, many of whom were working in trailers near a high hazard area of the refinery. Management of change procedures, as discussed in Chapter 10, should be used to evaluate the location, design, and occupancy of new buildings and trailers or revision to existing buildings or control rooms.

In most PHRAs, the facility siting review will usually be based on a checklist [11,18,23], as shown in Table 7.8, supplemented by review of a specialized study, if available. For larger facilities with many hazardous processes, an overall site facility siting study should be developed with results from each of the individual PHRAs, as needed. If developed, the site facility siting study is often then scheduled as a separate study for review on the same frequency as site PHRAs, typically five years. When an overall study has been completed, subsequent PHRAs should document any changes that needed to be reviewed as part of the site study when it is due for revalidation. Examples might be installation or removal of processes, building changes, or different building occupancies.

When in-depth evaluation of facility siting in addition to or as a substitute for checklists is needed to meet regulatory, industry, or company

TABLE 7.8

Example Facility Siting Checklist Topics [Modified from 11 and 23]

Spacing and layout of process equipment

Location, storage, and handling of chemicals

Location, occupancy, and construction of control rooms

Location, occupancy, and construction of site buildings (permanent, temporary)

Location and construction of possible public receptors

Location and construction of barricades

Location and construction of safe havens (permanent, temporary)

Location and adequacy of emergency equipment (PPE, showers, eyewash, etc.)

Location and adequacy of containment

Electrical classification

Emergency and contingency planning

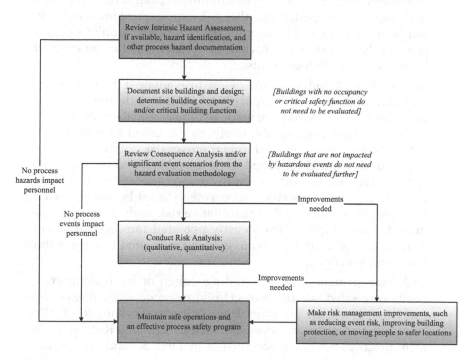

FIGURE 7.9
The steps used when performing facility siting analyses.

requirements, the steps shown in Figure 7.9 [23–25] are often followed by the PHRA team or a specialized team meeting specifically for this purpose, including:

- **Hazard review** – Review the site intrinsic hazard assessments or equivalent documentation to determine if process hazards are present that can impact facility buildings and personnel. If no process hazards are present, then no further analysis is required.

- **Building documentation review** – Develop a list of site buildings, control rooms, trailers, and other temporary or portable structures. Document intended occupancy, including transient location of workers in a building, critical safety functions, such as temporary safe havens or emergency equipment, and basic building design features. If buildings have no occupancy or critical safety function, they can usually be eliminated from further review.

- **Consequence review** – Compile different hazardous events that have been identified in all site PHRA consequence analysis and facility siting studies and determine if they impact site buildings (and in some cases, off-site receptors). This can be done based on specific spacing criteria or through use of consequence modeling depending on the type of event being evaluated. If buildings are not impacted by hazardous events, they can be eliminated from further review. If buildings are impacted, then either risk management improvements can be made, if needed, to improve the protection for building occupants or the next step, risk review, is conducted to further evaluate the impact of hazardous events.

- **Risk review** – Conduct a risk review to evaluate if the risk associated with potential event impact is tolerable or if design or operating improvements need to be made to reduce risk. Risk reviews can initially be completed qualitatively, as discussed in Section 7.10, but in some cases detailed quantitative risk analysis may be required. If risk is determined to be tolerable based on applicable risk criteria, then no further improvements are needed. If risk criteria are not met, then risk improvements must be made to meet tolerable risk requirements. Quantitative risk analysis may be conducted, for example, when a costly potential improvement, such as relocation of a control room, is being evaluated to ensure that relocation is needed.

- **Risk management** – When building design or location improvements are needed to better protect building occupants, critical safety functions, or off-site receptors from hazardous events, the PHRA team must identify and evaluate various options. Risk management practices are discussed further in Table 3.3 and Section 7.10. Examples might be to change building occupancy, provide explosion barriers, improve building construction, or further reduce the risk of the hazardous event being evaluated.

The results of a facility siting review vary with the types of hazardous events that are possible. For toxic events, the evaluation may focus on the ability of people to safely evacuate, the availability of appropriate personal protective equipment, design and location of temporary or permanent safe havens, emergency response, and medical treatment. For fire/explosion events, the specific damage to buildings must be evaluated to determine if the building is damaged or destroyed, causing harm to building occupants. Facility siting documentation often includes consequence contour plots on facility maps, as shown in Figures 7.5 and 7.6, to help evaluate the consequences to personnel and equipment at the site.

7.8 Inherently Safer Processes

The purpose of Inherently Safer Process (ISP) reviews is to identify opportunities to permanently eliminate or reduce process hazards rather than continuing to manage or control them through various safeguards [27]. The primary methods of ISP are discussed in Chapter 3. Processes using ISP designs are discussed in Chapter 5. ISP reviews are often conducted from detailed checklists that further develop the methods in Table 3.2, depending on the organization's policy for PHRAs and capital projects. Sample review questions could include:

- Are there opportunities to limit the use and/or inventory of hazardous materials?
- Can higher flash point materials be used to replace lower flash point materials?
- Can water-based systems be used to replace flammable materials?
- Can lower toxicity materials be used?
- Can lower vapor pressure materials be used?
- Can refrigeration be used to reduce the vapor pressure of hazardous materials?
- Can lower temperature/pressure operating conditions be used?
- Can lower pH or lower concentrations be used?
- Can piping headers be simplified to avoid confusion?
- Can piping headers be designed to only allow flow to one location rather than several?

The greatest benefit from ISP reviews occurs in the design or early project stages for new processes, equipment, or operations because safety can be significantly improved without the cost of additional or less-effective safeguards (see Chapter 5). ISP reviews may also be conducted for commercial processes to identify improvement opportunities, but costs associated with ISP changes at this point may be prohibitive.

7.9 Protection Layers (Layers of Protection)

A primary purpose of the PHRA is to evaluate the effectiveness of existing engineering and administrative controls and to identify if additional protection layers are needed to meet process risk criteria and help ensure safe and reliable operations. Major safeguards may be identified in the Consequence Analysis and detailed identification and evaluation of safeguard effectiveness occurs as part of the hazardous event evaluation reviews. Safeguards should be evaluated for effectiveness based on:

- **Specificity** – is the safeguard action effective for the purpose for which it is being used? Has the design basis (e.g., for a pressure relief device) been documented that demonstrates effectiveness for the intended application?

- **Independence** – is safeguard action independent of the basic process control system, other process equipment or safeguards, and human action?

- **Dependability** – is the safeguard reliable so that it will actually work for the event scenario if needed? Is redundancy (e.g., multiple of the same type of sensors) desirable? Is diversity (e.g., different types of sensors or control systems) desirable? Is separation (e.g., different locations of sensors or control inputs/outputs) desirable?

- **Auditability** – can the safeguard be inspected or tested to validate readiness?

- **Integrity** – was the safeguard properly installed and is it being maintained correctly?

Risk analysis, as discussed in the next section, is also used to help quantify the risk reduction benefits associated with multiple protection layers.

Different types of safeguards are available to help prevent and mitigate potentially hazardous events depending on the specifics of the process hazards and design. A hierarchy of risk management strategies for safeguards is discussed in Chapter 3, as shown in Table 3.3. Inherently safer process solutions to eliminate hazards are always preferred if practical. Use of passive, active, and administrative safeguards should be used in that order if possible. In many cases, short term administrative safeguards may be needed until the ultimate engineering safeguards can be installed.

Use of the Bow Tie diagram [33] and other methods for visualizing and evaluating protection layers is discussed in Chapter 3. Although the design and effectiveness of all safeguards must be documented and evaluated, the design of critical safeguards, such as alarm design and management, SIS, pressure relief, and containment, must be thoroughly reviewed and documented to ensure that appropriate reliability and effectiveness are obtained.

The PHRA team must evaluate these requirements for existing safeguards, specify the requirements for new safeguards, and ensure that practices are in place to ensure proper installation, maintenance, and testing of all safeguards. Layer of protection analysis (LOPA) may also be conducted, which is discussed in Section 7.10.

7.10 Risk Analysis

The potential result of the various PHRA methodologies is a set of recommendations for improving the safety of the process being evaluated. Recommendations must be well-written and supported by sufficient analysis to clearly describe what needs to be done and why. Risk analysis [21,28] is often completed to (1) evaluate event consequences to help determine if recommendations are needed, (2) better understand the risk reduction benefits of the recommendations, and (3) help justify the cost and other resource requirements associated with completing the recommendations. Risk reduction focuses on reducing hazardous event frequency and/or consequences, and sufficient protection layers must be provided to meet risk tolerance criteria. Risk analysis helps ensure that recommendations are not made solely on the potential consequences associated with hazardous events, unless required by specific regulatory or organization guidance. In many cases, higher overall risk may result from lower consequence, more frequent events when compared to worst case, very low frequency events. Risk criteria are set by organization guidance or, in some cases, based on regulation [29].

Depending on the scope of the PHRA and the resulting recommendations, qualitative risk analysis based on use of a risk matrix, as discussed in the next section, may be sufficient. In other cases, more quantitative risk analysis methods may be required to ensure that costly recommendations are justified in terms of the risk reduction benefits. Recommendations should also consider if interim actions are necessary to help improve safety when longer term projects are needed to fully implement recommended improvements.

7.10.1 Qualitative Risk Analysis

A qualitative risk matrix, as shown in Figure 7.10, is often used to determine the adequacy of available safeguards in the hazardous event evaluation methodology and to help develop recommendations for additional safeguards. Deviations in a HAZOP that lead to consequences of interest, such as toxic releases or fires, are often evaluated using the organization's risk matrix to evaluate the frequency and severity of the consequences. The risk ranking determined from these values will help the PHRA team evaluate if recommendations are needed to improve or add safeguards to achieve an

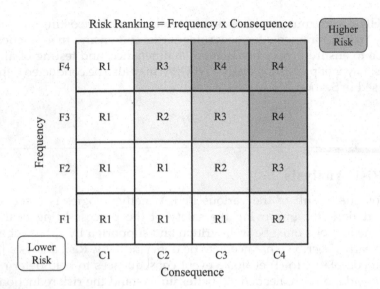

FIGURE 7.10

A example of risk ranking using a risk matrix.

acceptable risk category. In Figure 7.10, for example, a frequency of F3 and a consequence severity of C4 results in a R4 risk ranking. Based on review of the definitions provided in Table 7.9, this would require the team to make recommendations to achieve at least an R2 risk ranking, typically

TABLE 7.9

Example Risk Matrix Definitions [Modified from 1]

(a) Consequence severity	
C1 Minor	No or minor injuries
C2 Moderate	Moderate injuries (restricted work day)
C3 Major	Serious injuries (lost work day)
C4 Catastrophic	Fatalities or multiple serious injuries
(b) Event frequency	
F1 Extremely unlikely	>1 in 10,000 years
F2 Very unlikely	1 in 1000 to 10,000 years
F3 Unlikely	1 in 100 to 1000 years
F4 Likely	<1 in 100 years
(c) Risk ranking	
R1 Acceptable	Improvement opportunity
R2 Tolerable	Recommendation optional, improvement opportunity
R3 High	Recommendation required to reach R2 or better
R4 Very high	Recommendation required to reach R2 or better

by adding or improving safeguards that reduce the event frequency (see Figure 3.2).

The risk matrix can also be used to review all recommendations to confirm that they are justified and contribute to reduction in process risk if they have not already been evaluated as part of the hazardous event evaluation. More detailed risk analysis may be required for some recommendations to help quantify risk, especially if recommendations are very costly or the risk benefits are hard to estimate. Different sizes of risk matrices (e.g., 3 × 3, 4 × 4, or 6 × 4) are used by different companies, based on differing definitions of frequency and consequence categories.

The process for using a qualitative risk matrix includes:

- **Consequence severity**, based on review of Table 7.9a. Extensive consequence tables can be developed using injury, property damage, environmental effects, public impact, cost, or other organization criteria. For example, an event could be classified as C4 if a fatality could occur, but also could be C4 without injuries based on significant environmental or other impacts, as defined by organization guidance. Unmitigated, worst case consequences are generally used, but it is also possible to evaluate mitigated consequences based on successful action of some or all available safeguards, if desired. Separate risk matrices are sometimes developed depending on the type of consequence being evaluated (e.g., injury, public, and environment).

- **Event frequency**, based on review of Table 7.9b. The frequency is determined from estimates of the initiating failure rate (such as a pump seal leak) and the effectiveness of available safeguards (such as pressure relief) in mitigating the potential consequences, based on team experience or available failure rate tables [17]. The Bow Tie diagram discussed in Chapter 3 is a good way to visualize the barriers that are in place and that either work or fail leading to the specific event consequence being evaluated.

- **Risk ranking**, based on using the risk matrix and the rankings provided in Table 7.9c. The risk category is used to determine if PHRA recommendations are needed to improve process risk for the event being evaluated.

The criteria for defining risk categories and for when recommendations are required to improve risk are established based on organization or regulatory guidance. If the risk category for the event is low, the PHRA team may choose to identify and document improvement opportunities, such as additional safety improvements or minor action items, which are often documented separately from the PHRA report. The PHRA team should also consider the impact of poor OD on the risk evaluation, as discussed in Chapter 4, if appropriate.

7.10.2 Layer of Protection Analysis

Layer of protection analysis (LOPA) is commonly used to semi-quantitatively estimate process risk by developing an order-of-magnitude estimate of event frequency for a specific event scenario and consequence [30–32]. LOPA is based on barrier analysis, as discussed in Chapter 3, and provides a rule-based methodology for evaluating if the event frequency meets criteria developed by an organization. The major steps for applying LOPA are shown in Figure 7.11 and involve:

- **Scenario development** – The scenario is developed by applying a Hazardous Event Evaluation (HEE) methodology, such as HAZOP, to determine a specific initiating event failure (cause), such as a specific equipment failure (e.g., pump failure) or human error (e.g., open wrong valve), which leads to a specific hazardous event (consequence). Only one cause–consequence pair at a time can be evaluated by LOPA. For example, failure of a transfer hose could lead to release of ammonia during truck unloading, resulting in a toxic exposure hazard.

- **Target frequency (F_{target})** – If a risk matrix is used to assign a risk rating as part of the HEE, the risk rating can be used to determine if LOPA is required or optional. LOPA is usually not completed for all scenarios but is completed for the highest risk rankings based on organization criteria. Some companies may also specify that the highest consequence severity levels should also be evaluated using LOPA. Based on the risk or consequence rating, a target frequency (F_{target}), such as 1 event in 10,000 years (10^{-4}), is established using organization criteria.

- **Initiating event frequency (F_{init})** – The frequency of the initiating cause (F_{init}) is determined from frequency data based on organization or industry data tables. The failure frequency for a transfer hose, for example, may be one failure in 10 years (10^{-1}).

- **Enabling condition (EC_i)** – In some cases, the initiating event frequency can be adjusted with modifiers to reflect the actual time at risk. For example, if a hazardous process is operated only for one month a year, the frequency could be reduced by 1/12. Since LOPA primarily relies on order-of-magnitude, conservative calculations, the initiating frequency in this case would be reduced by 10^{-1} to reflect the actual time at risk versus continuous operation.

- **Conditional modifiers (CM_i)** – Conditional modifiers may also be used, such as probability of ignition of a flammable release, probability of personnel in the area, or probability of fatality. Since estimates of these probabilities can be very difficult, these consequence modifiers should be used with caution.

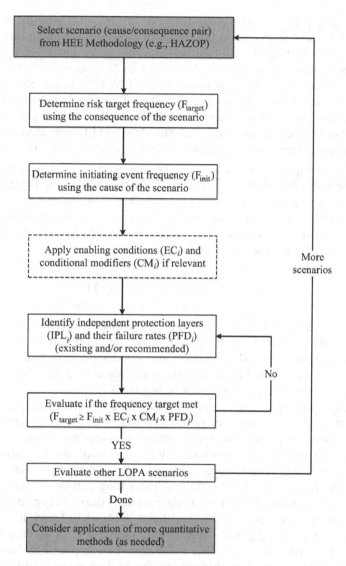

FIGURE 7.11
The steps for LOPA.

- **Independent protection layers (IPL)** – Independent protection layers (IPLs) are safeguards (barriers) that prevent the initiating event or mitigate the consequence to a lower severity. For use in LOPA, IPLs must be *applicable* to the specific cause–consequence scenario being evaluated in terms of preventing (or, in some cases, mitigating) the consequence, *independent* of both the cause and other safeguards, and *auditable*, usually as part of the facility mechanical integrity program. The probability of failure on demand (PFD$_i$) of IPLs is used to evaluate the reliability and potential benefit of the IPL, designating the IPL with a safety integrity level (SIL). Highly dependable IPLs may have PFDs of 1 in 10 years (SIL 1, 10^{-1}), 1 in 100 years (SIL 2, 10^{-2}), or more rarely, 1 in 1000 years (SIL 3, 10^{-3}). As the reliability increases, the cost and testing requirements also increase, so in some cases, several SIL 1 IPLs may be preferred to higher level IPLs. See ANSI/ISA Standard S84 [32] for additional discussion of safety instrumented systems and safety integrity levels.

- **Target frequency evaluation** – The overall predicted frequency for the cause–consequence pair scenario is evaluated using Equation 7.1. This is compared to the target frequency to determine if the number of available IPLs are sufficient or if one or more recommendations are required for additional or more reliable IPLs to meet the organization risk tolerance criteria.

$$F_{\text{target}} \geq F_{\text{predicted}} = F_{\text{init}} EC_i CM_i PFD_i \tag{7.1}$$

A variety of different equipment failures (cause–consequence pairs) and layers of protection can be quickly evaluated using LOPA, though to be effective, LOPA requires a qualified leader and good sources of failure rate data. LOPA can also be useful for semi-quantitatively evaluating the risk reduction achieved by addition of new IPL safeguards.

A simple example is provided in Table 7.10 for reactor overpressure and burst from a runaway polymerization due to loss of cooling water. The primary independent protection layers to prevent overpressure include a safety instrumented system that automatically injects an inhibitor solution to stop the runaway reaction and a pressure relief valve to vent pressure from the reactor. The failure rate of the cooling water loop can be determined from available data sources, which for a control loop failure, is typically 10^{-1}. The two safeguards are very reliable and have PFDs of 10^{-1} and 10^{-2}. The predicted frequency $F_{\text{predicted}}$ for a reactor burst from loss of cooling water for this scenario is therefore 10^{-4}. This value must be compared to the organization LOPA criteria, which in this case is 10^{-4}, to determine if additional safeguards are needed. Since the target frequency is met, no recommendations are required for this example.

7.10.3 Quantitative Risk Analysis

Quantitative risk analysis (QRA) [28] is outside the scope of this book. The main goal of a QRA is to identify and evaluate the consequences and associated risks of fire, toxicity, and explosion hazards to on-site personnel. In some cases, it can require complex methodologies, such as Fault Tree

TABLE 7.10

Example LOPA Evaluation

Scenario:
Reactor burst from runaway polymerization due to loss of cooling water resulting from a control loop failure ($F_{init} = 10^{-1}$)

Available Safeguards/IPLs:
1. Safety instrumented system to automatically inject shortstop inhibitor, which is independent of the basic process control system ($PFD_1 = 10^{-1}$)
2. Pressure relief valve ($PFD_2 = 10^{-2}$)

LOPA Analysis:

Target frequency – company criteria (F_{target})	10^{-4}
Loss of cooling – control loop failure (F_{init})	10^{-1}
Enabling conditions (EM) – none	1
Conditional modifiers (CM) – none	1
SIS failure rate (PFD_1)	10^{-1}
PRV failure rate (PFD_2)	10^{-2}
Predicted frequency ($F_{predicted}$) = $(10^{-1})(10^{-1})(10^{-2})$ =	10^{-4}

Since $F_{predicted}$ (10^{-4}) equals F_{target} (10^{-4}), no recommendations are required.

Analysis discussed in Section 7.5, and can be very time consuming and costly to complete. Because of this, QRA is most often used when recommendations can be very costly to implement, such as moving the location of a control room to outside of a potential blast zone, or when dealing with difficult to analyze scenarios. An example QRA that shows the predicted risk contours for a facility is shown in Figure 7.12. Depending on the analysis, these risk contours, depicting risk of fatalities/year, can be used to show the societal risk, the risk for someone outside, or the risk for someone inside a building (e.g., a frame structure with masonry walls and a concrete roof), such as part of a facility siting study. In all cases, qualified QRA resources should lead these studies with involvement of other specialized resources as needed.

7.11 PHRA Revalidation

PHRA revalidations are conducted after the initial or baseline PHRA has been completed. Revalidations are usually conducted every five years in the United States, due to the U.S. OSHA PSM regulation, but could be conducted more frequently if there have been many process changes or incidents. Revalidations can be conducted as a complete redo of the initial PHRA, a partial redo of certain sections, or an update and revision based on changes and incidents since the initial PHRA.

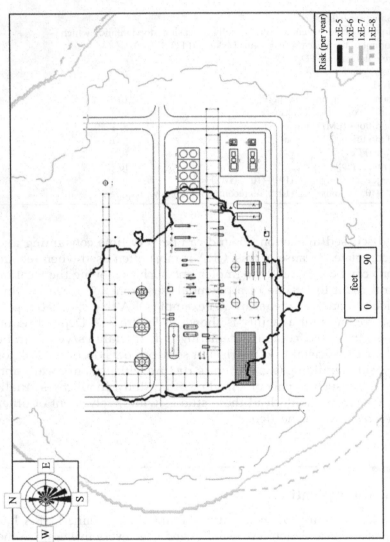

(QRATool© contours, Baker Engineering and Risk Consultants, Inc. —Permission for use granted Jan. 2017)

FIGURE 7.12

An example of a quantitative risk assessment—Risk contours.

The steps involved in a PHRA revalidation [11] include:

- **Gather information**: The initial PHRA should be reviewed to determine if it is complete and meets current regulations, with any deficiencies noted. The status of existing recommendations should be checked to see if all have been completed. Other useful information includes: management of change records, incident reports, supplemental hazard studies (e.g., a site-wide facility siting study), and audit reports.

- **Set scope**: Based on the number of changes, deficiencies, and other factors identified in the document review stage, the scope of the PHRA revalidation as a redo, partial redo, or revise and update is established. Some companies may also require that a redo be completed every 2 to 3 revalidations to help ensure the PHRA is thorough and current after 10 or more years of process operation and change. Some key considerations for determining the scope of revalidation PHRAs are provided in Figure 7.2.

- **Conduct the revalidation**: Redo or partial redo PHRAs may be conducted as described in this chapter for the various parts of the PHRA, though the existing prior PHRA is often used to simplify the study. Revise and update PHRAs usually consist of team review of change documents, process incidents, and previous PHRA recommendations to ensure that PHRA sections are updated as needed. For example, some of the changes may involve installation of the new equipment, and the PHRA team should consider if these changes create new deviations, causes, or safeguards in the HAZOP tables. Similarly, process incidents, including incident recommendations, may lead to revision of the consequence analysis, HAZOP, and other parts of the PHRA.

Revalidation PHRAs must be conducted by teams following similar procedures as discussed in Section 7.2. A new PHRA report must be documented describing how the revalidation was conducted and include any new recommendations that have been developed.

7.12 Measures of Success

☑ Design and implement an effective process safety system for evaluating and managing process risks.

☑ Assess for process risks using comprehensive and accurate process technology/process safety information, including documentation on the material and process hazards, the process design basis, and the equipment design basis.

☑ Complete a thorough Process Hazard and Risk Analysis (PHRA) on processes handling hazardous materials and energies, including:

    o Properly chartering the PHRA so that the scope, timing, resources, and other expectations are clearly defined

    o Ensuring that an experienced, trained, and multi-disciplinary PHRA team is assigned, including qualified methodology leaders

    o Ensuring that the selected methodologies and checklists are adequate for the scope of the PHRA scope (e.g., what-if/ checklist, Hazards and Operability studies (HAZOP), human factors checklists, facility siting checklists, etc.)

    o Developing well-written and supported recommendations based on reducing process risk

    o Documenting a comprehensive PHRA report

    o Communicating the PHRA results to affected personnel

    o Tracking the recommendations until closure.

References

1. Center for Chemical Process Safety. 2008. *Guidelines for Hazard Evaluation Procedures,* 3rd ed. Wiley-AIChE, New York.
2. Center for Chemical Process Safety. 2007. *Guidelines for Risk Based Process Safety.* Wiley-AIChE, New York.
3. U.S. Chemical Safety Board. 1998. *Chemical Manufacturing Incident.* Report No. 1998-06-I-NJ. www.csb.gov.
4. U.S. Chemical Safety Board. 1998. *Catastrophic Vessel Overpressurization.* Report No. 1998-002-I-LA. www.csb.gov.
5. U.S. Chemical Safety Board. 2002. *The Explosion at Concept Sciences: Hazards of Hydroxlamine Case Study.* Report No. 1999-13-C-PA. www.csb.gov.
6. U.S. Chemical Safety Board. 2004. *Catastrophic Vessel Failure.* Report No. 2004-11-I-KY. www.csb.gov.
7. U.S. Chemical Safety Board. 2006. *Sterigenics.* Report No. 2004-11-I-CA. www.csb.gov.
8. U.S. Chemical Safety Board. 2009. *T2 Laboratories, Inc. Runaway Reaction.* Report No. 2008-03-I-FL. www.csb.gov.
9. U.S. Chemical Safety Board. 2011. *Pesticide Chemical Runaway Reaction Pressure Vessel Explosion.* Report No. 2008-08-I-WV. www.csb.gov.
10. U.S. Chemical Safety Board. 2010. *Case Study: Explosion and Fire in West Carrollton, Ohio.* Report No. 2009-10-I-OH. www.csb.gov.
11. Frank, Walter L. and David K. Whittle. 2001. *Revalidating Process Hazard Analyses.* AIChE, New York.

12. Center for Chemical Process Safety. 2010. *A Practical Approach to Hazard Identification for Operations and Maintenance Workers*. Wiley-AIChE, New York.
13. Center for Chemical Process Safety. 1999. *Guidelines for Consequence Analysis of Chemical Releases*. AIChE, New York.
14. Center for Chemical Process Safety. 1996. *Guidelines for Use of Vapor Cloud Dispersion Models*, 2nd ed. AIChE, New York.
15. Center for Chemical Process Safety. 2010. *Guidelines for Vapor Cloud Explosion, Pressure Vessel Burst, BLEVE, and Flash Fire Hazards*. Wiley-AIChE, New York.
16. Center for Chemical Process Safety. 1989. *Guidelines for Process Equipment Reliability Data*. AIChE, New York.
17. Center for Chemical Process Safety. Process Equipment Reliability Database. www.aiche.org/ccps/resources/process-equipment-reliability-database-perd.
18. (a) PHA LEADER. *www.absconsulting.com;* (b) PHA PRO. www.cholarisk.com; (c) PHA WORKS. www.primatech.com.
19. Card, Alan J. et al. 2012. Beyond FMEA: The structured what-if technique (SWIFT). *Journal of Healthcare Risk Management*. 31(4):23–29.
20. Clements, C. Curtis. 2012. *The Structured What If/Checklist Hazard Evaluation Methodology*. 8th Global Congress on Process Safety, Houston, Texas.
21. Rausland, M. 2011. *Risk Assessment: Theory, Methods, and Applications*. Wiley, New York.
22. Center for Chemical Process Safety. 2007. *Human Factors Methods for Improving Performance in the Process Industries*. Wiley-AIChE, New York.
23. Center for Chemical Process Safety. 2012. *Guidelines for Evaluating Process Plant Buildings for External Explosions, Fires, and Toxic Releases*, 2nd ed. Wiley-AIChE, New York.
24. American Petroleum Institute. 2009. *RP 752: Management of Hazards Associated With Location of Process Plant Buildings*, 3rd ed.
25. American Petroleum Institute. 2007. *RP 753: Management of Hazards Associated With Location of Process Plant Portable Buildings*.
26. U.S. Chemical Safety Board. 2007. *Refinery Explosion and Fire*. Report No. 2005-04-I-TX. www.csb.gov.
27. Center for Chemical Process Safety. 2009. *Inherently Safer Chemical Processes: A Life Cycle Approach*, 2nd ed. Wiley-AIChE, New York.
28. Center for Chemical Process Safety. 2000. *Guidelines for Chemical Process Quantitative Risk Analysis*, 2nd ed. AIChE, New York.
29. Center for Chemical Process Safety. 2009. *Guidelines for Developing Quantitative Safety Risk Criteria*. Wiley-AIChE, New York.
30. Center for Chemical Process Safety. 2001. *Layer of Protection Analysis: Simplified Process Risk Assessment*. AIChE, New York.
31. Center for Chemical Process Safety. 2013. *Guidelines for Enabling Conditions and Conditional Modifiers in Layers of Protection Analysis*. Wiley-AIChE, New York.
32. ANSI/ISA S84. 2004. *Functional Safety: Safety Instrumented Systems for the Process Industry Sector*.
33. Vaughen, Bruce K. and Kenneth Bloch. 2016. Use the Bow Tie Diagram to Help Reduce Process Safety Risks. *Chemical Engineering Progress* 112:30–36.
34. UK HSE. 2016. *Policy and Guidance on Reducing Risks As Low As Reasonably Practicable in Design*. www.hse.gov.uk.
35. Center for Chemical Process Safety. 2017. *Guidelines for Siting and Layout of Facilities*. Wiley-AIChE, New York.

Section III

Practical Approaches for Implementing Process Safety

8

Operate Safe Processes

Safe operation costs much less than unsafe operation.

Kenneth Bloch

Why We Need to Operate Safe Processes

Safe process operation is essential for reducing the process safety risks and for reducing incidents. To help provide training consistency and to help ensure personnel competency, administrative controls are established for everyone working with or supporting the hazardous processes. Because decisions can either directly or indirectly impact one or more phases in the equipment life cycle, it is essential that everyone has the operational discipline and conduct their operations consistently with established protocols. Hence, there are administrative procedures for designing, fabricating, installing, commissioning, operating, maintaining, changing and decommissioning the equipment. For example, engineers have management of change procedures, operators have operating procedures, and maintenance personnel have maintenance procedures. Emergency responders help reduce operational risks using emergency response procedures designed to reduce incident consequences. Other groups, including senior leadership, upper management, occupational health and safety, environmental, security, quality, purchasing, receiving and shipping, must have procedures recognizing the process safety hazards and risks in their day-to-day conduct to support safe operations. The procedures are written for each specific group, documenting the expectations for their conduct of operations—their expected role—for supporting safe operations, and for providing the standard by which personnel competencies can be evaluated.

Operate safe processes by ensuring that personnel
are provided appropriate procedures and training to
develop capability in operations and process safety.

8.1 Introduction

Every group within the organization plays a role in operating safe processes, from the groups directly managing the process hazards and risks, such as engineering, operations and maintenance, to the groups affected by the process hazards and risks, such as emergency response, occupational health and safety, environmental, security, quality, purchasing, receiving and shipping. However, it is senior leadership, whether corporate, regional or local, that will have the greatest impact on safe operations. Leadership that provides a nourishing safety culture drives the behavior of the organization, providing a working atmosphere to integrate the process safety efforts across all disciplines and process safety systems.

Inherent to operating safe processes, the training efforts across the organization depend on the effective communication between these different groups affecting or being affected by the equipment life cycle and the other process safety systems, such as:

- designing the process
- understanding the hazards and managing their risks
- operating the process
- maintaining process integrity and reliability
- changing processes safely
- managing incident responses and investigations, and
- monitoring process safety program effectiveness.

This chapter focuses on how to address the training and competency of those directly impacting the equipment life cycle (refer to Figure 5.1) and of those directly or indirectly impacting the effectiveness of the process safety system's Plan, Do, Change, Act (PDCA) life cycle (refer to Figure 3.7). Hence, how an organization can safely run their operations depends on the operational discipline of everyone from the shop floor to senior leadership. No one is exempt from their responsibility to manage their roles and responsibilities.

8.2 Key Concepts

8.2.1 Using Procedures

One of the main goals of administrative procedures is help reduce mistakes—to reduce the likelihood for human error. This can be achieved with a consistent and effective training approach on the expected behavior and actions which apply to everyone assigned to perform the same task or tasks. Personnel are expected to have access to and to follow these procedures, with effective training relying on well-written and up-to-date procedures, checklists and decision aids. For example, for personnel directly impacting the equipment life cycle, such as engineers, operators or mechanics, there are administrative management of change procedures for changing the equipment design, operating limits or inspections. Operators rely on written operating procedures that include a description of the steps for each operating phase, define the safe operating limits, and provide guidance on how to respond to process upsets. Mechanics and electricians have written maintenance procedures for their inspections and tests on the process equipment. In addition, safe work practices, permits and administrative procedures focusing on the hazards and risks associated with work activities in the process units, are written for and apply to everyone performing the tasks, including contractors. The safe work practice procedures and permits reflect how work is safely performed when hazards are present. The Center for Chemical Process Safety (CCPS) has provided guidance on developing, using, and ensuring that procedures are reviewed [1–5].

8.2.2 Providing Competency Training

If leadership at the facility takes too much for granted, thinking that their personnel should already know what the hazards and risks are, they may not recognize that their personnel may not have the basic knowledge of the hazards and do not have a full understanding of why their associated risks are dangerous. Personnel can perish when they lack the knowledge and awareness of the hazards and risks, and not teaching them the basic competencies increases the likelihood of and consequences from incidents. Building everyone's competency provides more organizational resilience for managing unexpected events, helping prevent serious incidents [6]. Keep in mind that training will range from basic awareness to detailed, skilled, and performance-based testing depending on the role and its tasks. All initial competency training must include basic awareness, such as an overview of the process and its associated hazards and risks. Depending on their level in the organization, specific written and performance-based evaluations must be required and documented. All personnel must have training documentation that verifies their knowledge, their skills and their abilities ensuring that they can safely carry out the duties and responsibilities specified in their procedures. Table 8.1 lists some of the general items that are used to sustain effective training programs. Developing process safety competency, no matter what discipline you work in, is discussed in further detail in Chapter 13.

8.2.3 Sustaining Safe Processes

The process safety procedures must be used, verified regularly, and sustained to ensure safe operations. It is only a matter of time when knowledge,

TABLE 8.1

Items for Sustaining the Effectiveness of Training Programs

Effective Training Programs
• Are designed and implemented for all personnel, including awareness training for personnel who will not be touching the equipment.
• Are designed and implemented to ensure that contractors are qualified to perform their tasks safely.
• Ensure that all personnel understand the overall process and the specific safety and health hazards associated with the process and their tasks.
• Ensure that the trained personnel are qualified to perform their tasks, including having the required knowledge, skills, and abilities to carry them out.
• Ensure that qualifications are verified through oral/written exams, skills demonstrations, simulator demonstrations, or a combination thereof.
• Ensure that training records are documented, including initial, and refresher training.

procedures, and training becomes outdated due to newer and better technologies, procedural changes, workarounds, or system improvements. When newer technologies change the processing materials, the processing conditions, or the equipment, they must be reflected in the procedures. Periodically scheduled audits of the practices performed in the field are used to verify that the procedures reflect the current practices. Because skills become dormant when specific tasks are not done very often, refresher training must be provided at a regular frequency to sustain both procedures and personnel competencies. In some cases, particular procedures may be legislatively mandated for annual reviews [7].

The capability of any organization depends on how well it successfully combines its physical assets, its capacity and its expertise in order to perform its core functions. When changes occur, whether driven by new technologies or organizational changes, a program must be in place to manage and update the existing procedures and to ensure personnel competencies. Thus, a facility's change management system, described in Chapter 10, must be in place to ensure that the changes are implemented across all disciplines [1,8–9].

8.3 Developing Procedures: Process Safety Systems

Effective process safety systems depend on senior leadership providing the resources to manage the organization's process safety program policies, standards and guidelines—the corporate's administrative controls. Senior leadership must visibly endorse its "Safety, Health and Environmental" policies since its SHE policy establishes the principles that apply to everyone in the organization. As was depicted in Figure 3.6, process safety systems are the second protection layer that directly impact every effort required to safely manage and operate processes. The administrative parts of each process safety system must be developed to address the design, construction, commission, operation, maintenance, change and decommission phases of the equipment life cycle. Thus, these administrative controls affect each protection layer: the process and equipment designs, the basic process control systems and operating procedures, the critical alarms, the safety instrumented systems, the active or passive physical barriers and the emergency responses.

8.4 Developing Procedures: Safe Operations

Safe process operations require administrative procedures for designing, constructing, commissioning, operating, maintaining, changing and decommissioning the equipment (see the life cycle in Figure 5.1), and for responding to emergencies, ensuring everyone's health and safety, environmental stewardship, facility security and contractor safety. These procedures must be tailored for and used by each specific group involved in the day-to-day operations of the processes. Without them, the operational risks are increased, potentially resulting in incidents with serious consequences to people, the environment, the community surrounding the facility and the business.

Engineering Procedures

Engineers from many different disciplines impact the process and equipment design. Engineers apply mathematics and discipline specific knowledge in many different engineering fields, such as:

- Chemical engineers – apply chemistry, material properties, heat transfer, solid and fluid transport phenomena, unit operations, and process control to safely manage the materials and energies
- Environmental engineers – apply material properties to safely manage their potential impact to the air, land and the water
- Mechanical engineers – apply material properties, heat transfer, and solid and fluid mechanics to safely manage the reliability of the equipment and its instrumentation
- Civil engineers – apply material properties, including building and supporting structure design, to safely manage the infrastructure of the facility, and
- Geotechnical engineers – apply physical, mechanical and chemical properties of the soil and the earth materials to the terrain-specific location properties to safely manage the facility's infrastructure (e.g., includes the potential for earthquakes).

When any of these engineering disciplines change the design of the process equipment or construct new facilities, the discipline-specific, written procedures should include the steps to ensure that current, acceptable codes and standards have been used and the steps to verify that the facility or equipment was fabricated and installed per the design specifications.

Operations Procedures

Probably the most important administrative control emphasized for safe operations are the process unit's day-to-day operating procedures. The procedures must be written in the language that the personnel can understand. The operating procedures may have legislative requirements such that they receive much attention during process safety program audits.

To help ensure safe day-to-day operations, the written operating procedures should include:

- the safe operating limits, including the consequences of deviations and the steps required to correct or avoid the deviation, and
- the steps for each operating phase: startup, normal operations, temporary operations, normal shutdowns, emergency operations and shutdowns, and startups following a turnaround or after an emergency shutdown.

The different types of safe operating limits were shown in Table 5.3, with normal operations performing at or near the process aim and deviations from the quality or safety limits initiating responses form administrative and/or engineering controls. The limits in Table 5.3 are illustrated in Figure 5.3 based on the protection layer example discussed in Chapter 3 and shown in Figure 3.3, when responses occur due to system pressure increases. The standard operating conditions and both the manual and the automatic responses to alarms and interlocks should be documented in the normal operating procedures. The emergency shutdowns and emergency response procedures are activated when critical process or equipment design limits are exceeded. As shown in Figure 5.3, depending on the pressure, different protection layers are activated.

In addition, the written operating procedures should include or refer to the following:

- the health and safety aspects, such as the properties and hazards of the materials and the safety measures used to prevent exposure (i.e., engineering controls, administrative controls, and personal protective equipment)
- the descriptions of the engineered safety systems and their functions (e.g., interlocks, detection, suppression systems, etc.)
- the emergency response actions taken if physical contact or airborne exposure occurs
- the steps for controlling the quality of raw materials to ensure that potential material incompatibilities do not occur, and

- the steps for controlling the storage and inventories of hazardous materials to ensure that the hazards assessments are still valid, such as its facility siting study, and, or that legislative threshold values are not exceeded.

A summary of the different types of procedures and operating modes used for the day-to-day operations is provided in Table 8.2.

TABLE 8.2

Special Considerations When Writing Procedures for the Different Types of Operational Modes

Operational Mode Procedure	Item Included in Typical Procedure
	Addresses hazardous conditions that may exist when operating the process.

1 Normal operations	Descriptions are consistent with the information documented in the process technology (Chapter 5)
	Specifies a process aim based on the design intent
	Specifies standard operating conditions for the process aim (Table 5.3)
	References equipment manuals, as needed
	Specifies required personal protective equipment (PPE)
	Includes checklists and decision aids for troubleshooting, as needed
	Addresses minor process upsets/deviations from the design intent, documenting the:
	   • Consequences of deviation
	   • Steps required to correct or to avoid deviations
	   • Alarms and set points for normal operation and shutdowns
	   • Other engineering and administrative controls/safeguards which help reduce the risk of exposure to the hazardous materials (e.g., gas detectors, emergency isolation valves, etc.)
	For continuous processes: defined as the "steady state" mode

2 Normal startups and shutdowns	Addresses hazardous conditions which may occur when transitioning to/from steady state operations.
	
	May need to address or develop conditional procedures, checklists and decision aids to address potential hazardous conditions, such as:
	   • Additional PPE required during the transition
	   • Special handover protocols after maintenance or turnarounds
	   • Special startup protocols after emergency shutdowns
	   • Special operations during winter (i.e., potential freezing issues when shutdown)
	   • Startups after curtailed operations (e.g., during reduced customer demand)
	   • For mothballing equipment ("abandoned in place ")
	For continuous processes: defined as the "unsteady state" mode

(Continued)

TABLE 8.2 (*Continued*)

Special Considerations When Writing Procedures for the Different Types of Operational Modes

Operational Mode Procedure	Item Included in Typical Procedure
3 Emergency operations and shutdowns	Addresses how to immediately control or place the process in a safe state in response to hazardous conditions occurring during an incident.
	Responses defined by safe operating limits , critical operating limits and/or critical operating parameters (Table 5.3)
	Specifies steps for safe shutdown by qualified operators
	Specifies additional PPE, if needed
	For reactors in continuous or batch processes: Addresses actions to prevent runaway reactions or on how to respond them once they occur
4 Startups after turnarounds or emergency shutdowns	Addresses how to start up the process after a turnaround or emergency shutdown which may have left the process in a different condition or state than the process would be in after a normal shutdown.
	Specifies additional PPE, if needed
5 Temporary or special, nonroutine operations or activities	Addresses hazardous conditions that may occur when the normal or emergency operating procedures do not apply.
	Specifies additional PPE, if needed
	This includes testing on existing processes with new materials, equipment and procedures and should be formally addressed when planning for and implementing changes. If reactive materials are part of the proposed changes, there must be formal reactive chemical review and accurate documentation of the chemical reactivity risks.
6 Additional procedures	Addresses the controls for the raw material quality.
	Addresses the controls for the hazardous material inventories.

Other day-to-day administrative procedure-based controls include the facility's safe work practices and permits system. Operators must understand the scope of the work activities in their area, such as hot work or confined space entries, and must assist when preparing their process unit for electrical isolations and line breaks.

Maintenance Procedures

Probably the second most important administrative control emphasized for safe operations are the critical equipment's maintenance procedures (see types of critical equipment noted in Table 5.1). If the critical equipment is not "fit for service" at normal process design conditions, special risk reduction controls must be implemented to ensure safe operation or the equipment must be removed from service. The maintenance procedures may have legislative requirements for how inspections and tests should be performed and documented. Thus, the procedures receive much attention when the equipment integrity part of the process safety system is audited.

To help extend the useful life of the equipment and sustain the process and equipment integrity, the written maintenance procedures should document the inspections and tests on the critical process equipment, including:

- how the test or inspection was performed
- when the test or inspection was performed
- who performed the test or inspection, and
- the qualifications of who performed the test or inspection.

The test and inspection documentation must include the following to ensure that the equipment is fit for service:

- the "as found" condition
- the actions taken to address any deficiencies (e.g., recalibrations, repairs, replacements, or measures taken to ensure safe operation), and
- the "as left" condition.

Maintenance personnel must follow the facility's safe work practices and permits before working in a process area, as well. The safe work practices focus on reducing task risks, such as hot work, electrical isolation, line breaks, fall protection, heavy process equipment lifting and excavation. These safe work practices apply to both facility personnel and contractors.

Emergency Response Procedures

Unlike the operating and maintenance procedures that help reduce the likelihood of an incident, emergency response plans and procedures are administrative controls designed to reduce the consequences of the incident. Emergency responses are designed to reduce the operational risks and are considered as the eighth protective layer which is illustrated in Figure 3.6. The procedures for emergency responders are designed to ensure that personnel do not harm themselves when responding to process safety events. The presence of hazardous materials and conditions may be obvious, such as noxious fumes, fogs or smoke. However, in other situations, the hazardous nature of the released material may not be immediately apparent, as with odorless but poisonous and/or flammable vapors and liquids. The emergency procedures must address responder accessibility and the types of PPE for the situation, such as respirators for toxic releases or turnout gear for thermal hazards.

Occupational Safety and Health Procedures

Every facility handling hazardous materials and processes has occupational health and safety procedures that are designed for safe operations: protecting personnel from potential exposure to hazards in the work place. Not only do these procedures include managing the chemical hazards, such as toxicity and flammability (process safety hazards), they include mechanical electrical, thermal, pressure, gravity, ergonomic, biological, radiation and noise hazards. In particular, the occupational health and safety group includes industrial hygienists who create procedures for proper PPE selection and use and safety professionals who create procedures from safe worksite assessments. These administrative procedures describe the types of engineering controls, such as guarding for rotating equipment, insulation on hot surfaces or noise absorption barriers, as well as the required PPE, such as hart hats, safety glasses and ear plugs.

The safety procedures include the general safe work practices (SWP) and associated Personal Protective Equipment (PPE), such as hot work, electrical isolation (e.g., lockout/tagout), confined space entry, and opening process equipment or piping (e.g., line breaks). All SWP procedures define what type of PPE is required for the work (Table 5.2).

The permits complement the SWP procedures, addressing hazardous conditions that may occur when performing the work in areas with hazardous materials. Permits are designed to ensure effective communications on the hazards and risks and with the handovers between the operations personnel and the personnel performing the work in hazardous areas (i.e., those in control of the equipment have placed it in a safe state before the work begins). SWPs include the shift turnover procedures that convey the current equipment, unit, and facility status to the oncoming shift. The operating logs should have sufficient information documented, including the acceptable limits based on the standard operating conditions. Table 8.3 includes the typical activities addressed by SWPs, procedures, and permits.

Environmental Procedures

The environmental group focuses on ensuring that procedures address applicable air, land and water permits, helping guide the facility on these regulatory limits, especially on how to notify regulatory agencies if permit limits are exceeded. The facility's environmental procedures address safe operations by addressing "loss of containment" issues impacting air, land, or water. Examples include emitting volatile organics above their permitted limit, improper disposing of hazardous wastes, or exceeding effluent discharge limits to a river. These procedures may describe how a facility manages its volatile organic compound (VOC) emissions through a Leak Detection and Repair (LDAR) program, how hazardous wastes are managed in temporary storage locations, or how manifests are used when shipping hazardous

TABLE 8.3

Typical Activities Addressed by Safe Work Practices, Procedures, and Permits

Safe Work Practices, Controls, and Permits
• Hot work/fire permit
• Safeguard "bypass" procedures (e.g., interlock bypass, fire protection impairment, etc.)
• Electrical equipment isolation
• Mechanical equipment isolation
• Ignition source control
• Vessel entry (e.g., confined space entry)
• Opening process piping and equipment (e.g., line breaks)
• Pressurized cylinder gas storage, movement and use
• Excavation permit
• Electrical/high voltage permit
• Elevated work/fall protection
• Roof access
• Hot tapping of lines or equipment permit
• Explosives/blasting agent use
• Cranes/lifting over process equipment permit
• Hydroblasting permit
• Powered areal platform permit
• Scaffolding construction and use permit

materials. Some companies may include a tolerable risk matrix specifically addressing acceptable and unacceptable environmental risk based on the consequence and its likelihood. Some companies may add environmental risk criteria into their process hazards analysis procedure, requiring that no environmental residual risk remains after the scenarios have been discussed.

Security Procedures

Facility security is essential for safe operations, ensuring that visitors and deliveries are controlled upon entry to the facility. Vehicular traffic is often restricted in some areas, especially those with the potential for flammable atmospheres that could ignite due to a running engine. Other areas of a facility may be restricted to "authorized personnel," only. These procedures address the engineering controls, such as gates, monitoring cameras and fences with warning signs. Due the types and quantities of hazardous materials stored and handled at a facility, there may be required regulatory security measures and protocols that must be understood and followed. For example, in the United States, the Department of Homeland Security (DHS) requires vulnerability assessments through its Chemical Facility Anti-Terrorism Standards (CFATS) [10].

Supporting Group Procedures

Whether the materials have process safety hazards and risks or not, supporting groups are essential for a successful and safe operation. The support groups include quality, purchasing, receiving (warehousing) and shipping.

Each group has procedures dedicated to the success of their group-specific tasks, all which must be consistent with the expectations that the process units will be safely operated and that the materials entering and leaving the facility are handled safely. Each of these group's procedures and their impact on process safety risks are described briefly below.

The quality group writes procedures to help the business meet its customer expectations. This includes rigorous instructions for managing changes when the business is certified through the ISO 9000 quality management series [11]. Note that no business can survive in the long run if it does not consistently meet its customer's quality standards, and that no business will survive if it makes "quality" products unsafely (Figure 5.3). The quality procedures include Certificates Of Analysis (COA), which is an authenticated document issued by an appropriate authority that certifies the quality and purity of the material.

The receiving, warehousing and shipping groups support safe operations with procedures written to ensure that incoming raw materials and parts meet the process and equipment design specifications. Positive Material Identification (PMI) programs managed by receiving and warehousing include procedures using X-ray fluorescence (XRF) or optical emission spectrometry (OES) guns that analyze the metallic alloys, establishing the part's composition by reading the quantities by percentage of its constituent elements. The PMI program ensures that incompatible or substandard materials are not used during maintenance or turnaraounds. In most cases, the supporting group's procedures are incorporated and highlighted within the process unit's operating procedures.

Contractor Procedures

Contractors are expected to perform their work safely, adhering to procedures based on the standards and guidelines at the facility. They must follow safe work practices and procedures associated with their work, no matter whether their specialized work activities involve welding, removing or installing equipment and piping, entering vessels, working from lifts, lifting equipment, or excavating ditches or foundations. Safe operations depend on everyone performing their tasks safely, including those hired for special tasks in hazardous areas [12].

8.5 Developing Capability: Process Safety Systems

The capability of any organization depends on how well it successfully combines its physical assets, its capacity and its expertise in order to perform its core functions. With this definition, "process safety capability" is an organization's ability to combine an understanding of the magnitude of its process safety risks and its expertise to manage these risks, especially during emergencies. In other words, an organization's process safety capability could be defined by how resilient it is when stressed by an incident.

By developing process safety system capability, a facility will integrate its processes and equipment, including the engineering controls, with its administrative controls, and its resources to manage the process safety hazards and risks. A facility must have clearly established expectations and intents for each of its process safety systems—its core process safety system functions—with the following systems supported by engineering, operations, maintenance, or a combination of these groups:

- to design safe processes (engineering)
- to identify and assess process hazards (engineering and operations)
- to evaluate and manage process risk (engineering, operations, and maintenance)
- to operate safe processes (operations)
- to maintain process integrity and reliability (maintenance).

In addition, successful management of the process safety risks includes these process safety systems, as well:

- to change processes safely (everyone at the facility, including support groups)
- to manage incident response and investigation (emergency response and investigation teams)
- to monitor process safety program effectiveness (auditing teams).

It is evident that these process safety systems interact—managing process safety is complex—and that an organization must develop its capabilities for each group associated with each process safety system. Each group must understands its role in managing process safety with the organization providing support and resources to meet each group's outcome. The organization must clearly establish each group's outcome using the purposes for each process safety system listed in Table 3.4, with all phases of the equipment life cycle being addressed and understood by each group: its design, fabrication, installation, operation and maintenance phases (Figure 5.1).

8.6 Developing Capability: Safe Operations

After an organization has established its process safety system expectations, the capabilities of everyone must be developed to help ensure safe operation. As noted earlier, the different disciplines have specific skills that must be competently applied every day. This includes engineering, operations, maintenance, emergency responders when needed, as well as operations support personnel in safety, environmental, quality, purchasing, receiving, shipping and security. Contractors must be competent and must understand the facility- and process unit-specific hazards associated with the tasks they perform. Because procedures are developed to ensure consistently performed tasks for each group, they naturally become the common basis for training across the group to help ensure safe operations. This section briefly addresses the competencies expected from each of the groups which are used to build the company's capability for safe operations. Additional discussion on building your personal process safety capability is provided in Chapter 13.

Engineering Capability

The outcome of the engineering groups, whether the group is trained in chemical, safety, mechanical, electrical, civil, environmental, coastline or even offshore engineering, is to apply process, equipment and facility designs to safely manage the hazardous materials and energies and their associated risks. This includes storage and handling of hazardous materials, converting raw materials into products (i.e., process design, equipment design, equipment integrity design, and facility design) and ensuring that the decisions made in one engineering discipline do not adversely impact the process safety risks. A process support engineer working with the day-to-day operations must understand the process design as well as the equipment design, including how equipment integrity is maintained (i.e., its design, its expected operation, and how changes can be made safely). For example, pressure vessels are built to code, with specifications for the materials of construction and installed with proper supporting structures and foundations. These vessels must have sufficient instrumentation, relief system design, and an established equipment integrity program.

Engineering capability includes the discipline to follow the administrative procedures, such as those for managing engineering changes on the processes or equipment. The engineering groups must understand the codes and standards applicable to their discipline. When making changes, they must ensure that the design specifications for each phase in the equipment life cycle are communicated with other groups. Each of these disciplines must be properly resourced because they are often specialized, requiring both educational training and task competency.

Operations Capability

Although it may seem obvious, the outcome of the operations group process safety efforts focuses on safely operating equipment handling hazardous materials and energies. Poor operations, for whatever reason, simply affects the organization's bottom line. Operating capability hinges on following the operating procedures which define the safe operating limits and help ensure product quality. Production delays occur with both poor product quality and loss of containment incidents, affecting both internal and external customers.

The operators have extensive training expectations. They are subject to detailed auditing protocols to ensure that they have been trained properly, demonstrating specific competence in managing the process during each operating phase, including startups and shutdowns. Their capabilities extend to their ability to respond to unusual situations and how to troubleshoot issues without posing risks to themselves or others in the area.

Maintenance Capability

The maintenance group's process safety outcome focuses on the processing equipment's integrity, extending the equipment's useful life with routine maintenance and applying inspections and tests to determine that the equipment is still fit for service. Unexpected equipment failures cause production delays and incidents.

The mechanics and electricians have extensive training expectations and are subject to detailed auditing protocols to ensure that they have been trained properly. They must demonstrate specific competence in managing the hazards during their routine or emergency maintenance tasks. Their capability extends to their ability to respond to unusual situations and how they troubleshoot issues without posing risks to themselves or to others. The maintenance group capability hinges on well-developed inspections, tests and repairs, with discipline to follow the specific procedures designed to extend the useful life of the equipment.

Emergency Response Capability: Respond

The process safety outcome for emergency responders is to reduce risks by reducing the consequences of an event with prompt and effective responses. Emergency responders have extensive training expectations and are subject to detailed auditing protocols to demonstrate competence. Their capability extends to their ability to respond to unusual situations and how they troubleshoot issues, such as changing weather conditions or the possibility of domino effects (when nearby hazards may worsen the consequences), without posing risks to themselves or to others. Capability hinges on well-developed emergency response procedures that are understood by everyone at the facility, whether they are to respond by evacuating, sheltering in place, or donning PPE to address the emergency. Unfortunately, too many emergency responder fatalities have occurred when the responders were unaware of the hazards. For this reason, there are highly specialized hazardous materials emergency response procedures and qualifications designed to protect emergency responders. For example, in the United States, there are the hazardous waste operations and emergency response (HAZWOPER) requirements in addition to those expected for covered processes [7,13].

Occupational Health and Safety Capability

The outcome for the occupational health and safety group is to prevent injuries to personnel, an essential, complementary effort to process safety. Health and safety groups determine the workplace hazards, understand what engineering controls are used to protect workers, and establish the personal protective equipment and the safe work practices for workers when they are performing their tasks (see Table 5.2). Health and safety personnel develop the procedures describing the safe work practices and permitting systems, the administrative controls used to reduce operational risk. These professionals have extensive training expectations and are subject to detailed auditing protocols to demonstrate competence. Capability expertise includes, but is not limited to, regulatory compliance, fire protection, ergonomics and emergency response. Building their capability in the United States, for example, includes an understanding and application of exposure Threshold Limit Values (TLVs) recommended by the American Conference of Governmental Industrial Hygienists (ACGIH). Occupational safety and health professionals can be accredited as a Certified Industrial Hygienist (CIH) or as a Certified Safety Professional (CSP) [14–16].

Environmental Capability

The outcome of the environmental group is closely aligned with the process safety group's outcome: to prevent or mitigate loss of containment events that lead to spills or releases to the environment. From the process safety perspective, spills or releases of toxic or flammable materials may result in additional consequences beyond harm to the environment, such as fatalities, fires, and explosions. Environmental engineers must be able translate environmental legislation into facility-specific procedures, such as monitoring programs and procedures addressing permit limit deviations. The permit limits define the quantities of materials which can be released into to the air, the water or to the ground. Air permits for hazardous air pollutants include limits on volatile organic compounds (VOCs), which are often monitored with Leak Detection and Repair (LDAR) programs [17]. Hazardous waste permits are often legislated for "cradle to grave" responsibility by the organization, with special protocols for handling, storage, and disposing of toxic and flammable materials [18]. For this reason, spills of hazardous materials that do not result in a toxic release, fire, or explosion are still monitored by both environmental and process safety professionals, because these loss of containment "near misses" (or "near hits") could have been worse.

Security Capability

The outcome of a facility's successful security efforts combines both physical security risk reduction, such as effective fence line monitoring and facility accessibility, and electronic security risk reduction, such as effective firewalls to prevent loss of proprietary information. Because the security efforts require specialized training, both in designing effective physical fence line and electronic/Internet/computerized barriers, organizations tend to rely on specifically trained security organizations to manage the accessibility of people (i.e., employees, visitors, and contractors), to ensure controlled flow of materials into and out of the facility (e.g., delivery of raw materials, equipment and associated parts, etc.), and information technology (IT) expertise to prevent electronic security breaches. Capability for security personnel includes the discipline to follow procedures addressing facility access to employees, visitors, and contractors, as well as monitoring vehicle entry and departures, deliveries, and shipping. Capability for electronic security requires the discipline of everyone accessing the organization's intranet and servers to adhere to the company's IT procedures.

Supporting Group Capability

The primary process safety-related outcome from all facility and business supporting groups focuses on ensuring that all of the support procedures and activities are designed to safely manage the quality, procurement, logistics, shipping, and receiving of critical materials. These supporting groups complement operations and maintenance, with their impact on process safety risks including meeting:

- the equipment design specifications (i.e., through purchasing and receiving)
- the raw material and product quality (i.e., purchasing, receiving, and shipping), and
- the product monitoring and auditing criteria (i.e., quality).

Each group has procedures dedicated to the success of their group-specific tasks, all of which must be consistent with managing the facility's process safety risks. Capability of these supporting groups extends to meeting the expectations of those operating and maintaining the equipment in the process units at the facility, as well.

Contractor Capability

Because the size of the facility and its role in the organization may range from a one-site facility to a multisite/multinational company, the types and number of external resources—contractors—needed to operate safely will vary. These contractors support the facility with specialized tasks that cannot be resourced at the facility or within the company. Contractor outcomes include supporting safe operations across the many disciplines, such as:

- providing current process and equipment design technologies (engineering)
- meeting fabrication and installation specifications (construction)
- ensuring robust equipment testing, inspection, and repairs or replacements (maintenance)
- supporting emergency response (when needed), and
- securing the facility location and its operations (security).

Contractor capabilities when supporting process safety efforts focuses on them competently managing equipment integrity test and inspections, maintenance, repair, turnarounds, and major process changes.

8.7 Measures of Success

- ☑ Design and implement an effective process safety system for operating safe processes.
- ☑ Provide training to all of the operating groups and ensure that they are staffed with experienced and capable personnel.
- ☑ Ensure that operating procedures, safe work practices and permits are developed, authorized, and followed.
- ☑ Ensure that operating personnel have the appropriate tools to perform their tasks safely, with comprehensive controls for tool inspections and suitability for use.
- ☑ Schedule authorized reviews to keep procedures accurate with current practices.
- ☑ Schedule refresher training, as needed, to sustain personnel capabilities.

☑ Ensure and document that contractors have the training, knowledge and skills required for their tasks, including understanding:

- o The process hazards and risks,
- o The safe work practices and permitting systems,
- o The correct tools required for and their tasks,
- o Their tool inspections and suitability for use, and
- o How they must respond in emergencies, both specific to their work location and for the entire facility, if needed.

References

1. Center for Chemical Process Safety. 2007. *Guidelines for Risk Based Process Safety.* Wiley-AIChE.
2. Center for Chemical Process Safety. 1995. *Guidelines for Safe Process Operations and Maintenance.* Wiley-AIChE.
3. Center for Chemical Process Safety. 1996. *Guidelines for Writing Effective Operating and Maintenance Procedures.* Wiley-AIChE.
4. Center for Chemical Process Safety. 2006. *Guidelines for Mechanical Integrity Systems.* Wiley-AIChE.
5. Center for Chemical Process Safety. 2011. *Guidelines for Auditing Process Safety Management Systems,* 2nd ed. Wiley-AIChE.
6. Weick, Karl E. and Kathleen M. Sutcliffe. 2007. *Managing the Unexpected, Resilient Performance in an Age of Uncertainty,* 2nd ed. Wiley.
7. U.S. OSHA. 1992. 29 CFR 1910.119: *Process Safety Management of Highly Hazardous Chemicals.* www.osha.gov.
8. Center for Chemical Process Safety. 2008. *Guidelines for the Management of Change for Process Safety.* Wiley-AIChE.
9. Center for Chemical Process Safety. 2013. *Guidelines for Managing Process Safety Risks During Organizational Change.* Wiley-AIChE.
10. US DHS (Department of Homeland Security). Chemical Facility Anti-Terrorism Standards (CFATS). www.dhs.gov/chemical-facility-anti-terrorism-standards.
11. International Organization for Standardization. 2016. *ISO 9000 Series on Quality Management.* www.iso.org.
12. Center for Chemical Process Safety. 1995. *Contractor and Client Relations to Assure Process Safety.* Wiley-AIChE.
13. US OSHA. No date. 29 CFR 1910.120: *Hazardous Waste Operations and Emergency Response Standard (HAZWOPER).* www.osha.gov.
14. American Council of General Industrial Hygienists. 2016. ACGIH. www.acgih.org.
15. American Industrial Hygienist Association. 2016. AIHA. www.aiha.org.
16. Board of Certified Safety Professionals. 2016. BCSP. www.bcsp.org.
17. US EPA. 2007. *Leak Detection and Repair, A Best Practices Guide.* EPA-305-D-07-001. www.epa.gov.
18. US EPA. No date. *Hazardous Waste Guidance.* www.epa.gov.

9

Maintain Process Integrity and Reliability

Process and equipment integrity and reliability depend on the discipline to sustain all phases of the equipment life cycle.

Bruce K. Vaughen

Why We Need to Maintain Process Integrity and Reliability

Safe processes must be maintained to prevent loss of equipment integrity and reliability. The equipment has been designed to contain the hazardous materials, to prevent loss of control, to prevent loss of containment, and to reduce the consequences if control or containment is lost. Included in an effective maintenance program are written procedures and checklists, parts and supplies quality assurance programs, routine maintenance tasks (such as lubrication or gasket replacements), scheduled tests and inspections, and procedures for responding to the results of and fixing the issues identified during the tests and inspections. Sustaining equipment integrity and reliability requires an understanding of how the different phases in the equipment's life cycle impact or are impacted by the equipment's maintenance programs.

Maintain process integrity and reliability by ensuring that appropriate tests and inspections, preventative maintenance, and quality assurance programs are conducted.

9.1 Introduction

This chapter presents several key concepts for equipment life cycle maintenance programs, including identifying critical equipment inspections and tests and whether the programs are designed for equipment reliability or equipment integrity. A list of typical "critical equipment" used to handle hazardous materials and energies was provided in Table 5.1. Capable and competent maintenance personnel must be aware of and effectively implement the risk-based management system, including understanding the causes of equipment failure, developing, and implementing preventative maintenance (PM) programs, ensuring that the inspections and test results verify the equipment's fitness for service and that any identified equipment deficiencies are corrected. Similar to all the other process safety systems, for effective process safety programs, there must be a monitoring and auditing program for these maintenance programs. This chapter concludes with a case study of the maintenance-related issues which contributed to an incident.

9.2 Key Concepts

9.2.1 Inspection, Testing, and Preventative Maintenance Programs

A facility must have an effective inspection, testing and preventative maintenance (ITPM) program to maintain safe processes. ITPM programs are designed to prevent unexpected equipment failures which lead to process upsets, emergency shutdowns, and could potentially cause process safety incidents. If the equipment's integrity is thought of as the heart of a process unit, then the ITPM programs are the life blood for maintaining safe processes; they are an essential part of the facility's maintenance program. The planned and scheduled preventative maintenance activities are used to inspect the equipment's current condition or test its functionality, compare the inspection or test results to the equipment's design specifications, and if gaps are identified, then restore the equipment to its design specifications before it is recommissioned and placed back into service. Thus, equipment in use is validated for use – its fitness for service. If the equipment can no longer be restored, it must be removed and replaced. Although these planned maintenance activities are designed to help extend the useful life of the equipment, unexpected equipment failure will occur if the other aspects of the equipment life cycle are not managed effectively.

The elements that are expected in an effective ITPM program include the following [1–5]:

1) Maintenance procedures developed on current industry standards and guidelines for critical equipment and safeguards designed for safe operations (see Table 5.1 for a list of critical equipment)

2) Inspection and testing schedules with frequencies recommended by the manufacturer or determined through operating experience

3) Documentation of the equipment inspected or tested, when the inspection or test was performed (the date), the type of inspection or test, and of who performed the test (including qualifications or certification, if needed)

4) Documentation of the "as found" state, what action was performed to return the equipment to its design specifications (addressing deficiencies, if any), and what the "as left" state was before the equipment was placed back into service (i.e., evidence of the equipment's fitness for service).

Each of these elements is described in more detail below.

9.2.2 The Difference between Integrity and Reliability

Since there is a distinct difference between an "equipment integrity" program and an "equipment reliability" program, this section briefly describes the differences for integrity and reliability in context of the maintenance programs designed to sustain equipment, thus helping prevent loss of containment incidents from occurring and helping reduce the magnitude of the incident if it does occur. The definitions of "integrity" and "reliability" which can apply across all disciplines are as follows:

Integrity

Wholeness: The state of being sound or undamaged

Reliability

Dependable: Able to be trusted to do what is expected or has been promised

Likely to be accurate: Able to be trusted to be accurate or provide a correct result

For the purposes of this book, we will use the following definitions with our focus on the processing and the safeguarding equipment identified as a protection layer:

Equipment integrity

The aspect of the equipment to ensure that the equipment meets all of its design, fabrication, installation, commissioning, operation, and maintenance specifications, including any associated changes, such that the equipment retains its fitness for service through every phase of its life cycle.

Equipment reliability

The aspect of the equipment to perform under given conditions consistently *over time* meeting both its design specifications and its performance expectations.

Fitness for service

A systematic approach for evaluating the current condition of a piece of equipment in order to determine if the equipment item is capable of operating at defined operating conditions (e.g., temperature, pressure). [6]

Reliable equipment will run dependably and do what it is expected to do every time it is needed. Increased equipment reliability is achieved when every phase of the equipment's life cycle is sustained, from its industry-accepted design phase, through its quality fabrication and accurate installation phases, and when it is properly operated, maintained, and changed.

9.2.3 Ensuring Equipment Fitness for Service

Processing and safeguarding equipment sustain a long, useful life when inspection and testing programs are used to verify that they meet their design specifications—the equipment is fit for use. In other words, the equipment is trusted to do what it is supposed to do when it is supposed to do it. The equipment's health hinges on understanding what degrades the equipment's performance over time, including how both the environmental and operating conditions affect its useful life. Although equipment lifetimes differ due to their different uses and conditions, a general diagram depicting the equipment's "useful life" is shown in Figure 9.1. The likelihood that the equipment will fail when it is commissioned and at the end of its useful life when it is worn out is relatively high compared to its potential failure when properly maintained during its lifetime. Hence, due to the shape of the equipment's age versus failure potential curve, this diagram is commonly referred to as the "bathtub" curve. The critical failure times in the equipment's life is when it is most susceptible to failure—at its wear-in time ("birth" or "infant mortality" period) and when it is at is wear-out time (on its "death" bed and is worn out).

9.2.4 Maintenance Capability and Competency

Maintenance capability and competency depends on the written inspection, testing, and repair procedures that describe how the inspections and tests are performed, addressed, and documented. Employees and contractors performing these mechanical and electrical activities on the equipment

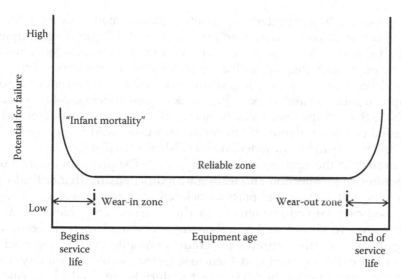

FIGURE 9.1
The "bathtub" curve depicting an equipment's reliability over its useful life.

are expected to develop and have the knowledge of the training, skills, and tools to perform their tasks. For example, if certified welders are required per the equipment or piping design specifications, they must be qualified and certified before they fabricate, install, or repair the equipment. To extend the useful life of the equipment, everyone's capability hinges on their discipline to follow their procedures, no matter which of the equipment's life cycle phases they impact.

9.3 Identifying Critical Equipment

Identifying the critical equipment for the inspection, testing, and preventative maintenance (ITPM) program is rather straightforward when the critical processing and safeguarding equipment have been identified (refer to Chapters 5 through 7). The critical equipment list will include all equipment which has been identified as a preventive or a mitigative protection layer (see Table 5.1). Typical processing equipment designed to manage hazardous materials and energies includes: vessels, such as storage tanks, pressure vessels, reactors, distillation columns, dryers, etc.; their associated piping and equipment, such as flanges, valves, pumps, compressors, etc.; their associated operation and control system instrumentation, such as the

BPCS, sensors, alarms, interlocks, safety instrumented systems (SIS), and emergency shutdown systems. Additional critical safeguarding equipment added to reduce the consequences of a loss of control include: active and passive physical protection, such as relief devices and systems, venting systems, dikes, flares, fire proofing, sprinklers, etc., and emergency response equipment, such as ambulances, fire trucks, foam systems, deluge systems, etc. Note that equipment located in electrically classified areas, areas where potential or known flammable materials or dusts could form an explosive atmosphere, should be included in an ITPM, as well [7–14].

To sustain the equipment's integrity, the ITPM program must be used to monitor the inspection and testing schedule, ensure that critical tools, materials, and replacement parts associated with equipment are available and used by qualified personnel. For this reason, companies include the purchasing, receiving, and warehousing personnel in their equipment integrity and quality assurance programs to monitor, verify, and track the use of critical equipment and their associated parts. The quality assurance program ensures that the correct materials are available during the equipment's scheduled inspections or tests. When unexpected equipment degradation or failures occur, procedures must be in place for temporary or emergency repairs.

At this point, we recognize that seven of the eight preventive and mitigative protection layers presented in Figure 3.6 may have equipment which can be identified as critical (see additional discussion in Chapter 5 on "critical equipment"). Although not all of the barriers will apply to all potential hazard/consequence scenarios, the barriers which *may have* critical equipment are as follows:

- Processing equipment (Barrier 1)
- Basic process control systems (BPCS, Barrier 3)
- Critical alarms with manual and automatic response (Barrier 4)
- Safety instrumented systems (SIS, Barrier 5)
- Active or passive physical barriers (Barriers 6 and 7), and
- Emergency response equipment (Barrier 8).

For each of these protection layers which have processing or safeguarding safety equipment identified through hazards assessments to help reduce the process safety risks, there should be an ITPM program in place on the critical equipment. As noted earlier, the ITPM program must have defined schedules for its inspections and tests, defined criteria for determining the equipment's fitness for service, and defined procedures that address deficiencies when they are detected. A robust ITPM program helps sustain the equipment through its useful life.

9.4 Identifying Causes of Equipment Failure

The basic goals of an effective ITPM program include understanding, addressing, and responding to the equipment's potential failure modes. Understanding and responding to potential failures before they occur extends the time between potential equipment failures, helping prevent unexpected shutdowns and increasing the equipment's reliability, integrity, and useful service life. The main objective is to provide operations with the longest, useful fitness for service life time as possible (see Figure 9.1).

One effective method for identifying potential equipment failures is through a formal root cause analysis (RCA) program on the critical equipment. RCAs are used to help identify what, how, and why a failure occurred so that steps can be taken to prevent future failures. The RCA process involves collecting failure data, charting the causal factors, and the identifying the root causes. Although the failure mechanism may be complex—no "one" root cause is the answer—RCAs are useful for understanding how to design the equipment's inspection and testing program.

An example of an RCA on a pump is presented in Figure 9.2 to help illustrate how an inspection and testing program can be developed around the pump. The failure conditions which lead to pump failure include poor installation and shaft misalignment leading to excessive vibration and premature bearing wear. Leaks occur when seals fail from poor installation, including using incorrect parts or incompatible materials, and from impeller or seal damage due to pump cavitation attributed to a blocked pump inlet or outlet. The consequences of the loss of containment range from operational and maintenance issues, such as process upsets or maintenance costs, to process safety incidents.

The reasons for unexpected equipment damage or deterioration include poor equipment design, fabrication, installation, operation, or maintenance and poor management of process and equipment changes. Hence, when design specifications do not clearly address the processing hazards, when construction does not follow design specifications, when the safe operating limits are not followed, or when incorrect or incompatible materials and parts are used, the equipment's integrity is jeopardized. In addition, more frequent scheduling of the ITPM program may be required when equipment is located outdoors or in corrosive environments. Since there are many ways that equipment can fail during its use, identifying the cause of failure requires a combination of design knowledge and operating experience to understand what to look for when performing the inspections and tests. Keep in mind that as new processes and equipment are being developed, the exact processing and environmental stressors on the equipment's integrity will not be known. However, most manufacturers and producers have

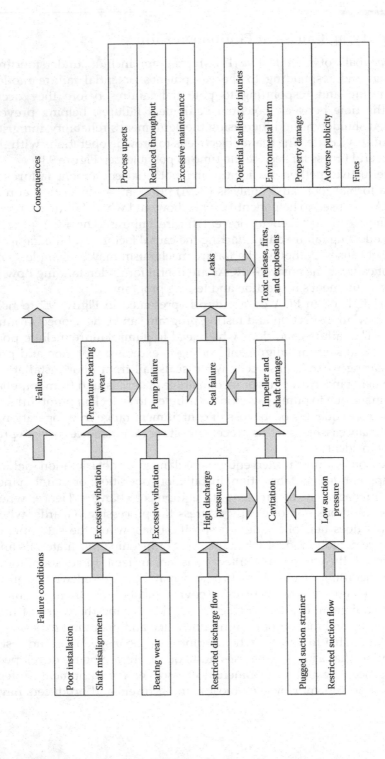

FIGURE 9.2
An example of a pump failure root cause analysis chart.

experience with similar equipment and processes, and, hence, can provide preliminary guidance on how best to maintain the integrity and reliability of the new equipment.

9.5 Developing an Effective Maintenance System

The most effective maintenance system to manage the equipment integrity and reliability uses a risk based approach for its ITPM program. Since equipment will eventually fail the longer it is service, the success of the maintenance program relies on understanding and preventing equipment failure. Once the causes of failure have been identified, preventative maintenance programs can be developed to improve the equipment's reliability. The rest of the organization then supports the equipment's integrity by sustaining the other phases in the equipment's life cycle. This includes a quality assurance program that establishes the requirements for those performing the maintenance and the administrative controls for the materials and parts used to maintain the equipment.

Equipment inspection, testing, repair or replacement must follow industry best practices or manufacture's guidelines and the materials and parts used must match the design specifications. The inspections and tests on the equipment are used to ensure equipment fitness for service: the equipment will perform its duties as expected. If there are equipment deficiencies identified during an inspection or test, there must be specific procedures to address and correct the gaps.

Keep in mind that the overall maintenance strategy will include coordinating the schedules between the operations and maintenance groups, as the equipment must be made available when its maintenance is needed. Hence, inspections and tests which do not require a process unit or utilities shut down are easier to schedule and perform. Inspections and tests that do require a shutdown must be coordinated with the production schedule to prevent or reduce business interruptions. Unscheduled maintenance, including emergency repairs, may cause schedule delays and adds undue stress to personnel across the organization.

9.5.1 Questions Posed When Developing the Maintenance System

For sites with processes and equipment handling hazardous materials and energies, a practical maintenance system can be developed by focusing on the answers to these questions:

1. **Why** is the equipment considered critical?
2. **How** will the critical equipment fail?

3. **What** type of maintenance needs to be done to prevent such failure?

4. **When** is the maintenance performed?

5. **Who** is qualified to perform the maintenance?

6. **Where** best is the maintenance performed?

The brief discussion on the goals of and potential responses to these six questions below should help clarify how they can be used to establish an effective equipment maintenance system.

Discussion on Question 1: Why is the equipment considered critical?

The list of critical equipment helps ensure a manageable maintenance system. This critical equipment list is based on the risk-based evaluation of the process equipment, since the maintenance group does not have the time or resources to inspect and test all of the processing equipment with the same effort and rigor. The risks and the associated risk reduction measures are evaluated through hazards and risk assessments, helping identify the critical equipment (see examples listed in Table 5.1).

Discussion on Question 2: How will the critical equipment fail?

Using the critical equipment list, a root cause analysis can be used to understand what factors influence the wear and tear on the equipment, and thus, how it will fail in the future. The failure-related causes include: the equipment's location, such as indoors, outdoors, or corrosive environments; extreme processing conditions, such as very high or cryogenic temperatures, very high pressures or vacuums, very high flow rates with the potential for erosion; and stream compositions which may affect corrosion rates. Although the starting point for understanding the potential failure mechanisms of the equipment is from the manufacturer's guidelines, the real story on its failure mechanisms that occur in the field at a specific location in a specific process. Hence, operating experience provides invaluable inspection and testing information which should be used when developing or improving the ITPM programs.

Other factors that can cause premature and unexpected equipment failure include incompatible materials—the wrong materials of construction—used when fabricating, installing, or fixing the equipment. Not all pipe is created equal, as has been proven with many loss of containment incidents. As will be noted below, if the equipment integrity, including its quality assurance, is not sustained by other groups, the ITPM program will be doomed once the process begins operating.

Discussion on Question 3: What type of maintenance needs to be done to prevent such failure?

The answers from Question 2, how the critical equipment can fail, determine what kinds of inspections and tests are needed. The first step when selecting the type of ITPM program is to follow the industry-wide operating experience and guidelines. Some tests are nondestructive, such as inspecting the wall thickness of a pressure vessel, whereas other may simply be scheduled replacement of a part that has been in service for an extended period, such as replacing a pump's o-ring seal after a defined number of operating hours. Sometimes, there may be many options for developing the process- and equipment-specific inspections and tests; other times, there may be only one option. The different types of maintenance include parts replacement, simple repairs, and complete equipment replacement.

Discussion on Question 4: When is the maintenance performed?

One of the greatest challenges facing an ITPM program is when to perform the inspection or test on the equipment, especially since there may be several schedules from which to choose. These scheduling choices include: one recommended by the equipment manufacturer; one accepted and recommended in the industry; or one developed from historical, site-specific reliability data. Depending on the type of equipment, there may be a legislated requirement for a maximum interval between inspections, as well, such as an annual pressure vessel inspection. Once the maintenance schedule has been determined, the facility must have a maintenance management program combining both the ITPM program with routine maintenance, such as lubrication schedules. Maintenance software packages used today can automatically issue and track the status of the work orders.

Discussion on Question 5: Who is qualified to perform the maintenance?

Qualified personnel include operators who can perform routine visual checks during their daily rounds, mechanics who can perform routine equipment lubrications, or electricians who can perform routine electrical checks. There are specialized ITPM programs, such as vibration studies, Infrared (IR) analyses on welding, Leak Detection and Repair (LDAR), and Ultrasonic Thickness (UT) testing programs. The qualifications depend on the type of inspection or test being performed, such as specialized welding certification.

<u>Discussion on Question 6: Where best is the maintenance performed?</u>

Maintenance working areas must be suitable and sufficient for proper repairing or refurbishing of the equipment. Depending on the type of ITPM program that is chosen, the locations for the inspections or tests may be at the equipment in the field or at dedicated maintenance locations on- or off-site. For example, the annual relief valve "pop test" requires special testing apparatus in the maintenance shop or with a contractor off-site. Since weather conditions, such as rain or ice, and the work environment, such as dirty, dusty, or congested areas, play a role when executing the inspection, test, or routine preventative maintenance task, each facility will have to decided where best to perform the required maintenance.

The six questions posed above help when selecting the equipment and logistics for the inspections and tests. Once these parts of the maintenance program are determined, the procedures for analyzing, interpreting and responding to the results must be designed. This includes documenting the tracking and analyzing of the "as found" conditions, tracking the work performed to correct any deficiencies, and documenting the "as left" condition after the work is performed. If the condition is not as expected, then there should be a formal review of the existing ITPM program to determine if the inspection or test is adequate, or if an accelerated equipment deterioration rate is detected in the field to determine if the inspection or testing frequency should be shorter.

Keep in mind that not all equipment should be under the scrutiny of an ITPM program designed to manage the process safety critical equipment. A cost-effective maintenance strategy often applied to common, easily repaired or replaced equipment is the run-to-failure approach. As an example, a water pump used to transfer potable water from one storage tank to another storage tank could be considered in a run-to-failure program since it:

- will not cause a hazardous process safety event if the pump catastrophically fails
- could be installed in parallel with a second (spare) pump
- could have the spare pump inventoried for easy replacement in the field.

Although this is a business decision (as all equipment needs a bona fide PM program for reliable operations), the unexpected loss of the non-critical equipment should not pose problems that can't be addressed quickly.

9.5.2 Equipment Risk Based Inspection Programs

Effective maintenance systems have an optimized equipment inspection approach based on the hazards and associated risks of the processes. These programs use a Risk Based Inspection (RBI) approach, providing a focused, more cost-effective maintenance system, thus helping a facility maintain the safety and reliability of its operations [1,2,15–19]. Some of the potential benefits of an RBI maintenance program include fewer inspections, fewer or shorter shutdowns and longer production run lengths without compromising process reliability, process integrity and process safety performance.

Recall that the engineering group selects the process and equipment designs used to manage the hazardous materials and energies, the hazards analysis team identifies and addresses the risks, and the maintenance and reliability groups identify the critical equipment failure modes and develop the ITPM program. The information from each of these groups is combined and used to develop appropriate inspection, testing and preventative maintenance tasks, to prioritize the ITPM schedule, to ensure that competent personnel perform the tasks, and to ensure timely and proper responses to the results, as needed. When the processing and safeguarding equipment integrity is maintained, both the likelihood and the consequence of an event should be as low as reasonably practicable (ALARP) (See Chapter 3). In summary, these general steps are used when establishing an RBI program:

1) Understand the process design and the equipment design (Chapter 5)
2) Understand the hazardous materials and energies associated with the process (Chapter 6)
3) Assess the risks of the equipment failure (Chapter 7)
4) Identify the equipment damage mechanisms and equipment failure modes (this chapter), and
5) Implement an ITPM on the critical equipment (this chapter).

Please refer to the literature for more discussion on how RBI maintenance programs can be designed to meet the expected frequency and documentation requirements and industry Recognized and Generally Accepted Good Engineering Practices (RAGAGEP) [1,2,15–19].

9.5.3 Equipment Reliability Programs

Reliability Centered Maintenance (RCM) programs focus on the maintenance phase of the equipment's life cycle depicted in Figure 5.1. The definition of an RCM program is as follows:

> A systematic analysis approach for evaluating equipment failure impacts on system performance and determining specific strategies for managing the identified equipment failures. The failure management strategies may include preventative maintenance, predictive maintenance, inspection, testing, and/or one-time changes (e.g., design improvements, operational changes) [6].

RCM programs include the predictive and preventative maintenance programs, with scheduled equipment inspections and tests used to verify that the equipment is still functioning in the way it was designed. Hence, if the equipment does not pass its test, then equipment repair or replacement must be scheduled. As noted earlier, industrial best practices and operating experiences are combined when determining the types of equipment failure modes and the types of ITPM program needed for the process unit.

It is important to note that the processing equipment and safeguarding equipment have distinctly different reliability curves from which to develop their ITPM programs. Although the processing equipment is used daily, with its reliability checked during each startup, the safeguarding equipment *is not expected* to be in use during normal process operations. Safeguarding equipment includes the computerized control systems, the critical alarms, the process permissives and interlocks, safety instrumented systems (SIS), active and passive designs, and emergency response equipment. Hence, when the safeguarding equipment is needed, unlike normal processing equipment, the safeguard must not suffer from the potentially high failure rate when it is needed ("on demand"). This distinct difference in the bathtub curves between normal and on-demand equipment is portrayed schematically with the safeguard equipment reliability curve in Figure 9.3. Although the processing equipment may have a wear-in zone, as is depicted in Figure 9.1, critical safeguarding equipment *must not have* a wear-in zone.

Although there may be equipment reliability databases used to predict the equipment's useful life, there are many factors which affect its lifetime. Not all equipment follow the general bathtub curve shown in Figure 9.1, but each equipment does have its own curve, usually based on its field-specific operating experience. These factors include whether the equipment is located

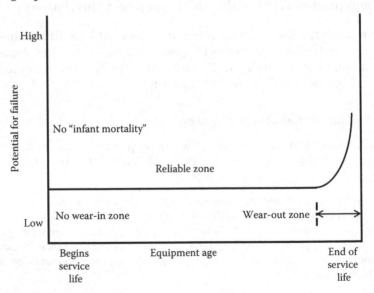

FIGURE 9.3
An expected reliability curve for critical safeguarding equipment.

indoors or outdoors, whether it is subjected to processing temperature or pressure extremes and if it is being operated within its design specifications. Note that for new, novel processes and equipment being implemented from lab or pilot-scale experiments into a full production scale, not even experienced engineers can predict the lifetime performance of the new processing equipment since the new equipment's "true" processing and operating conditions have yet to be tried on the full-scale equipment.

Keep in mind that preventing equipment failure is less stressful on an organization than having to respond during emergencies when the preventive engineering barriers fail. Thus, preventative maintenance includes lubrication, visual inspections, and simple parts replacements (e.g., gaskets, o-rings, etc.). For example, some pumps must be cleaned out when switching between product grades to avoid product contamination issues, with inexpensive gaskets replaced before placing the pump back in service. The drawback with preventative maintenance programs is that parts or equipment that may have some useful life are removed. In other words, parts or equipment which still function are being replaced. If the parts or equipment are expensive, such preventative maintenance programs may be expensive and wasteful.

Preventative maintenance programs can be complemented with predictive maintenance programs which inspect the equipment's deterioration or degradation rate and help extend the useful life of the equipment. Predictive maintenance programs test and inspect specific equipment conditions to help predict its remaining life time. By finding ways to nonintrusively monitor the equipment's condition, the focus is on looking for signs of impending failure to increase the reliable operational time of the equipment. When degraded performance is discovered, then equipment maintenance can be scheduled proactively before it fails. These proactive tests and inspections include vibration analyses, nondestructive thickness tests and infrared heat monitoring. More sophisticated predictive maintenance programs include continuous feedback loops, as well. These predictive and nondestructive tests provide information on the equipment's current state, independent of when it was fabricated and installed or whether it has been used much or not.

In summary, equipment reliability programs fit into the equipment life cycle by focusing on the state of the equipment currently being used and in service. Effective predictive and preventative maintenance programs include routine maintenance, inspections, and tests, and proper responses to the equipment's condition upon analysis of the ITPM results. The ITPM program must ensure that the equipment is fit for use, especially if the equipment is critical for process safety, noting that critical safeguarding equipment must not fail upon demand. A schematic of how these reliability programs fit into the equipment life cycle is presented in Figure 9.4.

Equipment reliability is often measured by its performance, recognizing that the equipment will eventually wear out. Note that the definition of reliability includes two specific criteria: meeting both its design specifications

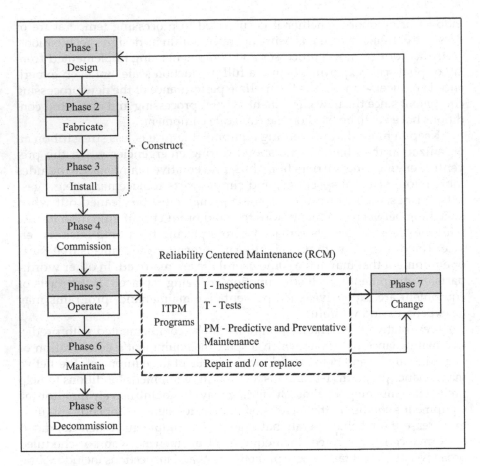

FIGURE 9.4
An equipment reliability program.

and its design expectations. The former refers to the equipment's inherent design through its commissioning (Phases 1 through 4, Figure 5.1), whereas the latter refers to the equipment's operability (Phase 5, Figure 5.1). If there are many unexpected equipment failures, especially those which lead to loss of containment and incidents, the facility should review its operational discipline on its equipment reliability program.

9.5.4 Equipment Quality Assurance Programs

The Quality Assurance (QA) programs focusing on critical equipment include several distinct activities for each of the equipment's life cycle phases. These programs must be written with facility-specific standards or guidelines and have clearly identified personnel responsible for each of the facility's QA activities. Combined with the inspection, testing and preventative maintenance programs—the facility's reliability programs, quality assurance programs complement a facility's overall maintenance strategy to ensure the integrity of the equipment handling hazardous materials and energies.

The construction management QA program helps verify proper equipment fabrication and installation per the design specifications or manufacturer's instructions; a training QA program helps verify proper equipment operation; and a maintenance QA program helps verify proper scheduling, analyzing, and responding to the equipment's inspections or tests. The maintenance QA program also verifies that proper parts or materials are used during replacements or repairs. The initial startup, operational, and maintenance QA programs complement the change management QA program, which is used to verify that changes to the equipment are properly evaluated and addressed before commissioning the change and then handing the equipment back to operations. The final goal for all these QA programs is to detect for and correct flaws before repaired or replaced equipment is placed back into service. Thus, the equipment is then fit for use for safe operations meeting its fitness for service specifications.

There are many equipment-specific QA programs which can be used to ensure that the equipment is fit for use. One particular program, a Positive Material Identification (PMI) program used to verify existing and incoming metallurgy, has gained popularity due to significant incidents that occurred when incorrect materials of construction were used [20,21]. Facilities handling hazardous materials and energies need to develop special receiving, warehousing, and distribution systems, including contractor or vendor supplied equipment, to ensure that all materials, spare parts, and spare equipment will be suitable when used in the field. Facilities must also verify that the equipment complies with established safety codes and standards.

9.5.5 Equipment Integrity Programs

How does the equipment integrity program relate to the equipment reliability and quality assurance programs? Essentially, an equipment integrity program oversees the entire life cycle of the equipment—the "wholeness" of the equipment used to control the hazardous materials and energies. The facility's equipment integrity program includes the entire equipment life cycle. An equipment integrity program includes:

- Trained and qualified resources—people and tools—for inspection, testing, and certified equipment repairs
- Quality assurance programs for ensuring quality parts replacement using recognized engineering practices
- Inspections and tests responding to correct any deficiencies when found, and
- Thorough records documenting the equipment's performance, including its "as found" and "as left" results from the inspections and tests.

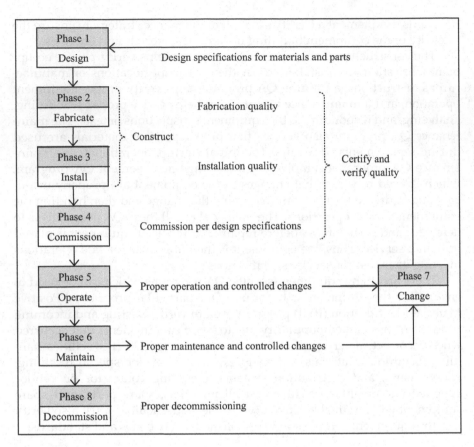

FIGURE 9.5
An equipment integrity program.

In other words, an equipment integrity program ensures that the equipment has not been compromised from its design intent during its entire lifetime, including when repairs are made or when changes are implemented, such that the equipment will be reliable and will perform as expected when it is needed. An equipment integrity program encompasses all phases of the equipment's life, as shown in Figure 9.5.

9.5.6 Combining Programs

Expanding upon the equipment's life cycle, the foundation for an equipment integrity program is formed when a facility combines its equipment reliability and assurance programs across all of the equipment's life cycle phases [22]. The goal of an equipment integrity program is to maximize the useful life of the equipment in service. Thus, the facility establishes an

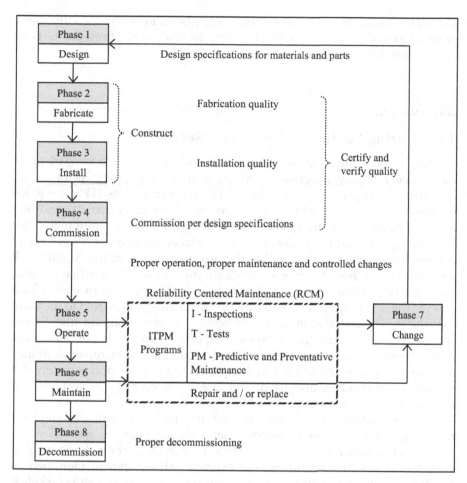

FIGURE 9.6
An effective equipment maintenance program.

effective maintenance program by performing and responding to its ITPM programs using qualified personnel and quality components when designing, fabricating, installing, operating, maintaining, changing, and removing equipment handling hazardous materials and energies, as is depicted in Figure 9.6.

Unfortunately, inadequate or postponed equipment tests and inspections have led to many loss of containment incidents that have caused fatalities, facility damage and environmental harm. For example, in January, 2014, a storage tank leaked and the containment dike (which had not been adequately inspected) allowed the hazardous material to enter the Elk River, contaminating the drinking water of up to 300,000 residents in nine counties in West Virginia, USA [23]. Other examples of loss of containment incidents have been reported due to inadequate leak testing on piping and related equipment (either before being placed in service or after many years of extended wear and tear) or due to inadequate pressure tests or weld inspections [24,25].

9.6 Ensuring Equipment Fitness for Service

Although performing regular inspections and tests is a crucial part of an equipment integrity program, ensuring that the equipment is fit for its intended use depends on how the facility responds to its ITPM program results. When gaps are detected between the equipment's current condition and its design specifications (the expected condition), the equipment's fitness for service must be verified before it is placed back into service. Critical equipment designed to ensure that the process risks meet the ALARP risk level must meet their design specifications, must be available when needed, and must be capable of being operated at a specified performance level for a specific time. The equipment's quality assurance program is used to measure that the initial commissioning meets design, construction and installation specifications to ensure that adequate inspections and tests are performed during operations; necessary calibration, adjustment and repair activities are planned, controlled and executed; quality repair parts and maintenance materials are available and used; and overhauls, repairs and tests do not increase operational risks.

It is important for a facility to identify the equipment designed for handling and storing the hazardous materials and energies, as well as the equipment designed as the safeguards to help manage the risks associated with the processing conditions. For example, all equipment identified in HAZOPs as independent protection layers (IPL) including safety critical systems must have scheduled testing to ensure that they will operate "on demand" (refer to the safeguarding equipment performance curve shown in Figure 9.3).

9.7 Addressing Equipment Deficiencies

The facility must analyze, understand, and respond to equipment failure modes which are detected outside acceptable limits; then it must document repairs, replacements, recalibrations and the equipment's condition before it is returned to service. An effective equipment integrity program will review its test inspections and results, promptly address conditions that can lead to equipment failure, examine the results to see if there are broader, systemic issues, and investigate chronic failures using a structured methodology (see Section 9.5.2). The equipment must have specific maintenance procedures, personnel must have training, and the construction (fabrication and installation) must be per design.

After any maintenance activity, the equipment must be verified that it is ready to be placed back into service. The decisions after these predictive or preventative maintenance steps include whether to continue using the equipment ("ok as is"), to rebuild it (use new parts), or to replace it (remove it from service). By definition, this is the proof that the equipment will meet its performance expectations, with the ITPM program detecting for potential equipment deterioration as it is being used. These validations are used to help prove the equipment's fitness for service before placing it back in service [26].

9.8 Monitoring the Equipment Maintenance System

It is important to select and monitor the processing and safeguarding equipment designed to manage the hazardous materials and energies. Note that critical equipment is the "tip of the iceberg" of all the equipment at a facility—this equipment *must have* robust maintenance programs. Monitoring the maintenance systems with equipment reliability studies across the facility builds understanding between the engineering-related demands, the production-related demands, and the maintenance-related demands, keeping expectations aligned across each group. Each group must balance time constraints and resources using key maintenance metrics [3], such as:

1) overdue Preventative Maintenance (PM) efforts on critical equipment

2) delayed or no response to required actions from ITPM results, and

3) down time due to frequent equipment-related failures.

Although these metrics are direct indicators of the health of a facility's maintenance program, there are some indirect indicators that may help, such as expected equipment breakdowns (e.g., the failure is part of normal day-to-day operations). By carefully monitoring and troubleshooting unexpected equipment failures, even those without severe consequences, there may be systemic issues which need to be addressed. Trending the failure data can provide information on where the business is going. See Chapter 12 for more discussion on monitoring ITPMs.

9.9 Case Study

This chapter concludes with a case study complementing reports on an incident in Amuay, Venezuela [27–30]. This case study is described in greater detail in this section using the following parts:

1) The consequences
2) The perceived hazards
3) An analysis of equipment failures
4) Another view of the root cause
5) Understanding the systemic weaknesses

The ultimate failure of the protection layers originally in place to manage the equipment's life cycle led to the catastrophe.

Amuay Discussion Part 1 – The consequences

In August 2012, flammable material from a leaking olefins pump accumulated in a large flammable vapor cloud, found an ignition source, and then exploded, killing 47 people and injuring at least 135 more. In addition, the blast destroyed a National Guard barracks and damaged more than 200 homes in the nearby community.

Amuay Discussion Part 2 – The perceived hazards

This analysis is based on assuming that the facility perceived no significant hazards with a small leak: the leak would disperse and could not accumulate and produce a significant flammable vapor cloud. This assumption relied on the normal prevailing wind speed and direction at the location of the refinery. Thus, a small vapor cloud area would be localized around the pump, as is depicted in Figure 9.7 (the wind speed and direction are shown with an arrow). Assuming that this normal direction and speed is sustained, any small leaks of flammable vapors would be dispersed such that the potential area covered with a vapor cloud above its lower explosive limit (LEL) would be small. This small cloud would be isolated near the leak, with dispersions of the vapors beyond this area not presenting a significant risk to personnel on site or the surrounding community, no matter how long the pump remained leaking.

However, the wind direction and speed changed, such that the heavier-than-air olefin vapors formed a large, flammable vapor cloud that dispersed in another direction (see example dispersion depicted Figure 9.8). This cloud drifted across the refinery's fence line, across a road and accumulated in congested areas off-site. Days before the explosion, people at nearby businesses and the surrounding residents had complained of

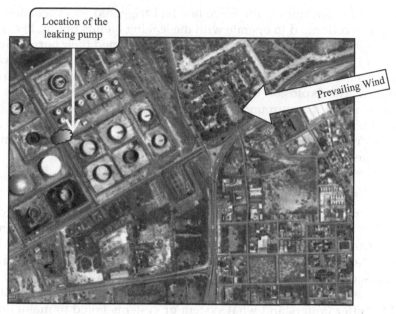

FIGURE 9.7
Amuay, Venezuela: Several days leading up to August 25, 2012 [Modified from 29].

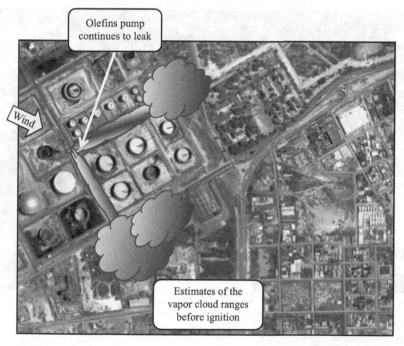

FIGURE 9.8
Amuay, Venezuela: During the day, August 25, 2012 [Modified from 29].

the odor. LEL detectors at the fence line had triggered an evacuation, but the refinery continued to operate with the leaking olefins pump. Personnel in the neighboring businesses and community resumed their normal duties in spite of the odor. On August 25th, the vapor cloud ignited, possibly when it eventually reached the nearby highway and a car on the road drove through it. The explosion caused fatalities, injuries and significant structural damage in the surrounding community. The fires continued to burn days later, as shown in Figure 9.9 and Figure 9.10. The refinery was shut down for months while the investigation was being held and the damage property replaced or restored.

Amuay Discussion Part 3 – An analysis of equipment failures

One lesson we can take away from this incident can be obtained by analyzing the equipment failure using a pump root cause failure analysis chart (see RCA example in Figure 9.2). With the scenario described above, a leaky pump resulted in a flammable vapor cloud which ignited and caused fatalities, injuries, environmental harm, property damage, and business interruption. The answers to the following questions can be used to better understand what system or systems failed to maintain the pump's integrity:

Olefins pump location

FIGURE 9.9
Amuay, Venezuela: The evening of August 25, 2012 [Modified from 29].

FIGURE 9.10
Amuay, Venezuela: Two days later, August 27, 2012 [Modified from 30].

1. Phase 1: Was the pump designed to manage the hazardous materials and energies?
2. Phases 2, 3, and 4: Was the pump fabricated, installed, and commissioned per its design specifications?
3. Phase 5: Was the pump operated per its design specifications?
4. Phase 6: Was the pump's ITPM program created and sustained?
5. Phase 7: Were changes after commissioning controlled?

The pump lifetime integrity will be jeopardized if any if these programs are weak. Some program weakness, such as operating, maintaining, changing or commissioning the pump, may have contributed to the pump's failure, as is depicted in Figure 9.11.

Did operating personnel hear any noises due to the excessive vibration? Did maintenance personnel receive a work order to investigate for the cause and fix the pump? Did Amuay personnel have a pump vibration monitoring program? It was only a small leak at the olefins pump; was this just the daily routine for maintenance and operations? Were normal operating conditions adhered to for the process handling hazardous materials and energies? Was a small leak considered normal operations at any of the many process pumps? Was the pump's integrity sustained? As we have noted before, poor operational discipline increases the operational risk when the program is not sustained (see Figure 4.3). The integrity of critical equipment depends on everyone.

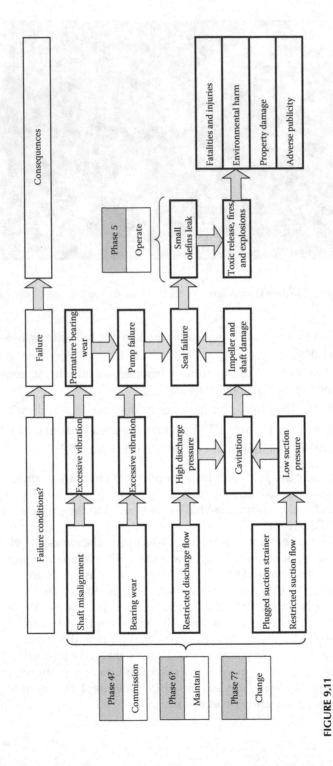

FIGURE 9.11

A root cause analysis on the pump failure at Amuay.

Amuay Discussion Part 4 – Another view of the root cause

The official report noted that a bolt needed to secure the pump to its foundation was missing [27]. The missing bolt may have contributed to excessive vibration and catastrophic failure of the piping and subsequent release of the flammable olefins. While most of the questions noted in part 3 above were not answered in any of the reports, the fundamental facts cannot be questioned: there was a loss of containment, a significant explosion occurred, and there were fatal consequences. The foundations for process safety, the safety culture and leadership, operational discipline, and process safety systems, must be in place to ensure effective management of process safety.

Amuay Discussion Part 5 – Understanding the systemic weakness

Equipment integrity relies on inspection, testing, and preventative maintenance (ITPM) programs that are implemented, analyzed and used. Ensuring that the equipment is fit for its intended purpose involves more personnel than those in the maintenance group. Everyone who impacts the equipment's life cycle must understand their important role in maintaining equipment integrity. These personnel may have a direct impact, such as designers, fabricators, installers, operators, and those making changes, or an indirect impact, such as managers controlling engineering, operational, or maintenance resources and tools, or personnel purchasing, receiving and warehousing spare parts and materials.

The basis for designating critical equipment hinges on the types of hazards and risks that must be addressed. For highly hazardous processes, specialized equipment designs will be warranted, whether the equipment is used for the process operations or as a safeguard. Although not all of barriers apply to all potential hazard/consequence scenarios, the barriers which may have critical equipment include seven of the eight protection layers identified in Chapter 3 as depicted in Figure 3.6:

- Processing equipment (Barrier 1)
- Basic process control systems (BPCS, Barrier 3)
- Critical alarms with manual and automatic response (Barrier 4)
- Safety instrumented systems (SIS, Barrier 5)
- Active or passive physical barriers (Barriers 6 and 7), and
- Emergency response equipment (Barrier 8)

For each of the barriers which have critical process safety operating and/or safeguarding equipment there must be an ITPM program in place. Remember, it is the small neglected items which can pose great risks. Figure 19.12 provides a perspective on the Amuay incident with acknowledgement to Benjamin Franklin's "For want of a Nail" introduced with protection layers

The Historical View   A View from Amuay

For Want of a Nail
For want of a nail the shoe was lost.
For want of a shoe the horse was lost.
For want of a horse the rider was lost.
For want of a rider the message was lost.
For want of a message the battle was lost.
For want of a battle the kingdom was lost.
And all for the want of a horseshoe nail.

For Want of a Gasket
For want of a gasket the seal was lost.
For want of a seal the pump was lost.
For want of pump the olefins were lost.
For want of the olefins the containment was lost.
For want of containment control was lost.
For want of control process safety was lost.
For want of process safety lives were lost.
And all for the want of a pump's gasket.

FIGURE 9.12
Equipment reliability—pay attention to the details.

in Chapter 3. It is likely that a combination of failures and weaknesses contributed to the Amuay incident, especially due to the complexity of the refining process. This, in this case, lack of maintenance operational discipline, combined with lack of safety leadership and culture, could have led to a flammable vapor cloud which ignited and caused the catastrophe. This is the similar conclusion presented in the Bhopal case study described in Chapter 11 [31].

9.10 Measures of Success

- ☑ Design and implement an effective process safety system for maintaining process and equipment integrity.

- ☑ Provide training to maintenance groups and ensure that they are staffed with experienced and capable personnel (see discussion in Chapter 8).

- ☑ Ensure that maintenance procedures, safe work practices and permits are developed, authorized, and followed.

- ☑ Schedule authorized reviews to keep procedures accurate with current practices.

☑ Schedule refresher training, as needed, to sustain personnel capabilities.

☑ Ensure that maintenance personnel have appropriate tools to perform their tasks safely, with comprehensive controls for tool inspections and suitability for use.

☑ Ensure regularly scheduled equipment inspections, tests, and preventative maintenance programs, focusing resources on process equipment and safeguards critical for safe and reliable operations.

☑ Ensure that the equipment continues to meet good engineering design practices and standards by thoroughly documenting the equipment inspections, tests, and preventative maintenance efforts, including:

  o Documenting the "as is" and "as left" conditions before and afterwards, and

  o Tracking until closure of the follow-up actions proving that the equipment is fit for service before being placed back into service.

☑ Ensure that a maintenance quality assurance program is in place for maintenance materials, such as gaskets, bolts, lubricants, for spare parts or equipment, and that all personnel involved in the design, procurement, and storage of the maintenance materials understands the equipment design specifications.

☑ Ensure and document that contractors have the training, knowledge and skills required for their tasks (see discussion in Chapter 8).

References

1. Center for Chemical Process Safety. 2006. *Guidelines for Mechanical Integrity Systems*. Wiley-AIChE.
2. Center for Chemical Process Safety. 2007. *Guidelines for Risk Based Process Safety*. Wiley-AIChE.
3. Center for Chemical Process Safety. 2010. *Guidelines for Process Safety Metrics*. Wiley-AIChE.
4. U.S. OSHA. 1992. 29 CFR 1910.119: *Process Safety Management of Highly Hazardous Chemicals*. www.osha.gov.
5. UK HSE. 2012. *Maintenance Procedure*. Technical Measures. www.hse.gov.uk.
6. Center for Chemical Process Safety. 2016. *Process Safety Glossary*. www.aiche.org/ccps/resources/glossary.
7. American Petroleum Institute. 2012. *API Recommended Practice 500: Classification of Locations for Electrical Installations at Petroleum Facilities Classified as Class I, Division 1 and Division 2, Third Edition*. www.api.org.

8. American Petroleum Institute. 1997. *API Recommended Practice 505 (R2013): Recommended Practice for Classification of Locations for Electrical Installations at Petroleum Facilities Classified as Class I, Zone 0, and Zone 2, First Edition.* www.api.org.
9. UK HSE. 2015. *Hazardous Area Classification and Control of Ignition Sources.* Technical Measures. www.hse.gov.uk.
10. British Standards Institution. 2009. *Explosive Atmospheres. Classification of Areas. Combustible Dust Atmospheres.* BS EN 60079-10-2:2009.
11. Energy Institution. 2005. *Model Code of Safe Practice Part 15: Area Classification Code for Installations Handling Flammable Fluids,* 3rd ed. (based on IP 15, 4th ed. expected July 2015). www.energyinst.org.
12. International Electrotechnical Commission. 2009–2015. *IEC 60079 Series Explosive Atmosphere Standards,* webstore.ansi.org (Includes references to: British Standards Institution. 2009. *BS EN60079-10-1:2009, Explosive Atmospheres—Part 10-1: Classification of Areas—Explosive Gas Atmospheres* and British Standards Institution. 2015. *BS EN60079-10-2:2015, Explosive Atmospheres—Part 10-2: Classification of Areas—Explosive Dust Atmospheres.*)
13. National Fire Protection Agency. 2012. *NFPA 497: Recommended Practice for the Classification of Flammable Liquids, Gases, or Vapors and of Hazardous (Classified) Locations for Electrical Installations in Chemical Process Areas.* www.nfpa.org.
14. National Fire Protection Agency. 2013. *NFPA 499: Recommended Practice for the Classification of Combustible Dusts and of Hazardous (Classified) Locations for Electrical Installations in Chemical Process Areas.* www.nfpa.org.
15. UK HSE. 2005. *Safety Implications of European Risk Based Inspection and Maintenance Methodology.* Research Report 304. www.hse.gov.uk.
16. UK HSE. 2015. *Risk Based Inspection (RBI)—A Risk Based Approach to Planned Plant Inspection.* SPC/Technical/General/46. www.hse.gov.uk.
17. American Petroleum Institute. 2009. *API Recommended Practice 580: Risk-Based Inspection,* 2nd ed. American Petroleum Institute. www.api.org.
18. American Petroleum Institute. 2008. *API Recommended Practice 581: Risk-Based Inspection Technology,* 2nd ed. www.api.org.
19. Geary, W. 2002. *Risk Based Inspection—A Case Study Evaluation of Onshore Process Plant.* HSL/2002/20. www.hse.gov.uk.
20. American Petroleum Institute. 2010. *API Recommended Practice 578: Material Verification Program for New and Existing Alloy Piping Systems,* 2nd ed. www.api.org.
21. U.S. Chemical Safety Board. 2006. *Positive Material Verification: Prevent Errors During Alloy Steel Systems Maintenance.* Bulletin 2005-04-B.
22. Vaughen, Bruce K., John Nagel, and Mathew Allen. 2011. An approach to integrate plant reliability efforts with a mechanical integrity program. *Process Safety Progress* 30(4):323–327.
23. U.S. Chemical Safety Board. 2016. *Freedom Industries Investigation.* Final report to be issued September 28, 2016. www.csb.gov.
24. U.S. Chemical Safety Board. 2009. *Allied Terminals, Inc., Catastrophic Tank Collapse.* Report No. 2009-03-I-VA. www.csb.gov.
25. U.S. Chemical Safety Board. 2014. *Catastrophic Rupture of Heat Exchanger, Tesoro Anacortes Refinery Anacortes, Washington.* Report 2010-08-I-WA. www.csb.gov.

26. UK HSE. 2006. *Plant Ageing: Management of Equipment Containing Hazardous Fluids or Pressure.* Research Report RR509. www.hse.gov.uk.
27. Venuzuala. 2013. *PDVSA, Evento Clase A Refineria de Amuay.* Gobiemo Boivariano de Venuzuela, Ministerio dei Poder Popular de Petroleo y Minera. www.pdvsa. com/interface.sp/database/fichero/publicacion/8264/1632.PDF.
28. Pearson, T. 2013. *Venezuelan Report: Refinery Disaster Caused by Intentional Manipulation of Gas Pump Bolts.* venezuelanalysis.com/news/10013.
29. RiskMgmtGroup. 2012. *Amuay UCVE Event August 12 2012.* www.slideshare. net/RiskMgmtGroup/amuay-ucve-14177498.
30. Bodzin, Steven. 2012. *What will Venezuela learn from its Amuay refinery explosion? Christian Science Monitor.* www.csmonitor.com/World/Americas/Latin-America-Monitor/2012/0827/What-will-Venezuela-learn-from-its-Amuay-refinery-explosion-video.
31. Vaughen, Bruce K. 2015. Three decades after Bhopal: What we have learned about effectively managing process safety risks. *Process Safety Progress* 34(4):345–354.

10

Change Processes Safely

Change is a constant for all systems: physical equipment ages and degrades over its lifetime and may not be maintained properly; human behavior and priorities usually change over time; organizations change and evolve, which means the safety control structure will evolve.

Nancy Leveson

Why We Need to Change Processes Safely

Understanding and managing change is probably the most difficult issue when managing safe and reliable processes. This is due, in part, to the relationships between each of the process safety systems and that *many different people* are required to manage each of the process safety systems. Thus, the real challenge in managing change is how best to communicate the changes to others who may be affected by the change. Everyone must understand how a change impacts the process safety risks and how the changes affect the other process safety systems. We still suffer from major incidents today due, in part, to ineffective change management systems—evidence to the difficulties when managing them. Although equipment, processing, and systemic changes are often driven by new technologies and our continuous improvement efforts, they may be driven by findings from other efforts, such as process hazards and risk assessments, investigations, reviews, or audits. If the direct or indirect effects of these changes on other equipment, processes, or systems are not thoroughly understood, authorized, and communicated effectively, significant incidents may occur. Process safety and reliability relies on an effective change management system.

Change processes safely by ensuring that process, equipment, system, and organizational changes are evaluated and authorized and that operational readiness reviews are conducted.

10.1 Introduction

This chapter is deliberately shorter than the other chapters in Parts II and III of this book, not because it is not important but because changes may either directly or indirectly impact *every other process safety system*. A change may not be significant, or may not matter at all to the other process safety systems, however, unless you verify with those managing the other systems, you most likely will not know if it does impact them. For example, incident investigations have revealed that well-meaning purchasing agents, during business-wide cost reduction efforts, could save expenses by substituting less expensive carbon steel piping for stainless steel required to handle the corrosive process fluids and conditions. Unfortunately, what the cost-conscious purchaser failed to recognize—and did not share with others—was that stainless steel was required. The operations and maintenance personnel were unaware of the different material of construction; the engineers were unaware that the design specifications were no longer being met. Thus, once in the field, the undetected carbon steel piping failed due to unanticipated, severe corrosion causing a subsequent loss of containment.

Many of the worst disasters described elsewhere in this book were due, in part, to ineffective change management systems. Changes were made to the original equipment design and specifications without understanding either their short- or long-term impact on the safety and reliability of the process. Often the thought processes of and decisions made by those making the changes at the time were sound based on their limited understanding of its impact elsewhere. It is only through the incident investigation when we discover that others essential to managing process safety were unaware of the change or unaware of consequences of the change. As noted when researching and analyzing incidents, it is clear that everyone consciously managing safe and reliable operations makes the best decisions they can every day; they do not deliberately set out to do harm to themselves, to others, to the environment, or to property [1,2].

Instead of focusing on how to design and implement an effective Management of Change (MOC) system, given that most companies recognize how important managing change is, this chapter explores approaches on how better to understand and verify whether a proposed change has direct, an indirect, or no effect on the other process safety systems. In particular, history has shown that subtle changes made over time on process equipment or operating and maintenance procedures have slowly and adversely affected the equipment's integrity, resulting in quite unanticipated and sometimes catastrophic consequences. When working with hazardous processes and energies, these unintended consequences may cause significant incidents, leading to fatalities, injuries, environmental harm and property damage. Unfortunately, many significant process safety incidents have occurred when the MOC system was ineffective, such as Flixborough, Mexico City, Bhopal, Pasadena, and Buncefield [2–8].

Since changes can occur at all phases in an equipment's life cycle, it is essential that designers, fabricators, installers, operators, and maintainers of the processes and the equipment understand the reasons for the change and how their responsibilities and tasks may change.

10.2 Key Concepts

10.2.1 Defining Changes

Change can be simply defined as "something different" than what exists now. However, this simple definition has been far from easily managed when dealing with processes handling hazardous materials and energies. Although the drivers for change can vary from using new or improved process technologies to addressing gaps identified through incident investigations, reliability studies, audits, or new legislation, when "something different" happens, it must be understood and managed correctly. For the purposes of this book, any one of the following definitions for process safety-related changes may apply:

1. To replace or substitute someone or change, modify, or combine roles and responsibilities (i.e., change people or the organizational structure, both corporate and local)
2. To replace or substitute something (i.e., change chemicals, materials, equipment, or process technologies), or
3. To revise or vary from a routine or a task (i.e., change operating procedures, maintenance procedures, or process safety management systems and procedures).

With these definitions in mind, it is worth noting that not all changes are created equally, nor do all changes have similar effects. Some companies even have definitions for "replacement in kind," recognizing, for example, that when the new equipment is designed to meet the design specifications of the original equipment, a rigorous MOC review is not needed. Hence, when companies can adequately define changes, separating minor changes from major ones, their process safety risks associated with changes can be effectively managed.

10.2.2 Managing Changes

Another concept essential for safe and reliable operations is simply that a company must effectively manage its changes. As noted by Leveson, Bloch, and many others: Change is inevitable—staffing will change, equipment will age, and there will always be continuous improvement efforts across all levels of and systems in the organization. Many companies have developed systems to manage changes to their staffing, technologies, and processes. However, they may not fully appreciate how a disciplined MOC system not only helps reduce process safety risks, but helps improve its operations, reliability, quality, customer satisfaction, and overall business performance. The costs associated with safety far outweigh the costs of having a severe process safety incident. Effective change management systems ensure that all changes with process safety impact are recognized, that all changes to process and equipment information are documented, and that all personnel directly affected by the change understand the change and are trained on how to manage it once it is implemented.

10.2.3 Communicating Changes

The third concept, often lost when people struggle with the details of their company's MOC system, is *why* they have to go through the whole management of change process itself. The reason is quite simple. We fill out all the forms, get the proper authorizations, and then confirm that all the process safety technology information reflects the change to verify with others that our proposed change will not adversely affect them and our operational risks. The key is to talk to others across the other disciplines to make sure that what we change does not miss any important connections between groups. MOC systems are designed and implemented to provide consistency across the company when evaluating, authorizing, and documenting changes, to ensure and to verify that the changes to process hazards or risks are properly managed. MOC systems are designed to effectively communicate these changes *before the change is implemented*.

From a cook's perspective when preparing a satisfying meal and working with many different ingredients, think of your proposal as a new recipe. Do we add salt? Do we add sugar? How much do we add? If we make a mistake in these materials or the amounts, will we have a delicious cake? The consequences were quite evident when an author's eight-year-old son surprised his family with a hot and fresh creation (Disclaimer: mom helped with the hot oven, only, not in preparing the mix). The kitchen sure smelled great! However, one very salty bite into the delicious-looking cake revealed that salt had mistaken for the sugar. To any one working with unlabeled, granular, white powders, they both look the same. Could you make a similar mistake when taking a round, white, and unlabeled gasket from the maintenance bin? After all, it is the right diameter and it does fit... Is this gasket a bonafide replacement in kind? Does it meet the design specifications?

Why do we have long, extensive, and seemingly excessive MOC checklists? It is because somewhere, sometime ago, someone working with hazardous materials and energies did not ask the specific line to the right person and verify that the change did or did not impact them. The change was not effectively managed and communicated across the groups. And someone got burned.

10.3 Case Study

The incident that occurred at the Buncefield petrol depot in the United Kingdom illustrates how a seemingly small change may result in a significant incident when an effective management of change system is not in place. The release occurred when one of the Buncefield storage tanks was overfilled on December 11, 2005, resulting in a vapor cloud that ignited causing a massive explosion and fire. The depot had been in operation since 1968, distributing fuels to London and to the south-east of England and was the fifth largest fuel distribution site in the United Kingdom at the time of the release, explosion, and fire. Offices and warehouses surrounding the depot suffered major damage, nearby residents were evacuated, and the fire closed the busy M1 motorway twice during the five days that it burned. Fire and satellite images showing the black smoke from the fire that reached London during this incident are shown in Figure 10.1. The fire engulfed 20 large storage tanks, with 25 fire engines, 20 support vehicles, and 180 firefighters needed to extinguish the blaze. Although the fire brigade water curtain protected tanks on eastern part of site, the bunds (dikes) surrounding the burning tanks failed to contain all the contaminated liquid run-off (a mix of firefighting water and escaped fuel), such that large amounts of run-off escaped from the site and contaminated ground and surface water.

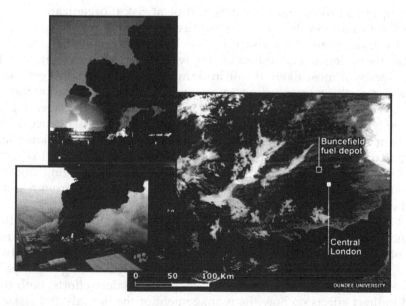

FIGURE 10.1
Images of the Buncefield incident [6–8].

Among several issues discovered by the Major Incident Investigation Board (MIIB), a major management of change system discipline failure had occurred *a year before*, in 2004, when an independent high level switch (IHLS) on the storage tank was replaced. In addition, there were operability and reliability issues with both the tank's level gauge and the IHLS that went unaddressed after this change. At the time of the incident, the operators did not know the tank level, they do not know exactly how much petrol was being added (due to poor pipeline feed system design), and had poor communications during shift handover. With these combined conditions, the new operators did not know that the tank was at a dangerously high level. The petrol overflowed during their shift through the vents at the top of the tank, formed a vapor cloud that reached an ignition source, causing the explosion and fire. The MIIB wrote that an adhered-to management of change "typically would have included an engineering assessment of the benefits and disadvantages of any such change, and a consideration of what changes in procedures (e.g., in testing) would be necessary as a result," concluding that the Buncefield operations had failed to implement an effective management of change process. For additional findings from the investigations, please refer to the literature [6–8].

10.4 Understanding Organizational Changes

A company's process safety efforts will be at risk if organizations do not adequately address changes in its required staffing or reporting structure. Ideally, there is an optimal staffing level needed to minimize the operational risks. Too many resources are wasted (and expensive). Too few resources will most likely result in dangerous short cuts to "get the job done," directly corresponding to an overall increase in the operational risks. When the process safety staffing is insufficient (this number depends on the types of risks and potential magnitude of the consequences), there may not be enough people with the right skills, experience, or knowledge to manage the company's process safety systems and tasks. This section explores some of the issues and approaches when managing organizational change from a process safety perspective, compiled from resources in the literature [9–11].

Often the driver for organizational change is financial. Such "cost cutting" reductions may include one or more of the following: reducing the number of people, reorganizing departments, combining/sharing tasks between groups, and outsourcing tasks to contractors. When assessing the impact of a proposed organizational change on its process safety efforts, both direct and indirect effects on how the management of the hazards and risks will be sustained must be understood. Depending on the extent and number of changes, the best option may include making slow, deliberate changes over a long time rather than abruptly changing everything at once. In addition, too many changes occurring at the same time may distract people, adversely affecting the effectiveness of the process safety systems they are managing. Whether slowly or quickly, all organizational changes should be realistic, reflecting thorough and methodical plans.

When the headcount is reduced, the company may outsource more of its work to contractors or may have to engineer more automated manufacturing controls. Some of the elements that should be considered during an organizational change include the following:

- understanding how the changes affect the staffing levels and tasks
- understanding how the changes affect the management of the process safety systems
- consulting with staff and contractors before, during and after the change
- providing training and adequate supervision for staff with new or changed roles, and
- ensuring that key roles, responsibilities, and tasks are identified and successfully transferred.

It is essential that the organization does not underestimate the training burden, the time needed to train, and how the processes will be operated during the training sessions.

In addition, there are items that should be considered when evaluating competencies for those managing the process safety systems who have been assigned to the new roles with new responsibilities. Some these items include the following:

- identify gaps in process safety skills and knowledge
- identify how these gaps will be addressed
- select suitable methods and trainers required for training
- plan for and allocate sufficient time needed for those to be trained
- set clear competency criteria, and then
- verify that the training meets minimum requirements for the role.

The reader is encouraged to review the literature for useful checklists specifically addressing competency evaluations [11,12]. Additional discussion on developing personal process safety competencies is provided in Chapter 13.

This section concludes with an example summarizing how a facility's overall operational risk increased, in part, due to poor management of an organizational change when both the operational and maintenance staffs were reduced during an economic downturn [12]. The example: An economic market collapse forced customers to stop placing orders and the business' sales drastically declined. The company responded prudently by curtailing production and making drastic budget cuts across all groups. These cuts did not guarantee longevity but did improve the odds that the business would be able to survive the economic downturn. Included in the budget cuts was a significant reduction in staff, with the hardest loss of staffing affecting both the production and the maintenance groups. As the economy slowly rebounded, the business responded by slowly rehiring operations personnel and, thus, slowly ramping production back up. However, the business delayed its maintenance group hiring as it regained its foothold into the market. Unfortunately, senior management did not address the increase in the process safety risks due to delayed and unperformed equipment integrity tests and inspections. Essential equipment responsible for helping manage the risks associated with the hazardous materials ultimately failed, causing a fatality. The increase in the overall operational risk after this inadequately managed organizational change is represented on a risk matrix in Figure 10.2 and an "optimal" staffing level chart shown in Figure 10.3. Point 1 reflects the normal operational risk at the facility's optimum staffing level, respectively; Point 2 reflects the higher operational risks upon resumed operations without adequate maintenance support.

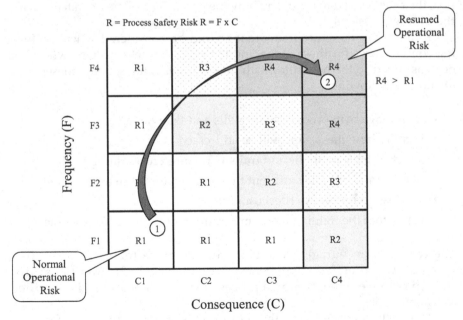

FIGURE 10.2
The effect on the process safety risk after an organizational change [Modified from 12].

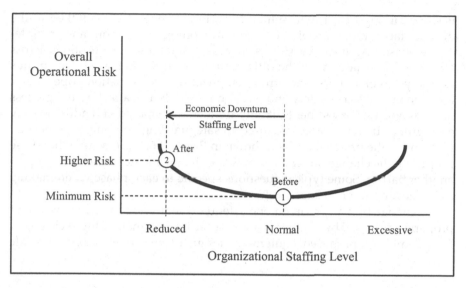

FIGURE 10.3
The effect on the operational risk after an organizational change [Modified from 12].

10.5 Understanding Technology Changes

An MOC system is designed to identify the impact on the process hazards and risks with changes to the process technology. The MOC system also is used when temporary changes are performed, such as testing different equipment, processing conditions, or procedures. When new, modified, or temporary changes are proposed, they may affect the risks to personnel and equipment due to changes in the toxic, flammable, explosive, or reactive process hazards. Depending on the scope of the change to a process handling hazardous materials and energies, the MOC protocols may require a Process Hazards Analysis (PHA) as a part of the MOC review. Additional guidance is provided in Chapter 7 on the different types of PHA methodologies that can be used during these reviews.

Since changes may be proposed at any phase in the equipment's life cycle, it is essential that those proposing the change understand how the change may impact other groups responsible for the equipment in the other life cycle phases. If the change affects another phase, it is essential that the issues are addressed during the MOC before implementing the change. Although a change may directly affect one group and is obvious to those making the change, the change may also *indirectly* affect another group responsible for the equipment in other phases, and, hence, *may not be obvious* to those making the change. One way to visualize the connections between these phases is shown in Figure 5.1, where the life cycle's change phase is linked to the design, fabrication, installation, commissioning, operation and maintenance phases. An effective MOC system is designed to ask questions across all groups associated with the different phases in the life cycle. The groups typically involve personnel from engineering, construction, operations, maintenance, process safety and environmental. Issues affecting the process hazards and risk assessments, if any, are then understood and addressed by each group. If changes are proposed to safeguarding equipment associated with any of the protection layers shown in Figure 3.6, it is essential that those proposing the change understand how the change in one barrier may impact another barrier. Some typical questions specific to each phase when making a change are shown in Table 10.1.

A change by one group on one safeguard may need to be reviewed and properly addressed by another group responsible for another life cycle phase. For example, a proposed equipment design change may impact the safe

TABLE 10.1

Typical Questions Posed During a Management of Change (MOC) Review

Life Cycle Phase	Question Specific to Equipment Life Cycle Phase
1 Design	Are there changes to the process technology or processing conditions? Are there changes to the equipment specifications?
2 Fabricate	Are there changes to the fabrication specifications?
3 Install	Are there changes to the installation specifications?
4 Commission	*Before handover and startup:* Have the new design specifications been verified? Have the new fabrication and installation specifications been verified? Have the new or updated operating procedures been written? Have the new or updated maintenance procedures been written? Has the training been completed and documented across all groups, as needed? Has the communications been completed and documented across all groups, as needed?
5 Operate	Are there changes to the safe operating limits? Are there changes to the operating procedures?
6 Maintain	Are there changes to the tests and inspections? Are there changes to the maintenance procedures?
7 Change*	What is the project's scope and proposed change? Does the change affect the process hazards or risks? Does the change require a process hazards analysis (PHA)?
8 Decommission	Will the change affect hazards or risks when the equipment is removed?

*Note: The MOC protocols usually start with the project's scope and proposed change (Phase 7).

operating limits or the types and frequencies of tests and inspections. The link is illustrated in Figure 5.1: the design change (Phases 1 and 7) is linked to operations (Phase 5) and maintenance (Phase 6). Or process changes, such as increased temperatures or pressures during operations (Phase 5), may exceed the equipment's safe operating limits (its design, Phase 1). In addition, delayed maintenance tests and inspections (Phase 6), a change from the equipment's ITPM schedule (Phase 7), may cause unexpected equipment failure during normal operations (Phase 5). The Buncefield case study presented in Section 10.4 provided an example how the process safety risks increased when maintenance was neglected. Therefore, it is essential for safe and reliable operations that proposed changes to the process or equipment address potential impacts on the other life cycle phases.

10.6 How Change is Managed

Since this book is written with a different approach on how to effectively manage process safety, it is only appropriate to actually define *what* a Management of Change (MOC) system is at this point:

> A management system (used) to identify, review, and approve all modifications to equipment, procedures, raw materials, and processing conditions, other than replacement in kind, prior to implementation to help ensure that changes to processes are properly analyzed (for example, for potential adverse impacts), documented, and communicated to employees affected [13].

All changes begin with an idea, a reason for change. The new concept (the idea) must be written such that the reason for the change is clearly defined and can be effectively shared with others. The technical basis (the scope) for the change must be clear and must be compared and evaluated to the current, if not original, design. However, before initiating the change management system (and having to fill out all the burdensome forms and struggle to get all the authorizations), the proposed change should be evaluated for any potential hazardous effects. What this really means is this: if you work with hazardous materials or energies, you will have to follow some formalized change management system to ensure that you do not put yourself or your colleagues at risk to injury or death.

The requirement that everyone proposing a change must use the MOC system is based on the results of the hazards identification and risk analyses. In particular, are there toxicity, flammability, explosivity, or reactivity hazards involved? If so, then perform an MOC; if not, then follow the normal work order system. If an MOC is required, the next step may or may not require a formal PHA (Chapter 7). The technical basis for the change includes any changes to raw materials and additives, product specifications, by-products, waste products, design inventories, processing conditions, procedures, equipment, materials of construction, instrumentation and control systems. The final step for most MOC systems due to the changes in the technical basis may include updating the equipment files, the P&IDs, and any associated operating or maintenance procedures.

Probably the most difficult aspect of any change management system is how best to communicate the change across the different groups involved in or indirectly affected by the change. People with knowledge and skills in process safety, engineering, operations and maintenance must be involved from the beginning to the end. For this reason, multidisciplined teams are required to review and authorize the work through each stage in the project's life cycle. By adding the major groups typically represented in the MOC approval stages and mapping the project life cycle on the equipment life cycle (Figure 5.1), we can visualize in Figure 10.4 the important communication links between the different phases of the equipment's life that must be addressed during the change.

One of the most challenging links in the past has been commissioning phase, when the project is handed over from the engineering group to the

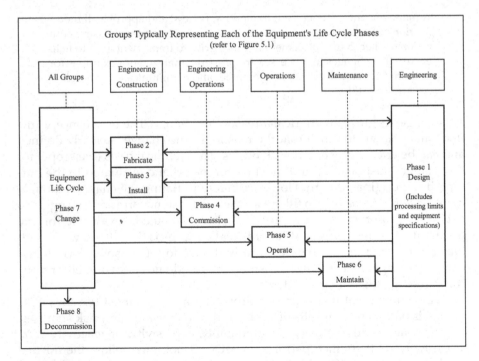

FIGURE 10.4
Communication links essential for effectively managing change.

operating and maintenance groups. Many incidents have resulted when a process is started back up after a change was implemented without adhering to the steps in a structured MOC system. For this reason, there must be an authorized operational readiness review to ensure and verify that all the prestartup work has been completed before the handover is performed and its operation is ready to go. For specific equipment-related changes, the readiness review must include reviewing and updating the operating and maintenance procedures *and* the training for operating and maintaining personnel before the handover.

Businesses handling hazardous materials and energies can operate safely and reliably when they have an effective change management system. Although the concept is really quite simple, it is the application of the MOC system which is not. To safely manage changes to a process handling hazardous materials and energies, a typical change management system is designed, in general, with the following steps listed below [adapted from 4, 14–16]:

1. Obtain approval for the scope of the proposed change to the hazardous process
2. Understand and document the changes to the process hazards
3. Design phase
   a. Ensure understanding of the changes to the process hazards
   b. Evaluate and assess the changes to the process hazards and risks
   c. Finalize design
   d. Identify construction resources
4. Construction phase
   a. Ensure communication of the hazards and risks
   b. Procure equipment per design specifications
   c. Fabricate per design specifications
   d. Install per design specifications
5. Commissioning phase
   a. Ensure communication of the hazards and risks
   b. Review and update operating and maintenance procedures, as needed
   c. Inform and provide training, as needed, to those affected by the change
   d. Verify that construction meets design
   e. Perform the operational readiness or pre startup safety review
   f. Handover responsibility to operations/maintenance
6. Complete the updates to the affected process technology

The company's MOC program must document that:

1. the changes were assessed for their effect on the process hazards and risks
2. the changes were thoroughly reviewed across all groups, and
3. the changes affecting the other process safety systems were addressed *before restarting the process*.

Some of the principles addressed and documented in an effective MOC system for safely managing changes to processes handling hazardous materials and energies include:

- involving every group directly or indirectly affected by the change
- holding reviews and authorizing approvals during each step
- describing the technical basis for the change
- describing the consequences to safety, health and the environment
- performing an appropriate process hazards analysis
- understanding and addressing the impact on the other equipment life cycle phases
- training and updating procedures for operating and maintenance personnel
- verifying that the design specifications were met before startup, and
- completing all updates to the process technology that changed.

Thus, most companies have developed their own MOC procedures with MOC-specific forms that must be followed and used when changes are proposed, reviewed, authorized and implemented. The MOC checklists may be exhaustive and, especially for small projects, include many questions that simply do not apply to the project. Granted, it may feel like it takes longer to fill out all the paper work, hold all the meetings, and get all the appropriate approvals than it takes to physically make the change in the field. However, please remember that these checklists reflect our history—when these items are not asked and verified properly for processes handling hazardous materials and energies, potentially fatal consequences may occur elsewhere. We must not forget what happened lest we get burned again. Safe and reliable operations must be sustained for processes managing hazardous materials and energies by applying what we have learned from our past (Chapter 11).

Some topics found on a typical MOC form are noted in Table 10.2. Authorization teams noted at specific stages in the MOC process include, at

TABLE 10.2

Typical Topics in a Management of Change (MOC) Checklist

Topic	Checklist Item*
1 Type of MOC	Permanent
	Temporary
	Emergency
2 Project scope	Area(s) affected
	Reason for change (the technical basis for change)
	Author
3 Technical requirements	Basic design review

	Hazards and risk assessments (if applicable):
	    Process hazards analysis (PHA)
	    Facility siting review
	    Human factors review
	    Occupational safety and health review
	    Environmental review

	Materials review
	Process control review
	Instrumentation review
	Equipment integrity and reliability review
	Product quality review
	Document updates (see Topic 7)
4 Operations requirements	Procedures review
	Operations personnel training
	Communications
5 Maintenance requirements	Procedures review
	Maintenance personnel training
	Communications (includes purchasing/warehousing)
6 Commissioning requirements	Operational readiness review or pre startup safety review (PSSR)
	Authorizations to startup with change
7 MOC closure requirements (as needed)	Equipment documentation update
	Process flow diagram and flow sheet update
	Piping and instrumentation diagram (P&ID) update
	Instrumentation and electrical drawing update
	System software back up and documentation update
	Inspection record and documentation update
	If temporary MOC: verification that the process has been returned to its original state after the specified temporary change period is over (e.g., when the test is completed)

(Continued)

TABLE 10.2 (*Continued*)

Typical Topics in a Management of Change (MOC) Checklist

Topic	Checklist Item*
8　People often involved in or resourced during the project (MOC Team members)	MOC champion
	PHA leader
	Process engineer
	Operations – supervisor
	Operations – operator
	Maintenance – supervisor
	Maintenance – mechanic
	Maintenance – instrumentation and electrical technician
	Equipment integrity and reliability technician or engineer
	Process safety technician or engineer
	Occupational safety and health technician or engineer
	Environmental technician or engineer
	Process control technician or engineer
	Laboratory/quality technician or engineer
	Purchasing/warehousing technician
9　Authorizing managers (depends on the project's scope and phase)	Process safety
	Occupational safety and health
	Environmental
	Engineering
	Operations
	Maintenance

*Note:　Typical MOC forms have check-boxes and fields for information entry, including physical/ electronic signatures for the MOC file documenting/verification management system (hard copy and/or electronic).

minimum, representatives from the engineering, operations, maintenance, process safety and environmental groups. In most projects, there are items that must be completed before the handover, such as updating procedures and training operators, and items that can be completed after the handover, such as updating the red-line P&IDs. These follow-up actions are clearly identified in the company's MOC system. However, to officially close an MOC, all of the follow-up items must be completed. This includes closing the often-delayed process technology updates.

In summary, a successful and effective change management program ensures that everyone directly or indirectly affected knows about and understands what the change is and how it impacts them. Therefore, everyone must be aware of and understand the proposed design or change and its associated process safety hazards and risks before resources are authorized. When changes are made to equipment, everyone must understand and be trained on to how to construct, operate, and maintain the new or modified equipment. This includes reviewing and updating the affected operating and maintenance procedures. All the design specifications of the project must be verified before authorizing the handover to operations. Everyone must be aware of how the changes have affected the process technology documentation, and verify that the documentation—accessible to all—correctly reflects any technology changes after the change has been implemented. It is important to remember that process and equipment changes without a disciplined change management program have resulted in significant incidents with fatalities, injuries, environmental harm and property damage.

10.7 Measures of Success

☑ Design and implement an effective system for changing processes safely, especially when managing changes to the:
  - o Staff, including organizational and personnel
  - o Process technology and design basis, including chemicals, chemistries, and processing conditions
  - o Equipment technology and design basis, including specifications for their design (i.e., materials of construction), fabrication, installation, operation and maintenance, and
  - o Administrative controls, including operating and maintenance procedures.

☑ Ensure that experienced, trained, and multi-disciplinary teams are assigned to manage changes throughout all phases of the change.

☑ Ensure that all appropriate changes are reviewed, authorized and documented, including:
  - o Using checklists to ensure a comprehensive review of the change
  - o Including every group directly or indirectly affected by the change
  - o Describing the technical basis for and scope of the change
  - o Describing the consequences to safety, health and the environment by the change
  - o Performing an appropriate level process hazards and risk assessment
  - o Addressing the impact, if any, on other equipment life cycle phases
  - o Updating procedures and training for operating and maintenance personnel

o Verifying that the design specifications were met before startup (i.e. a pre-startup operations review before resuming operations)

o Recording and documenting the review, and

o Completing all updates to the process technology documentation that was changed.

References

1. Vaughen, Bruce K. and Tony Muschara. 2011. A case study: Combining incident investigation approaches to identify system-related root causes. *Process Safety Progress*. 30:372–376.
2. Bloch, Kenneth. 2016. *Rethinking Bhopal: A Definitive Guide to Investigating, Preventing, and Learning from Industrial Disasters*, 1st Edition. Elsevier Press.
3. Center for Chemical Process Safety. 2008. *Incidents That Define Process Safety*. Wiley-AIChE.
4. Garland, R. Wayne. 2012. An engineer's guide to management of change. *Chemical Engineering Progress*. 108(3):49–53.
5. Kletz, Trevor A. 2009. *What Went Wrong? Case Histories of Process Plant Disasters and How They Could Have Been Avoided*, 5th Edition. Butterworth-Heinemann/IChemE.
6. UK HSE. 2011. *Buncefield: Why did it happen?* Competent Authority Strategic Management Group (CASMG). www.hse.gov.uk.
7. UK HSE. 2008. *The Buncefield Incident 11 December 2005: The final report of the Major Incident Investigation Board, Volume 1*. Buncefield Major Incident Investigation Board (MIIB). www.hse.gov.uk.
8. UK HSE. 2008. *The Buncefield Incident 11 December 2005: The final report of the Major Incident Investigation Board, Volume 2*. Buncefield Major Incident Investigation Board (MIIB). www.hse.gov.uk.
9. Center for Chemical Process Safety. 2013. *Guidelines for Managing Process Safety Risks During Organizational Change*. Wiley-AIChE.
10. UK HSE. n.d. *HSE Human Factors Briefing Note No. 11, Organisational Change*. www.hse.gov.uk/humanfactors/topics/orgchange.htm.
11. UK HSE. 2003. *Organisational change and major accident hazards. HSE Information Sheet. Chemical Information Sheet No CHIS7*. www.hse.gov.uk/humanfactors/topics/orgchange.htm.
12. Center for Chemical Process Safety (CCPS). 2016. *Guidelines for Integrating Management Systems to Improve Process Safety Performance*. Wiley-AIChE.
13. Center for Chemical Process Safety. 2016. *Process Safety Glossary*. www.aiche.org/ccps/resources/glossary.
14. Center for Chemical Process Safety. 2007. *Guidelines for Risk Based Process Safety*. Wiley-AIChE.
15. Center for Chemical Process Safety. 2007. *Guidelines for Performing Effective Pre-Startup Safety Reviews*. Wiley-AIChE.
16. Center for Chemical Process Safety. 2008. *Guidelines for the Management of Change for Process Safety*. Wiley-AIChE.

11

Manage Incident Response and Investigation

There's no harm in hoping for the best as long as you're prepared for the worst.

Stephen King

Why We Need to Manage Incident Response and Investigation

Imagine yourself walking through the thermal cracking area in your refinery on a calm sunny afternoon. You step around the corner and see an operator motionless on the ground near the coker. There is no visible vapor cloud and your personal hydrogen sulfide monitor is not alarming. Do you fight, flee, or freeze? What are you trained to do? Do you know where the proper emergency response gear is and how to don and use it? Do you locate the wind sock, run upwind, and find the nearest emergency alarm? Or do you simply stand there and hope that someone else comes along soon who knows what to do? Emergencies are terrifying. Thus, it is essential that you know what to do, whether you are to "fight" or to "flee" from the scene. If you "freeze" you most likely will be at the same dangerous risk level as if you try a rescue without being trained or if you do not know how to safely reach an alarm to notify those who are trained. Unfortunately, emergencies occur at our facilities today in spite of the progress we have made in managing our process hazards. And, without effective incident investigations, we would not have made such progress in the first place. What we learn from our disasters must be shared with others so that they will not suffer; what is shared by others must be understood by us so that we do not suffer, as well. We manage incident responses—the emergencies—safely to prevent us from placing ourselves at greater risk, and we investigate incidents to learn what happened, to apply, and to share this new knowledge to help prevent the incidents from happening again.

Manage incident response and investigation by ensuring that
personnel are trained how to respond during emergencies
and that incidents are thoroughly investigated.

11.1 Introduction

Like the knight playing chess against the dragon, we offer a different
approach for effectively managing process safety risks by combining the
emergency response and incident investigation elements into one of our
eight process safety systems. Although the first rule of how to respond in an
emergency is *to not get into one*, the emergency responses are planned and in
place *before* you need to respond to the emergency. Effective emergency management saves lives, protects the environment and property, and helps reas-
sure the surrounding community that the facility is well managed in spite
of the incident. We can only presume that the unsympathetic traveler in the
fable shown in Figure 11.1 provided counsel when the poor lad needed help.
An organization handling hazardous materials and energies should not wait
for an emergency to figure out how best to respond—by then, it is too late.

The focus of this process safety system is on how to prepare for, respond
to, and learn from emergencies when the administrative and engineering
controls fail to contain the hazards. Effective risk reduction occurs when the
response, recover and investigate phases are combined with the planning,
changing and sustaining phases. Although beyond the scope of this book,
a facility's emergency response plan should also address other potential

A boy bathing in a river was in danger
of being drowned. He called out to a
passing traveler for help, but instead
of holding out a helping hand, the
man stood by concernedly and
scolded the boy for his imprudence.

"Oh, sir!" cried the youth, "pray help
me now and scold me afterwards."

-Aesop

FIGURE 11.1
Aesop's fable of the drowning boy. Image courtesy of Sarah Vaughen. With permission.

hazards, such as geological, meteorological or biological. In some cases, the hazardous materials may pose security issues, as well, requiring additional measures to address potential acts of sabotage or terrorism.

This chapter begins by introducing three key concepts which set the foundation for successfully managing incidents: how an organization defines its incidents, how an organization defines emergencies, and how an organization needs to tie six phases to effectively manage its emergency responses and incident investigations. This chapter concludes with an approach for better understanding incidents, using a case study to show how the bow tie diagram can be used to illustrate both event-related and systemic-related weaknesses which ultimately led to the incident.

11.2 Key Concepts

Earlier in this book, we described how an organization identifies its hazards and understands its risks before developing the process safety systems that are used to manage its hazards and risks. In context of this chapter, the hazard and risk information is used to plan for and respond to emergencies. We have identified six phases for managing incidents and briefly describe how an organization can have a successful and effective incident investigation when it identifies and addresses weaknesses in its process safety foundations. Unfortunately, if the hazards of the materials and their associated risks are not understood by those at the facility, by those who are responding to the emergency, or by those who are curious about the ensuing commotion, fatalities may result. For example:

1) In 1982, hundreds of responders and people died when the contents boiled over at a burning storage tank and flowed downhill. *The incident (Tacoa, Venezuela)*: An explosion blew the top of a large oil storage tank at an electric company, resulting in a fire in the tank. Eight hours later, the oil boiled over, overflowed the tank dike and flowed down a steep hillside (the location of the tank allowed for gravity feed to the equipment below). Forty firefighters and 113 spectators died when they were caught in the downhill flow of burning oil [1].

2) In 2014, emergency responders and people in the surrounding community died when ammonium perchlorate stored inside a burning warehouse unexpectedly exploded. *The incident (West, Texas, USA)*: The ammonium nitrate storage facility caught fire and exploded, destroying most of the facility and leaving blast crater nearly 100 feet (30 m) across and 10 feet (3 m) deep. Fifteen people, including

10 members of the town's volunteer fire department, were killed and about 200 were injured. The volunteer fire firefighters were unaware of the explosive hazards of the ammonium nitrate [2].

Thus, effective emergency responses occur when an organization has defined their responses using their understanding of the potential hazards and risks.

11.2.1 The Six Phases for Managing Incidents

Since managing emergencies and their subsequent investigations are linked, we have combined them and identified six phases when effectively managing them. The first two phases describe how an organization prepares for and responds to an emergency. The second two phases are performed at the same time: how the facility recovers from and investigates the incident. And, the last two phases use other process safety systems for managing and sustaining the changes.

The six phases require a strong foundation in management leadership and organizational operational discipline at all levels to be effective. These phases, depicted in Figure 11.2 and described in greater detail below, are as follows:

1. Planning	for incidents	Phase 1
2. Responding	during the incident	Phase 2
3. Recovering	from the incident	Phase 3
4. Investigating	after the incident	Phase 4
5. Changing	after the incident	Phase 5
6. Sustaining	the changes after the incident	Phase 6

11.2.2 Effective Investigations

Although there are many approaches which can be used when investigating incidents, there is no "one size fits all" approach that works in all cases. It should not be a surprise that one approach will not work on the different types of incidents worth investigating, whether the incident is labeled as a "near miss" or has resulted in multiple fatalities, severe environmental impact, extensive property damage and significant business interruption. There are pros and cons to each of the different investigation approaches, with the types of tools ranging from those identifying "root causes" to those identifying "systemic causes" (additional details between these tools are provided below). However, the approach in this book focuses on the weaknesses in one or more of the three foundations that ultimately led to the incident (Figure 1.3). Hence, the better we understand the *complex and combined effects* of these foundational weaknesses, the better we will understand what happened, and the better we will be able to address the weaknesses that ultimately led to the incident (see discussion in Section 11.8).

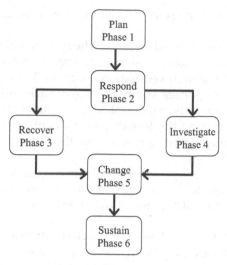

FIGURE 11.2
The six phases when managing emergencies and investigations.

11.3 Classifying the Types of Responses and Incidents

When an organization defines its criteria for an "emergency" and for an "incident," it has established a common language for everyone in the organization. This classification is often based on the organization's risk tolerance and may have specific threshold limits which must be attained or exceeded for specific responses. Emergencies can range from small, localized fires only requiring a single person and a fire extinguisher to large fires requiring multiple emergency response personnel and equipment. The size and types of consequences from the emergency then, in part, help classify the type of incident and vice versa. Approaches for defining emergencies and incidents are discussed in this section.

11.3.1 Defining Emergencies

Emergency responses are distinguished from the normal, routine day-to-day operator or control system responses when slight, expected deviations from the standard, safe operating conditions occur (Chapter 5, Table 5.3). Thus, an "extra" response is needed to respond to processing conditions exceeding the normal, safe operating limits. The extra responses may either be operator actions to an alarm combined with automatic actions, such as interlocks, that move the process into a safe state or shut the process down. At the risk of oversimplifying responses, the basic difference between a normal and an emergency response is the sense of urgency—emergencies require a much quicker response.

Since the type of response depends on the type of incident (as is described below), an organization should design its emergency response program to be consistent with its incident investigation program. Thus, "smaller" incidents will have "smaller" responses, whereas "larger" incidents will have "larger" responses. In addition, it is important to understand that the speed of the response is influenced by the loss of process control and subsequent loss of containment of the hazard. When we use the following definition for the process safety emergency:

> A sudden, unexpected loss of containment event involving a hazardous material or energy that requires an urgent response,

we recognize that the event is "unexpected" during normal operations (assuming every barrier is effectively in place) and that an immediate response is required to control the event.

Fortunately, more often than not, there are the events which could have resulted but do not result in a loss of containment, harm to people, harm to the environment, or even property damage. A localized loss of containment event is usually addressed only by operations or maintenance personnel in the area. Events impacting a broader area may activate additional personnel such as the site's emergency response team with much larger events activating the site's emergency response plan (discussed in more detail in Section 11.4). An illustration for this range of event responses is shown in Figure 11.3. In all cases, the magnitude of the response is directly related to the severity (or potential severity) of event's consequences, with faster responses expected for the larger events. Another way of looking at

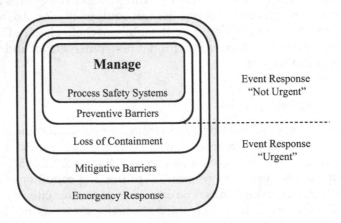

FIGURE 11.3
The sense of urgency when responding to events.

this range is from being (1) not as urgent, proactive, "fire prevention" actions to being (2) urgent, reactive, "fire-fighting" actions.

11.3.2 Defining Incidents

An organization needs guidance on what constitutes an incident and, more importantly, what events could potentially lead to an incident. Without this guidance, it will be difficult to anticipate, plan for, safely respond to and effectively investigate incidents. Some incidents are easy to define: when images of billowing black smoke and foggy toxic dispersions make the headlines and result in fatalities, injuries, environmental harm and significant property damage, the organization involved has certainly had an "incident." In particular, during the summer of 2010 on the restaurant monitors in China, one of the authors watched—with much sadness—live streaming of the oil bubbling into the Gulf of Mexico below what used to be the Deepwater Horizon deep sea oil platform at the Macondo well [3]. Eleven people lost their lives when the flammable gases released into the platform from the deep well ignited. Such loss of containment events are examples of catastrophic incidents (Chapter 1, Table 1.1).

In this book, we will define a "process safety incident" as follows:

> A combination of conditions and events that result in toxic releases, fires, explosions, or runaway reactions that harm people, damage the environment and cause property and business losses.

Please note that this definition specifically addresses a *combination* of conditions and events—not a *series* of conditions and events—since chemical processes are complex (see Section 11.8). In addition, we define a "near miss" process safety incident as follows:

> A combination of conditions and events that *could have* resulted in a toxic release, fire, explosion, or runaway reaction that *could have* harmed people, damaged the environment and caused property and business losses.

In both cases, safe and effective operations are influenced by the strength and condition of each of the three process safety foundations.

Many companies use a risk based process safety approach, focusing on the severity òf the actual or potential consequences. As noted for catastrophic incidents, these high severity events result in fatalities, irreversible personnel injuries, irreversible environmental harm and significant property damage and loss. Companies having potentially catastrophic incidents are usually regulated due to their potential impact (i.e., they are "covered processes"). On the other hand, low severity incidents may result in small or reversible personal safety or health effects, environmental damage,

TABLE 11.1

General Guidance for Classifying Incidents

Process safety event or incident type		Consequence or severity for loss of containment of hazardous materials or energies	
1	Catastrophic or Severe (Note 1)	High	Fatalities, irreversible personal injury or illness
			Irreversible environmental harm
			Loss of facility
			Off-site impact, off-site emergency response (Note 2)
2	Serious or Critical	Medium	No fatalities, irreversible personal injury or illness
			Significant environmental clean-up
			Major property damage
			On-site impact, on-site emergency response
3	Minor or Marginal		Reversible occupational illness or injury
			Limited environmental impact (e.g., permit violation)
			Limited property damage
			Local area-specific, incidental response
4	Negligible	Low	Little personnel safety or health affect, but violates specific corporate or regulatory criteria
5	Near miss or Near hit		No harm, but if conditions had been different (e.g., occupancy, time, or distance), it could have been worse
6	Unsafe act		External observation of personal actions
	Undesirable act		Internal reflection on personal choices (e.g., at-risk behavior)

Note:
1) Significant incidents ("accidents") have severe consequences: fatalities, irreversible injuries, irreversible environmental harm, bad publicity (e.g., multiple news feeds through the internet; globally televised reporting), regulatory responses and fines, harm to business reputation (e.g., consumer boycotts) and profitability loss (e.g., worker's compensation, higher insurance rates, lost production, and lost market share)
2) Smaller facilities may need to resource off-site emergency responders for larger incidents due to internal resource limitations.

or equipment damage. Thus, a range of event severities are needed when an organization defines "incidents," with some typically used definition provided for reference in Table 11.1 [4,5].

11.4 Planning and Responding to Emergencies

This section discusses the first two phases shown in Figure 11.2: Phase 1, when planning for incidents; and Phase 2, when responding during the incident. The planning phase is used to anticipate the types of emergencies such that an effective emergency response plan (ERP) can be developed and implemented. The responses to an incident depend on what has happened, where it has happened, and how much is *still* happening.

11.4.1 Planning for Incidents

Planning ahead for incidents ensures that everyone knows how to recognize and respond to unsafe conditions and events. Like the drowning boy and the unsympathetic traveler illustrated in Figure 11.1, failing to plan for emergencies or failing to execute a plan when needed can quickly transform an event into a catastrophic incident. People must know what can happen and must know what to do. Thus, understanding what hazards and risks exist, developing plans to address loss of containment events, and training for those responding to the event are essential for reducing the consequences of the event once it occurs. The emphasis in this section is on how to plan for and manage emergencies, the large events which require a coordinated emergency response. Smaller events which could have resulted in an emergency response, sometimes referred to as near-miss incidents, are discussed in Section 11.10.

Each facility should have an Emergency Response Plan (ERP) or an Emergency Action Plan (EAP) that is activated when an emergency occurs. An ERP is a written plan that describes how everyone will respond during an emergency. If all of the emergency response planning protocols are effectively implemented (i.e., with an ERP), when an incident occurs people will know how to safely respond, will have well-defined roles and responsibilities, and will have the proper training to control the situation. The ERP is developed in the planning phase shown as Phase 1 in Figure 11.2. Although this book provides some general guidance on what is included in an ERP, the reader is encouraged to review many of the resources, both from the industry and regulations (Table 11.2). A summary of important aspects considered when developing ERPs is provided in Table 11.3.

The basic elements of an ERP include immediately ensuring that everyone on-site is accounted for, those responding to the emergency do so safely, those in the affected area are rescued and treated, and then controlling the situation to reduce additional property damage or environmental harm. The emergency procedures should include raising an alarm, defining an initial local response (such as providing first aid and using fire extinguishers), and summoning the emergency response team and additional resources, as needed. The additional resources may include medical, rescue, and firefighting services using well-trained professional emergency responders and medical service personnel. Emergencies are clearly identified with site-wide alarms that are understood by everyone. The emergency responders need to evaluate the situation, including establishing whether they are to "fight or flee."

Everyone should know the different sounds from the facility's emergency alarms, and what they are expected to do once the alarm is activated.

TABLE 11.2

Potential References when Developing an Emergency Response Plan (ERP)

1	Center for Chemical Process Safety (CCPS). 1995. *Guidelines for Technical Planning for On-Site Emergencies.* AIChE/Wiley. www.aiche.org
2	U.S. EPA. 1999. Chemical Emergency Preparedness and Prevention Office (CEPPO). *Use Multiple Data Sources for Safer Emergency Response, EPA-F-99-006.* www.epa.gov/ceppo
3	U.S. EPA. 2001. Chemical Emergency Preparedness and Prevention Office (CEPPO). *Chemical accidents from electric power outages, EPA 550-F-01-010.* www.epa.gov/ceppo
4	UK Health and Safety Executive (UK HSE). 2016. Site Emergency Plan (SEP). *Planning for incidents and emergencies.* www.hse.gov.uk
5	UK Health and Safety Executive (UK HSE). 2016. Site Emergency Plan (SEP). *Emergency response / spill control.* www.hse.gov.uk
6	National Fire Protection Association (NFPA). 2012. *NFPA 704: Standard System for the Identification of the Hazards of Materials for Emergency Response.* www.nfpa.org
7	U.S. OSHA. 2004. *Principal Emergency Response and Preparedness Requirements and Guidance.* OSHA 3122-06R. www.osha.gov
8	U.S. OSHA. 2016. *U.S. OSHA 29 CFR 1910.38, Emergency action plans.* www.osha.gov
9	U.S. OSHA. 2016. *Evacuation Plans and Procedures eTool.* www.osha.gov
10	U.S. OSHA). 2016. *29 CFR 1910.120, Hazardous waste operations and emergency response standard (HAZWOPER).* www.osha.gov
11	U.S. OSHA. 1992. 29 CFR 1910.119, *Process Safety Management of Highly Hazardous Chemicals.* www.osha.gov

TABLE 11.3

Aspects when Developing an Emergency Response Plan (ERP) [Modified from 47]

- Ensures that all types of emergency responses are based on written and implemented emergency response procedures.
- Describes the actions of everyone, including contractors, when an incident is occurring at the facility with a written and implemented Emergency Action Plan (EAP).
- Describes the actions for those managing the responses and to those responding to the emergency with a written and implemented Emergency Response Plan (ERP).
- Ensures that everyone knows the different types of emergency alarms and how to safely respond to them.
- Ensures that everyone knows how to recognize hazardous emergency situations, such as fires, spills or releases, and how to notify others of the emergency.
- Ensures that safe evacuation routes exist and that everyone knows where to go when the alarm sounds.
- Ensures that personnel can safely shut down the processes and the facility during the emergency, as needed.
- Ensures that all emergency responders are knowledgeable of the hazards, have been properly trained and have sufficient and reliable emergency response equipment and PPE required for a safe response.
- Ensures that everyone is accounted for during the emergency.
- Ensures that regular, realistic, emergency response drills are scheduled and critiqued (both table-top and "live" simulations).

Everyone needs to know what is happening, where it is happening and whether they need additional emergency PPE or not (and how to don it, if needed). They understand their role and what they are to do during the emergency. Those responding to rescue potential victims need to know where to look for the injured and how to safely enter and leave the affected area. Everyone knows where their main and alternate escape routes are, as some escape routes may not be usable, and know where their mustering points are—the preplanned facility safe zones for headcounts. Depending on the incident's hazards, the types of emergency responses include facility evacuations as well as sheltering-in-place (i.e., when toxic gases are released). When hazardous materials are released outdoors, people need to know where the wind socks or pennants are mounted to help determine their safe, upwind escape route.

An essential part of the ERP includes how communications with those on-site and those off-site will be handled, both internally through the organization and externally to the surrounding community and to the media. Everyone who enters a facility understands what clear visual and audible alarms and signals occur when an emergency occurs. Specific personnel are expected to be trained to safely respond to the event, whereas other people are expected to stop working, either relocating to temporary havens for sheltering-in-place or relocating to a safe mustering point outside of the affected area. In all cases, the ERP identifies the proper PPE for the emergency response, such as escape respirators or emergency response and turnout gear.

The ERP specifies potential safe locations for an emergency control center that will be used during an emergency response. Information located at this control center includes maps of the facility layout and the surrounding community, drawings of the process areas, the utilities, and fire protection systems, such as fire water or foam accessing points, drawings of where the emergency response equipment is and safety data sheets on the hazardous materials at the facility. Communication information located in the control center includes a current, up-to-date personnel contact phone list and, depending on the nature of the emergency, contact information of mutual aid responders and notification lists of both local and governmental agencies. The emergency shutdown plans and procedures should be available and, if needed, responders should be able to access current weather conditions (wind direction, precipitation, etc.) which may be used for modeling dispersions.

Once the loss of containment occurs, the first objective is to protect people, whether they are on-site, off-site, or the emergency responders. This includes all facility personnel, all contractors and all visitors at the facility during the emergency. For larger releases, the ERP is used to help the emergency response team quickly to identify the hazards associated with the release, understand potential scenarios and consequences, and define people's responsibilities during the emergency. If effective communication is not quickly established between everyone responding, whether they are in the affected area or have the potential to be affected by the release, efforts to prevent them from being injured or killed will be not as effective. The ERP summarizes where the process equipment handling hazardous materials are located, such that people understand the potential for nearby hazards which could make the event worse if the impacted area extends beyond the event's release area (also known as the "domino effect").

The operations group usually provides the first, immediate emergency response, such as safely shutting down the process and isolating hazardous material inventories. Both area operations and maintenance personnel are trained to respond, since they are most likely to be in the area when the event occurs. However, since events can occur anytime, such as during troubleshooting efforts or special contractor tasks being performed in operating areas, it is essential that everyone working in the area knows how to respond. If the loss of containment event can be isolated and contained quickly, they may not officially "activate" a facility's ERP. These events are discussed further in Section 11.10.

If the immediate localized response does not work, for example, if one fire extinguisher does not snuff out the fire, the facility's ERP is activated (also known as a "one and done" response). The ERP begins with designated incident-response personnel relocating to a central location for communications, a where all activities are coordinated by a designated incident commander. Specially trained emergency response teams may be required, including, but not limited to, internal emergency response teams, mutual aid emergency response teams from nearby facilities, hazardous material teams, external fire departments, and external medical responders. No matter whether an event warrants an urgent, fully developed ERP response or not, it is essential that the facility has the operational discipline to effectively execute the other process safety systems required to sustain their ERP. This includes performing drills with debriefing sessions to note what went well during the drill and to identify and address gaps in the responses. Effective monitoring of the response plans is described in Chapter 12.

11.4.2 Responding during the Incident

For effective emergency responses during an incident, everyone is able to recognize the hazardous conditions and how to safely respond to them, whether they should "fight or flee," or in some cases, whether they should shelter-in-place. Once the emergency is declared, the facility's ERP is activated with an incident commander given the authority to coordinate the different teams responding to the emergency and someone given the authority to coordinate both on-site and off-site communications. Everyone's first and most important response is reactive: to protect themselves, to rescue those impacted by the emergency, and to keep everyone else safe from the hazardous conditions. The initial responders should know, at minimum, how to administer first aid for minor injuries, and how to request for additional emergency assistance when there are serious injuries. The initial responders know how best to control the hazardous conditions and the affected area. At this point, it is worth noting that actual emergency responses should be rare if all of the other process safety systems have been effectively designed and implemented [6].

Once the ERP is activated, the facility coordinates the response efforts within the facility and with local community responders by establishing an emergency control center. This emergency control center is located in a safe zone, so that it can be occupied for the duration of the emergency. The center serves as the major communications link between the onscene incident commander and the teams responding to the emergency, as well as the link to corporate management. If needed, one facility representative is designated and trained to communicate to local community officials and the media, as well. The primary communications and back up communication equipment should be independent in case one of them is disabled by the emergency, with options including radios, telephones, cellular phones and the internet.

The emergency response scenarios anticipated in the ERP depend on the types of hazards, as well, including where the hazardous events could occur and how the event could impact adjacent process units or the surrounding community. A dangerous area, an exclusion or "hot" zone, is often defined to separate it from other safe or recovery areas and mustering points. In addition, safe passageways are designed to protect emergency responders entering and exiting the dangerous areas. For specific hazardous material-related responses, this safe passageway may be designated as a decontamination corridor or zone in between the impacted and the recovery areas [7,8]. The boundaries at the facility are determined once the nature of the emergency is identified. Although process safety emergency responses focus on potential toxic release, fire and explosion hazards, everyone is aware of and respond safely to other hazards, such as mechanical, electrical, biological or radioactive hazards.

11.5 Recovering From and Investigating Incidents

Once the emergency is over, the people have been rescued, accounted for and treated, the hazards are back under control and the scene has been safely secured, the recovery and investigation phases begin. This section discusses the next two phases when effectively managing incident responses and investigations: when recovering from the incident and when investigating after the incident. These two interconnected phases are depicted as Phase 3 and Phase 4 in Figure 11.2, respectively; whereas a recovery team provides short-term solutions, the investigation team provides longer-term solutions which address the incident's causes. The combined goal is to get the facility back up and running safely as soon as practical and to learn from the incident. The investigation concludes with proposed and subsequently implemented changes. Thus, it is essential that these changes are sustained long after the memory of the incident has faded. Since these phases occur in parallel after the incident, an effective process safety program combines resources from both its recovery and investigation efforts, with common steps for both of these efforts: (1) establish a team, (2) gather and analyze the information, and then (3) identify solutions.

11.5.1 Recovering from the Incident

Incident recovery efforts depend on how severe the damage is to property, the environment and the business. In general, larger incidents require larger responses, have larger impact and take longer for their recovery, if there is an option to do so. Unfortunately, some significant incidents which have killed people have also killed the company, such as those in Flixborough [9], Bhopal [9], and West Texas [2]. When the extent of damage has been evaluated, a business has to decide what to do: rebuild and resume operations or not. This section describes general steps used during the incident recovery phase when short-term, interim solutions are developed. Section 11.5.2 describes the steps used during the incident investigation phase when identifying longer-term solutions.

The general steps in an effective recovery effort, shown as Phase 3 in Figure 11.2, are listed below and are described in greater detail in this section:

1. Establish the team
2. Gather information
3. Analyze information
4. Identify solutions
5. Implement solutions
6. Document and share changes

Step 1 – Establishing the recovery team

The number of the recovery team members depends on what happened, where it happened and on the impact. The sooner a recovery team members are selected, the sooner the facility damage can be assessed and the sooner the recovery efforts can begin. Note that many of these team members will serve on the incident investigation team, as well. Since the incident investigation team is established soon after the incident (see Section 11.5.2), both the recovery and investigation efforts are depicted as parallel phases in Figure 11.2. The sooner the impact and damage is identified, the sooner safe operations can resume and the sooner the environmental clean-up, if any, can begin. Although roles of the team members will vary depending on the type and extent of the incident, the members should represent, at minimum, personnel from operations, maintenance, and engineering. Additional team members may include environmental response personnel specially trained to clean-up hazardous materials, such as hazardous materials responders in the United States [7,8].

Steps 2 and 3 – Gathering and analyzing information

The recovery team members should gather and analyze the information acquired at the same time with the incident investigation team (see Step 2 in Section 11.5.2). Both teams identify the property and environmental damage, including: what equipment was damaged, how badly it was damaged, and whether the equipment can be repaired or replaced. The recovery team, in particular, provides senior leadership with time estimates on what needs to be done and how long the recovery may take before resuming safe operations. The environmental team members assess what environmental hazards, if any, remain. If the incident area is contaminated by hazardous materials, special decontamination procedures may be needed.

Steps 4, 5, and 6 – Identifying, implementing, documenting and then sharing solutions

Once the information has been collected, the recovery team identifies and helps the incident investigation team develop interim, short-term solutions. These solutions should be aligned with the longer-term solutions being developed by the incident investigation team. The interim recovery efforts may include personnel rehabilitation, equipment repair or replacement, and environmental decontamination. Significantly damaged equipment or property may require demolition, re-design and reconstruction. Significant damage, therefore, may take months to implement due to equipment procurement lead times and the time needed for its fabrication and installation.

The recovery solutions should be implemented through the facility's change management system, as shown as Phase 5 in Figure 11.2. The steps for managing the changes during the recovery phase are described briefly in Section 11.6.1 and in more detail in Chapter 10. Some interim measures in the affected area may include signs and barricades, isolating or removing damaged equipment, and cleaning up the area. In some cases, the recovery team may implement its solutions before the investigation team has completed its study. Therefore, it is important for both teams to work in parallel (see Figure 11.2). Although there may be specific documentation requirements of the changes if they are not a "replacement in kind," keep in mind that something *will need to change* to help prevent a repeat incident. In all cases, an operational readiness review and approval procedure should be followed, addressing and validating proper equipment design, fabrication and installation of the repairs, rebuilds or replacements, with effective communications to all affected personnel [6].

11.5.2 Investigating after the Incident

The investigation effort depends on the actual consequences with larger incidents, in general, having larger teams. In some cases, incidents with significant numbers of injuries and fatalities, significant property damage, significant environmental harm, and significant off-site impact may be subject to third-party investigation teams (i.e., governmental teams). In all cases, evidence gathering and investigation techniques should clearly defined in the organization's written protocols. An effective investigation will identify the systemic and foundational weaknesses that contributed to or allowed the events to occur and will then address these weaknesses to help prevent the incident from happening again. This section describes the steps used during the incident investigation phase for identifying the solutions. Important aspects to consider when developing a process safety system for effective incident investigation are provided in Table 11.4.

TABLE 11.4

Aspects When Developing Incident Investigation Procedures [Modified from 47]

- Ensures that all types of incidents are investigated based on written and implemented incident investigations procedures.
- Ensures that investigations are started promptly after the emergency is over.
- Ensures that evidence is secured and collected soon after the incident and before the impacted area is disrupted.
- Ensures that qualified people lead the investigation.
- Ensures that witness interviews are conducted by trained investigators as soon as practical after the incident (depending on the consequences to the witness, not too soon and not too long afterwards).
- Ensures that the investigation team has members, including contractors, who may have been directly involved (if possible), who understand the processing unit and area technologies, and who understand the processing operations and/or maintenance tasks taking place at the time of the incident.
- Ensures that an appropriate and effective incident investigation methodology is used.
- Ensures that event-related protection layer/barrier failures, any associated process safety system weaknesses, and any foundational weaknesses are identified (refer to Figure 1.3).
- Ensures that the findings are addressed with recommendations and follow-up actions.
- Ensures that the actions are developed, are tracked until closure, and are effectively implemented.
- Ensures that a final incident report is prepared, issued and shared.

The general steps in an effective investigation effort, shown as Phase 4 in Figure 11.2, are listed below and are described in greater detail in this section:

1. Establish team
2. Gather information
3. Analyze information
4. Identify causes and solutions
5. Implement solutions
6. Document and share changes
7. Document investigation
8. Share learnings

In addition to the first six steps noted for the recovery team's efforts in Section 11.5.1, the incident investigation team is expected to formally document the investigation and to share its findings and learnings with others. A comparison of the steps between these teams is shown in Table 11.5. Details for effectively investigating incidents are discussed in the literature [e.g., 10–14, 47].

TABLE 11.5

Summary of the Steps for the Recovery and Investigation Teams

Step #	Steps When Recovering From an Incident	Steps When Investigating an Incident
1	Establish team	Establish team
2	Gather information	Gather information
3	Analyze information	Analyze information
4	Identify solutions (Note 1)	Identify causes and solutions (Notes 1 and 2)
5	Implement solutions (Note 1)	Implement solutions (Note 1)
6	Document and share changes (Note 1)	Document and share changes (Note 1)
7	n/a	Document investigation (Note 2)
8	n/a	Share learnings

Note:
1) Changes to the process and equipment handling hazardous materials and energies should be developed, reviewed, approved and implemented using the facility's change management program (e.g., a Management of Change (MOC) system). See Chapter 10.
2) Incident investigation methodologies and documentation expectations should be clearly defined in the facility's incident investigation procedure.

Step 1 – Establishing the investigation team

The incident investigation team should be activated promptly after the incident, should be properly sized, and should have qualified members trained in investigation and data analysis methods. The sooner the investigation team members are selected, the sooner the causes can be identified and the sooner effective solutions can be developed and implemented. Catastrophic incidents may trigger regulatory requirements that specify when investigations should begin. For example, for PSM-regulated facilities in the United States, the incident investigation team must be formed within 48 hours [48]).

Since not all events are catastrophic incidents, the investigation teams need to be properly sized, with the number of the members depending on what happened, where it happened and on the severity. In general, the team's size depends on the actual impact of the incident, with the number of team members directly related to the event's severity. Although roles of the team members will vary depending on the type and extent of the incident, the members should represent, at minimum, personnel from operations, maintenance and engineering. Additional investigation team specialists may include members specially trained in gathering and securing evidence, performing witness interviews, and evaluating for equipment failure mechanisms. Note that investigation team members may be a member of the recovery team, as well.

It is crucial that an organization establishes clear policies on the expected roles, competencies, training of and responsibilities for members on an investigation team. No matter what size the investigation team is, there should be a qualified leader trained on the investigation methodology and members bringing the following expertise who:

- are from the area where the incident occurred
- may have been directly involved in the incident, including contractors (if possible)
- know how to collect and organize the forensic data
- know how to conduct and document witness interviews, and
- know how to analyze and interpret the information.

Additional resources may be needed to help analyze the information gathered by the investigation team, such as programmers for computer control-related issues or material scientists to help the team understand material failure mechanisms.

Step 2 – Gathering the information

Forensic information and evidence, including physical data, electronic data and witness interviews, should be secured and gathered as soon as the emergency is over. It is essential that everyone understands what hazards may remain in the impacted area so that the physical evidence is gathered safely. Delays in preserving the incident site, collecting the forensic evidence and interviewing witnesses will make it more difficult to re-construct the event. Both physical evidence and personal recollections degrade over time. Loss of data integrity may occur with delays in securing physical or electronic evidence. And the timing for witness interviews must be prudently balanced: too soon after the incident the witnesses may still be in shock; too long after the incident, the witnesses may not be able to recall exactly what happened. As noted in Section 11.5.1, the forensic information is shared with the recovery team as they evaluate options for safely restarting the process, as well. This shared information includes what equipment or processes were damaged, how bad the damage was, and what hazards remain in the affected area.

Steps 3 and 4 – Analyzing the information to identify the causes and solutions

The information is analyzed to help identify the incident causes. Data analysis methods, briefly discussed in Section 11.7, range from simple and flexible tools to more-detailed and fixed-structured procedures. The goal of the analysis step is to interpret the evidence and develop an understanding of both direct and indirect causes. Direct causes include human actions or inaction and unexpected equipment or component failures. Indirect causes include latent conditions and weaknesses in the three foundations shown in Figure 1.3 that are essential for effectively managing a process safety program.

The results from the witness interviews can provide insights on the systemic weaknesses associated with human actions or inaction before and during the emergency. The gaps between the ineffective or missing process safety systems and what is expected are identified and used to generate recommendations. Although systemic weaknesses adversely affect the integrity of protection layers required to operate safe and reliable processes, keep in mind that these weaknesses may have influenced the operating environment at the time of the incident well before it occurred due to actions or inactions during the preceding months or years. The Bhopal incident presented in Section 11.9 is one of many cases where foundational and systemic weaknesses years earlier set the stage for the catastrophe.

Once the gaps have been identified, the team develops recommendations (the actions) needed to help prevent the incident from happening again. These solutions combine both administrative and engineering controls, such as improving procedures, changing the process, or changing the equipment. Usually these gaps focus on the missing or ineffective safeguards and systems at the time of the incident (see the different protection layers shown in Figure 3.6). However, the most effective investigation will identify underlying foundational weaknesses that contributed the incident. In other words, poor safety culture, poor safety leadership or poor operational discipline may have, in part, contributed to weaknesses in the process safety systems that are or should have been in place to safely manage the hazardous materials and energies.

Steps 5 and 6 – Implementing and documenting solutions

The solutions may be short-term and used by the recovery team, or they may be longer-term which could require significant time for process or equipment redesign, construction and installation. The solutions should be implemented through the facility's change management system, as shown as Phase 5 in Figure 11.2. The steps for managing the changes after the investigation phase are described briefly in Section 11.6.1 and in more detail in Chapter 10. Although there may be specific documentation requirements of the changes if they are not a "replacement in kind," keep in mind that something *will need to change* to help prevent a repeat incident. As noted with the recovery efforts above, an operational readiness review and approval procedure should be followed, addressing and validating proper equipment design, fabrication and installation of the repairs, rebuilds or replacements, with effective communications to all affected personnel [6].

Step 7 – Documenting the investigation

It is expected that the investigation team will document and communicate its effort (see Step 8, below). The incident report should describe what happened, when it happened, where it happened, and who participated on the incident team. The report documents the causes of the incident and the recommendations resulting from the analyses. If the hazardous materials or processes at the facility are subject to regulatory requirements, there may be specific documentation requirements, including the following:

- the date of the incident
- the date the investigation began
- a description of the incident
- the causes that contributed to the incident

- the actions resulting from the investigation, and
- the tracking and documentation of the actions until closure.

In some jurisdictions, there may be an incident investigation report retention requirement, as well. For example, for PSM-regulated facilities in the United States, a minimum of five years is required [48].

Step 8 – Share learnings

The final step is to communicate with others by sharing the results and recommendations of the investigation. Sharing these findings helps others learn and change without having them personally experience pain and suffering. Companies should review incidents that have occurred elsewhere, learning from and implementing changes to their processes and systems, as needed. The intent is to learn from the incident, understand how it applies, make changes, and prevent a similar incident from occurring again. As noted in 1905 by George Santayana: "Those who cannot remember the past are condemned to repeat it" [15]. In some jurisdictions there may be a requirement for the company to share its findings with all affected personnel, including contractors. The reader is encouraged to review and share findings from case studies and incidents [e.g., 1, 9, 16–21].

11.5.3 Combining Resources When Recovering and Investigating

When an incident requires a significant recovery effort, the same people needed to help get operations back up and running will be needed to help with the investigation effort. Whereas the recovery efforts focus on understanding the extent of the property damage, suggesting interim solutions such as repairs or replacements to safely resume operations (Section 11.5.1), the investigation efforts focus on understanding the causes and then proposing permanent changes to address them (Section 11.5.2). Thus, it makes sense to combine resources between these teams. In addition, the recovery and the investigative phases have similar steps, including identifying the team members, gathering and analyzing the information and identifying solutions. What distinguishes the two teams are the types of their recommendations. In general, interim changes are often identified and implemented relatively soon by the recovery team and the permanent changes are fully understood and implemented once the investigation team publishes its findings. These parallel phases were illustrated in Figure 11.2, and the parallel steps for each team are summarized in Table 11.5. Keep in mind that both interim and permanent solutions should be reviewed and implemented using the facility's management of change program which is briefly described in Section 11.6.

11.6 Changing and Sustaining after the Incident

This section discusses the final two phases depicted in Figure 11.2: Phase 5, when implementing the changes; and Phase 6, when sustaining the changes. The recommendations from the recovery and investigation teams are implemented using the facility's management of change system, and they are sustained by monitoring them through inspections, tests and audits. Operational discipline is essential for managing and sustaining these changes once the facility resumes operation, as prompt closure of the actions and thorough monitoring of the actions' effectiveness from then on will help ensure that the overall risk to the business has been reduced.

11.6.1 Implementing the Changes

Interim and permanent changes recommended by the incident recovery and investigation teams should be implemented using the facility's management of change program. These changes, whether administrative and engineering, should be properly reviewed and approved before being implemented. Interim changes measures include erecting barricades, posting signs, isolating or removing damaged equipment, and cleaning up the affected area. Depending on the extent of property damage, interim measures may include temporary equipment repairs or replacement, before permanent solutions have been identified. Proposed changes could affect the types of the raw materials, the operating parameters defining the processing conditions, the design and operation of the equipment, or the steps within the procedures. The proposed administrative changes will be reflected in the reviewed and updated procedures. Changes to the equipment should consider all life cycle phases: from the equipment's design (or redesign), fabrication, installation, commissioning and operation phases through its maintenance phase. Once the proposed engineering changes have been reviewed and approved, process safety information, such as the equipment files and P&IDs, should be updated and reissued. A final prestartup safety review—an "operational readiness" review—is then performed to help ensure a smooth handover to operations of the new and updated process and equipment changes [6]. In summary, the proposed measures must be implemented with the facility's management of change system (Chapter 10) and the equipment's integrity program (Chapter 9).

11.6.2 Sustaining the Changes

The changes resulting from an incident are sustained by regularly monitoring them to ensure that they continue to be effective in day-to-day operations. Keep in mind that these corrective actions were implemented to help prevent a recurrence by addressing gaps identified during the incident investigation. Unfortunately, over time we keep repeating our incidents since we suffer from both individual and corporate memory loss [9]. The further removed we are in time from *when the incident occurred*, the less we remember, and we forget how much the pain actually hurts us. The further away we are from *where the incident occurred*, the less we think about it in our day-to-day operations, as well ("it has never happened here"). Compounding our problems is the fact that we tend to forget why the changes were made in the first place, and then we jeopardize their effectiveness by reverting back to old habits and practices or by removing the improvements altogether.

Hence, strong operational discipline is required to prevent deterioration of our change's effectiveness. If the changes suffer from neglect or subsequent countermeasures, the actual operating risk at the facility may increase, in spite of the lower operational risk achieved with the change. This was illustrated earlier in Figure 4.3, where the *perceived* operating risk, based on the effective control measures implemented after the incident, does not represent the *actual* operating risk due to poor operational discipline over time. The facility holds onto its perceived risk, moving closer and closer over time to more risky operation as the controls in place degrade. Thus, the likelihood of having another severe incident increases as time elapses.

A case study describing the impact of a degraded process safety program and its three foundations over time is provided using the Bhopal incident in Section 11.9 below. In particular, an internal audit at Bhopal a few years before the tragedy had noted some of the weaknesses in the facility's operations, but no corrective actions were taken. Thus, it is crucial that we respond to and address gaps identified from investigations, reviews and audits to help reduce our operating risks.

One way to sustain the effectiveness of the implemented control measures is to add specific verification questions to the facility's internal auditing system. Specifically monitoring their effectiveness at regular frequencies is one way to ensure that the facility has the operational discipline to maintain the changes. This could include specifying that the incident investigation team's final report has as its last recommendation, due six months after all the other actions have been implemented and closed, a requirement for an internal audit on the state of all the closed recommendations. Thus, the controls implemented to reduce the likelihood of another incident and the process safety risk are checked at some point afterward to ensure that they are still effective and in place. In particular, this internal verification should become a part of the facility's auditing system, as will be described in Chapter 12.

11.7 Incident Investigation Methods

There are many methods used today to investigate incidents, ranging from simple and flexible tools to more detailed, fixed-structured procedures. Incident data analyses are used to help identify the incident's direct and latent causes, with today's efforts focusing more on the foundational and systemic causes. Traditional investigations assume a simple cause-and-effect approach (an "event tree") to help identify the causes, such as why-trees (or 5 why's), fishbone diagrams, root cause analyses, and fault tree analyses. Although these methods are briefly discussed in this section, the reader is encouraged to review more detailed resources which are available in the literature [11,12]. No matter which tool is selected, an effective investigation identifies the foundational weaknesses that contributed to or allowed the events to occur and then addresses these weaknesses with corrective actions (i.e., the key concept noted in Section 11.2).

11.7.1 Why-Trees

The simplest event tree model is the why-tree model. The why-tree approach is used to determine the root cause by repeating the question "Why?" to identify the cause-and-effect relationship of an issue or problem. Its formal origin can be traced back to the automobile industry, when the causes of defects in the car quality were identified and resolved. The premise is simple and can be easily adopted when identifying the causes of *simple incidents*. Each "why" question forms the basis of the next question. Since five is the typical number of times needed, the approach is often referred to as the "5-why" approach. Unfortunately, there are no specific rules about the types of questions or how long to continue the search for additional causes, such that the outcome depends upon the knowledge and persistence of the members of the investigation team.

11.7.2 Fishbone Diagrams

Often the answers to the 5-why cause-and-effect questions are combined with the fishbone diagram to map the answers across several categories. The fishbone is an event tree diagram used to visually depict multiple causes to the sequence of events, such as primary, secondary, contributing or latent causes within several distinct categories prone to failure. The six categories typically used in a fishbone diagram for process safety incidents are as follows:

- Management – in context of this book, this category includes the process safety program foundations shown in Figure 1.3 (safety culture and leadership; operational discipline; and the process safety systems)
- Environment – the operating environment, such as the process location (indoors or outdoors), the time of day, and the operating conditions (such as processing temperatures, pressures, etc.)
- Materials – the process materials (such as raw materials, intermediates, products, and waste streams, etc.), the utilities materials (such

as inert gases, steam, heat transfer mediums, etc.), and the maintenance materials (i.e., the materials of construction for replacement parts and equipment, such as gaskets, bolts, etc.)

- People – anyone involved with or affecting the process (includes management, engineering, operations, maintenance, contractors, and support personnel)

- Process – the administrative controls and the specific requirements for how the process is managed, such as specific programs within the process safety systems (i.e., operating procedures, maintenance procedures, and inspection, testing and preventative maintenance programs), and

- Equipment – the engineering controls, such as the processing equipment, the tools required to perform the task, engineered protection layers and safeguarding equipment (see Figure 3.6), maintenance equipment and tools, and utilities equipment and tools.

The answers from the 5-why questions based on these categories are populated on a diagram which looks like a fishbone, as is illustrated in Figure 11.4. The primary causes "point" to one of the categories with any secondary causes (which may be related to a weakness in a process safety system) "pointing" to the primary cause. The outcomes noted on the fishbone diagram depend on the knowledge and questions posed by the different members of the investigation team in each of the categories.

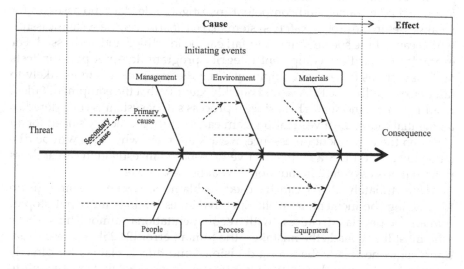

FIGURE 11.4
A fishbone diagram for depicting a cause-and-effect scenario.

11.7.3 Root Cause Analyses

One of the most popular incident investigation analysis tools based on the event tree model is a root cause analysis (RCA). The root cause can be defined as "an initiating event or failing from which all other causes or failings spring" [12]. RCAs can address failures in planning, in system management, and in communication across the organization. Similar to other event-related or chain-of-failure incident investigation approaches, RCAs can be used to determine the technical weaknesses, human factor issues, and the engineering and administrative controls which may have contributed to the incident.

An RCA uses a timeline to identify the sequence of causal factors for events that led up to the incident. This method is based on a traditional view of the three basic time-sequenced events in an incident: (1) its initiating phase, (2) its propagating phase, and then (3) its mitigating phase [11,22]. The RCA method is a "cause tree" approach—a *linear* event tree—which reconstructs the events leading up to the incident by selecting relevant causes, such as technical or human factor issues. It is a deductive approach looking backward in time to examine the preceding technical- and human-related events that caused the incident. The RCA approach focuses on component-based failures, such as a degraded equipment part or a missed procedural step, resulting in recommendations which include improving the reliability of the engineered safeguards, updating the administrative procedures, and training people. The reader is encouraged to review the literature for more information on predefined cause trees which provide more detailed information on identifying event-related causes [11,23].

Today, RCAs can be used to help identify systemic weaknesses, as well. When led by a trained and competent investigation leader who assesses for weaknesses in the process safety systems *as potential causes*, the systemic issues which caused the specific barrier to fail can be identified and addressed. For example, if a facility's equipment integrity program does not perform tests and inspections on critical equipment, then the equipment is more likely to fail unexpectedly. The RCA would quickly identify that the equipment failed, resulting in the incident. Then, since a process safety system is considered as a potential cause, the investigation team can identify the facility's need for an ITPM on the equipment (in essence, we ask another "why"). As will be discussed in Section 11.8, a more effective RCA-based investigation can address potential process safety foundational weaknesses, as well.

Unfortunately, RCAs can be used to blame someone or some group for causing the incident, as well. Effective investigations must not stop at human factor-related causes, with recommendations admonishing those who most likely made the mistake—the human error [24,25]. In these situations, someone "made the wrong decision," and thus is blamed for directly causing the incident. When the investigation concludes that people are to blame, any systemic causes which led to the "human error" will most likely not be addressed. Unfortunately, when an organization does not address systemic issues, another engineer, operator or mechanic in the same role will fall victim to a repeat incident caused by the unaddressed systemic-driven mistake and will cause yet another incident [26–28]. This reason, in part, is why an organization with a poor safety culture due to poor leadership focus on systemic problems will have recurring incidents.

11.7.4 Fault Tree Analyses

The Fault Tree Analysis (FTA) is the most commonly used technique for causal analyses in specific risk and reliability studies, applying the event tree model to better understand root causes. In particular, fault trees are excellent analyses to use when performing investigations on specific equipment-related failures (the failure being its "top event"). In addition, fault trees are applied proactively when designing safety instrumented systems (SIS), the last protection layer before a loss of containment as depicted as Barrier 5 in Figure 3.6. The reader is encouraged to review the literature for more details on SIS design using FTAs [29,30]. Also note that FTAs require a trained and competent leader and should only be used on a few selected scenarios due to the time and resources required for a thorough review.

In brief, a fault tree analysis is a top–down approach to failure analysis, starting with a potential undesirable event and then determining all the ways in which it can happen. The premise is based on a selected top event (an incident), with identified system faults that must occur for the top event to occur depicted with branches from the top event (a logic tree). The probability of the top event is determined by sequential and parallel or combinations of faults, with Boolean algebra used to quantify each fault's probabilities. The analysis proceeds by determining how the top event can be caused by individual or combined lower level failures or events, with the causes of the top event connected through logic gates. As will be described in the following section, the FTA approach can be combined with other analysis tools to help provide investigation teams with a visual tool to depict failures of protection layers that should be in place to reduce process safety risks (i.e., FTAs are incorporated into the bow tie diagram).

11.8 Applying Bow Tie Diagrams to Investigations

In this section we show how the bow tie diagram can be applied to help investigation teams understand the weaknesses in the process safety program foundations which led to an incident. This section describes how bow tie diagrams use the barrier model to depict barrier weaknesses and their event-related causes, and then we show how bow tie diagrams can depict the process safety system weaknesses which compromised the barriers. This section concludes by setting the stage for understanding how the barriers and the process safety systems were adversely affected by weaknesses in the process safety program foundations, ultimately leading to the catastrophe at Bhopal—the case study described in Section 11.9.

11.8.1 Searching for the Event-Related Causes

The simplest and most commonly used method when searching for event-related causes is the root cause analysis (RCA) approach discussed in Section 11.7. RCAs based on the barrier model focus on the events which led the incident, helping identify the barriers that have failed to prevent or mitigate the incident. Figure 3.12 shows how the Swiss Cheese Model is used to depict the missing or ineffective barriers identified in an RCA. Although this straight line (linear) pathway does not capture complexity (discussed in more detail below), it does provide the starting point for representing the multiple barrier failure pathways on the bow tie diagram.

11.8.2 Visualizing Barrier Weaknesses on the Bow Tie Diagram

The bow tie diagram introduced in Chapter 3 is used as a visual summary of the incident scenarios, helping incident investigation teams understand the hazards and how they should have been managed. Since multiple failure pathways can lead to loss of control, and multiple consequence pathways may be possible, the bow tie diagram helps visualize the missing or ineffective barriers. Thus, the bow tie diagram shows the chain of events or possible incident scenarios identified in the RCAs, starting with the fault tree leading up to the top event and an event tree leading up to the consequences (see Section 11.7 for discussions on fault and event trees). Causal factors are identified as the specific events which needed an effective barrier to prevent the hazard(s) progression from the left-hand side, through the knot of the bow, and escalating the consequences as the hazardous scenario proceeds to the right-hand side. With this image, the bow tie diagram helps visualize the control measures, including the management systems that should have been in place to reduce the overall process safety risk.

The bow tie diagram shows the *multiple linear pathways* that led to the incident, with both the preventive and mitigative barriers depicted as walls on each path (Figure 11.5). The ineffective or missing preventive barriers allow a *straight* path from the threat to a loss of containment (the knot of the bow), with another *straight* path to the consequence due to ineffective or missing mitigative barrier failures. An incident investigation that used a bow tie diagram approach was presented earlier in Figure 3.14. The investigation team identified causal factors that led up to the failure of the barrier, allowing the incident's consequences to escalate as its events evolved over time (from

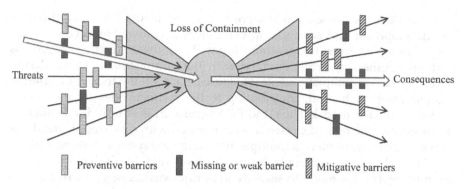

FIGURE 11.5
Mapping barrier weaknesses on the bow tie diagram—A linear path.

left to right). Additional discussion on the history of the bow tie diagram and its current uses when analyzing incidents is provided in the literature [31–36].

11.8.3 Searching for the Systemic Causes

Since the methods to better understand and identify the systemic causes in the chemical process industries are in their infancy at the time of this book's publication, this section can only provide the reader with a brief overview of concepts currently supporting the efforts to better model and thus identify the systemic causes. The key term to takeaway from this section is "complexity," which can be defined for chemical processes as follows:

1. Complex - not simple (i.e., cannot be modeled effectively with a linear event tree barrier analysis sequence)
2. Complexity - the complex man/machine/environmental interactions between the people, the process equipment and the operating environment that affect and are affected by deviations from the process and equipment design.

In other words, the performance of your chemical process, based in part on how well people follow the process specifications and maintain the equipment integrity, cannot be predicted using a simple linear event tree model. As discussed in Chapter 5, the process design specifies standard operating conditions and safe operating limits on the processing temperatures, pressures, flow rates, etc. and specifies the operating procedures as an administrative control when the process is started up, is running or is being shut down. In addition, the equipment design has specific operating conditions (Chapter 5) and specific inspection, testing, and preventative maintenance (ITPM) integrity programs (Chapter 9) to manage the process safety risks.

The RCA tools used today were developed in the 1960's, well before the development and advances in computers and distributed control systems that are common to operations today. Thus, these advances in process systems engineering (PSE), credited with the improvements in the design and operation of complex chemical facilities, have left the incident causation models woefully behind, contributing to the ineffectiveness of the simple linear event tree model. Thus, the RCA approach is simply inadequate to address systemic-related causes, as is evidenced with catastrophic incidents in recent years attributed to multiple, unanticipated *systemic* failures [28].

The advanced PSE controls are fundamental to the main concept supporting the approach to identify systemic weaknesses described with the normal accident theory [37]. The theory's basic premise is that process safety incidents are *inevitable* due to the complex, tightly coupled process safety systems that are in place to manage the process safety risks. Although we have implemented engineering controls—both preventive and mitigative protection layers—to reduce the system's vulnerability and failures by adding more complexity to the system, we actually have caused just the opposite effect, leading to an inevitable system failure and a process safety incident. The process safety systems are so intertwined that the failure of one process safety system—a systemic weak link—leads to the failure of the entire *process safety program*. By adding more complexity to the management of process safety risks, the complex, systemic interactions contribute to *new, unknown paths* for incidents. A good example is the trend for implementing more sophisticated, complex safety instrumented systems (SIS) with safety integrity level (SIL) ratings of 2 or 3. Not only does the facility's life cycle cost increase due to increased ITPM programs, the tests and inspections are often so complex that they may be compromised during the test, or just as troubling, people select to postpone the tests since there is insufficient time to do them properly [38].

Compounding the complexity when operating today's processes are the social communication networks between different groups within the business, within the facility, and within the process units. These different groups manage process safety, occupational safety, and health, environmental, quality, and security risks, as well as operational, environmental, regulatory, community, and financial risks [39]. The interactions between these social networks are still poorly understood. It is worth noting, however, that much progress has been made in other industries to better understand the role of complexity when investigating process safety incidents. These investigation teams have gotten better in identifying their industry's systemic issues [28,40]. The reader is encouraged to read more on the System-Theoretic method developments developed to help improve process safety performance. In particular, the System Theoretic Process Analysis (STPA) is designed to understand which "systemic safety constraints" have failed and which has been implemented with varying degrees of success in different industries including oil and gas production and chemical processes [40].

In summary, investigation teams that use the RCA approach to identify the *systemic causes* of why the engineering and administrative controls failed or were ineffective will be more successful in helping prevent another similar incident in the future. However, due to today's complex operating and control systems, systemic weaknesses create a work environment ripe for new, unanticipated "sequence" of events which inevitably result in yet another unpredicted incident.

11.8.4 Visualizing Systemic Weaknesses on the Bow Tie Diagram

Although the bow tie diagram has been used to map the gaps in the systemic causes which contributed to an incident using the linear model approach [41–43], it is difficult to visualize the non-linear path that actually occurred. In this section, we apply the systemic approach to the bow tie diagram with one major distinction: while the event tree path is a simple, linear illustration (Figure 11.5), the systemic approach offers a complex, non-linear illustration, as is shown in Figure 11.6. We have added the system failure as vertical lines which manage system-specific engineering and administrative barriers, with one system controlling a number of barriers designed to prevent or mitigate the incident. An example, assuming that a few critical safeguarding instrumentation and alarms have not been maintained (Barrier 4 in Figure 3.6), would be the loss of the equipment's integrity due to an inadequate ITPM program discussed in Chapter 9. A systemic failure would cause all barriers associated with the system being in a failed state, with multiple potential non-linear paths (e.g., paths 1 and 2 in Figure 11.6).

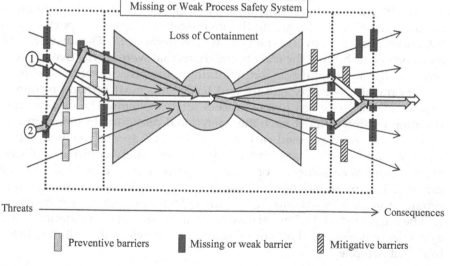

FIGURE 11.6
Mapping systemic weaknesses on the bow tie diagram—non-linear path.

Thus, with a common cause being an ineffective or missing process safety system, we can see there are opportunities for a path from the threats on left to the consequences on the right to be non-linear, as represented visually on the bow tie diagram. Note that a number of potential threat-to-consequence event tree scenarios due to the multiple, complex systemic interactions may help explain why PHAs using the HAZOP methodology typically do not address double-jeopardy scenarios since there are many more potential paths for each hazardous scenario (Chapter 7). This helps explain why hazards analysis teams cannot reliably anticipate all of the potential paths. The traditional HAZOP studies consume enough resources and time as is, with little incentive to require addressing additional, complex scenarios that only will add to the costs and duration of the study. This non-linear visualization on the bow tie diagram of how systemic failures led to an incident is demonstrated with the Bhopal case study in Section 11.9.

11.8.5 Searching for the Foundational Causes

Since the bow tie diagram does not look at one potential path but to all possible paths associated with the hazard, it lends itself to helping understand systemic weaknesses. While the ineffective or missing barrier led directly (linearly) to the event (Figure 11.5), the ineffective or a missing process safety system led to *multiple* barrier weaknesses (Figure 11.6), increasing both the likelihood and consequences of the event due to a complex, non-liner path to the event. The weaknesses in the systems were due to weaknesses in the other foundations.

Recall, process safety systems are one of our process safety program foundations (Figure 1.3), giving the investigation team an opportunity to ask another why: "Why did the system or systems fail?" The answers will focus on one or both of the *other* foundations that were ineffective in properly establishing, supporting, or maintaining the process safety system(s). The organization's safety culture and leadership must provide resources and the means to design robust process safety systems. The organization from top to bottom must have the operational discipline to maintain and sustain them. As was noted earlier in Section 11.7, an organization with a blame culture—poor safety leadership and poor safety culture—will stop investigating once someone or some group is found responsible for the incident. This reinforces our premise of the effectiveness of a strong safety program: *all three foundations must be strong for the program to be effective.* As will be illustrated in the Bhopal case study in Section 11.9, foundational and systemic weaknesses led to multiple barrier weaknesses that ultimately led to the catastrophe.

11.9 Case Study

Since much has been written about the Bhopal incident, we hope this case study complements them by illustrating how bow tie diagrams visually depict the systemic and foundational causes that led up to that fateful day in 1984. We set the stage by describing the sequence of events and the causal factors attributed to the incident at Bhopal [9,11,16,44–46]. This case study for the Bhopal incident is described in greater detail in this section using the following parts:

1. The consequences
2. The original design
3. Changes to the original design
4. Mapping the safeguards on the bow tie diagram
5. Understanding the foundational weaknesses

Our goal is to show how investigation teams can depict complex process incidents with a visual illustration of weaknesses in all three process safety program foundations. The bow tie diagram not only depicts the barrier failures, but also the process safety system failures due to weaknesses in the other two process safety program foundations. Although we know that the end result at Bhopal was catastrophic, if we had known then what we know now for safely managing hazardous processes, the event at Bhopal may have been prevented or may not have had such severe consequences. Thus, an effective process safety program at Bhopal could have reduced - or even prevented - the number of fatalities, the number of injuries, the environmental damage, the abandonment of the facility and the ultimate end of a company that was known for its innovation and safety. The lesson we wish for you to take-away after reading this case study: Effective process safety programs can be achieved only when all three process safety foundations shown in Figure 1.3 are strong.

Bhopal Discussion Part 1 – The consequences

"The worst disaster in the history of the chemical industry occurred in Bhopal, in the state of Madhya Pradesh in central India, on December 3, 1984. A leak of methyl isocyanate (MIC) from a chemical plant, where it was used as an intermediate in the manufacture of the insecticide carbaryl, spread beyond the plant boundary and caused the death by poisoning of more than 2,000 people. The official figure was 2,153, but some unofficial estimates were much higher. In addition, about 200,000 people were injured. Most of the dead and injured were living in a shantytown that had grown up next to the plant [9]."

"The vapor is highly toxic and causes cellular asphyxiation and rapid death. ... MIC is highly reactive in the presences of water and iron oxide, and it generates heat. In sufficient quantities, this heat may generate [the highly toxic] vapor... [11]."

"The incident was a catastrophe for Bhopal ... [with] significant damage to livestock and crops.... [over a decade later] upwards of 50,000 people remained partially or totally disabled [16]."

The impact of the catastrophe still affects the region today.

Bhopal Discussion Part 2 – The original design

The Bhopal incident took place within a pesticide facility operating in India licensed by an experienced company with its headquarters in the United States. The company was a global industry leader at the time, known for successfully managing the safety of its processes (although the term "process safety" was not identified as such at that time). The blueprint used for the design of the Bhopal facility came from an existing facility in the United States that had operated safely for almost 20 years. It was therefore reasonable to expect that the new manufacturing process to be built in India would be at least as safe as the original facility.

One of the intermediate chemicals in the process was methyl isocyanate (MIC), a toxic, reactive, volatile and flammable material that is a liquid at room temperature. MIC reacts readily with water, with an exothermic, runaway reaction occurring once the temperature gets too high. In addition, pure MIC is highly reactive with itself, forming an MIC "trimer," a stable, solid substance with physical properties similar to plastic. When in the presences of iron oxide (rust), the MIC to trimer reaction is catalyzed and significant deposits occur on the metal's rusted surface. Thus, the specification for piping and equipment expected to contain MIC liquid or vapor is stainless steel (the first safeguard, an inherently safer process design).

MIC leaks are very difficult to ignore since exposure to MIC induces tears and it produces an irritating odor at low concentrations. Therefore, small leaks of MIC are easily detected by both workers inside the facility and the surrounding community. The Bhopal facility was designed with a public address system to notify anyone who might be affected by a fugitive MIC release in the event of a loss of containment incident. As is well known due to the consequences of the incident at Bhopal, MIC is deadly.

Safeguards in the original equipment design included a refrigeration system on the MIC storage tanks to help prevent MIC liquid from vaporizing and to help prevent potential exothermic reactions if the pure MIC in the storage tank becomes contaminated. Another safeguard included in the design was a variable rate nitrogen purge system to help prevent potential flammable atmospheres of MIC vapors from being able to develop during maintenance and shutdowns.

Operations had several options to prevent unstable reactions in the MIC storage tank, including the ability to continuously spike the pure MIC storage vessels with 200–300 ppm phosgene. A high temperature alarm during normal operations was used to detect an unstable, potentially exothermic condition, as well. Once the alarm set point was triggered, operations had multiple administrative controls for responding to and managing the contaminated or unstable MIC:

1. Transferring the tank contents into an auxiliary reserve tank for additional cooling
2. Adding a solvent to the affected MIC tank to quench the exothermic reaction
3. Reprocessing the impure MIC back through the purification process
4. Neutralizing contaminated MIC in the vent gas scrubber section, and
5. Sending waste MIC into the flare tower for final destruction.

In addition, the MIC storage tanks were protected by a pressure relief valve that would automatically open at 40 psi when the unwanted exothermic reaction resulted in increased pressure. Once relieved, the MIC vapors could be directed into the vent gas scrubber or to a flare tower for destruction. As a last resort, a fire monitor discharge could be directed to the atmospheric MIC release point to quench any hot MIC vapor or liquid releasing into the atmosphere. Although these secondary control measures could reduce the consequences of an MIC release, the most effective safeguards were designed to prevent the conditions that might result in an overpressure scenario in the first place.

Bhopal Discussion Part 3 – Changes to the original design

The original process and equipment design for the Bhopal facility suffered from decisions and innovative problem-solving solutions that were made during its construction, when it was commissioned and once it began operating. Different hazards were inadvertently created as operating issues were addressed. These changes, in part, were necessary at Bhopal due to equipment reliability issues, beginning with the chronic failure of the MIC transfer pump. It was almost a year after the scheduled start-up before the commissioning team handed the process over to operations. Although the pump problem had not been solved, the team (including experts flown in from the United States) solved the pump transfer problem by using the Nitrogen system to push the material from the tank. The details for their solutions are described in more detail by Bloch [46]. The dire consequences of these changes manifested themselves into "normal operations" over time with unexpected, additional maintenance required to address new unanticipated issues arising from the previous solutions. Overall, each new decision to some degree defeated an engineering safeguard and replaced them with an additional administrative control subject to human error. Ultimately (and inevitably) during the decommissioning stage these decisions, each a hole or gap in the barrier, aligned and MIC was released into the community. Some discussion of these critical decisions during the facility's life is provided in this section, all of which to some degree defeated the safeguards specified in the original design.

During the construction at Bhopal, the inherently safer stainless steel specification for the MIC vapor transfer header system was changed to less expensive iron. The reasoning included the knowledge that during normal operations, the transfer header would never see MIC vapors. As a result, the process became very dependent on the nitrogen inerting for both flammability and reactivity protection during normal operations. Unfortunately, and unbeknownst to all making the decision, the process was changed to address the frequent pump shutdown by diverting MIC-laden vapors through the header transfer system. The rust on the vapor header carbon steel piping catalyzed the MIC into trimer which formed on the surface, eventually plugging the piping. To clean out the trimer, maintenance began to routinely wash out the piping with water. The problems addressed at one point introduced other problems due to poor understanding of the original process design and of the adverse consequences of the change.

Due to the chronic MIC storage tank transfer pump failures, the process had to be shutdown routinely to do pump maintenance. This included shutting the refrigeration system down, resulting in routine tank temperature increases as the ambient temperature heated the tank. Since there were no adverse consequences with this normal temperature rise, operations became conditioned to the nuisance alarm and subsequently ignored it. Later, process design modifications led to the loss of the high temperature alarm function and instrumentation reliability altogether. The fact that the temperature reading was supposed to detect an exothermic reaction which could lead to overpressurization of the tank was lost.

The frequent leaks from the MIC transfer pumps further compromised safety by conditioning the workers and the surrounding community to expect periodic MIC loss of containment incidents as the norm rather than the exception. Through this normalization of deviation process, after multiple MIC leaks in the facility and into the surrounding community, the foul MIC odors failed to elicit a defensive response by the people close to the facility. On the night of the incident, the public alarm system was shut off soon after the release was detected, even though the release had not been stopped.

Many of the changes on the process addressed the reliability issues associated with the MIC transfer pumps eventually manifested themselves into chronic operational and maintenance issues with the vent gas scrubber and flare systems. These issues were due in part to the routine maintenance needed to clean out the trimer plugging the vent header system. Both operational and maintenance workarounds were added to the normal operations, such that the scrubber was not on line and the flare had been disconnected on the day of incident. Unfortunately, an MIC exotherm occurred on December 3. Thus, MIC vapors formed, the tank's pressure increased and then MIC vapors were released through the pressure relief system. The inexperienced crew did not expect the release. Once the release was detected, they could not manage the vapors due to inoperative mitigative controls. More details on the decisions to not maintain the functionality and reliability of these mitigative systems have been provided by Bloch [46].

During the course of the struggling operations, several significant toxic exposure incidents occurred, one of which resulted in a fatality [46]. Unfortunately, the local leadership and the incident investigation team concluded that the fatality was the mechanic's fault, in spite of evidence that the process was no longer in control and that the required tasks to run and maintain the plant were placing the workers at undue risk. A formal, public complaint by the workers due to the poor working conditions only soured the relationship between the management and the workers. Over time, significant personnel turnover rates and inadequate training placed workers into operations and maintenance positions with little knowledge of the process, its hazards, and its safeguards. These new workers learned how things were getting done, not with how things were supposed to be done per the original design. The many changes to operations and maintenance procedures had become routine operations and maintenance tasks. When the inexperienced crew was hired to deinventory the process – the last production run at Bhopal – the reasons for the safeguards being there in the first place was lost. To them, it was just another routine day at the facility. A routine day that was anything but.

Bhopal Discussion Part 4 – Mapping the safeguards on the bow tie diagram

The preventive and mitigative safeguards for the original process design at Bhopal discussed in the previous section are mapped on the bow tie diagram shown in Figure 11.7. The preventive safeguards included the inherently safer process design specifications for stainless steel, the refrigeration system, the nitrogen purge, the phosgene spike (all represented as a part of Barrier 1, Figure 3.6), and the high temperature alarm (Barrier 4). As shown in Figure 11.7, the bow tie diagram also depicts the mitigative safeguards in place per the original design. These safeguards included the pressure relief system (Barrier 6) the vent gas scrubber system, the flare tower and the water curtain fire monitors (all a part of Barrier 7), the public warning system (Barrier 8).

Since the path for the combined systemic failures on complex processes is difficult to predict (it is non-linear), we introduce the bow tie diagram at this point to help us visualize this unanticipated and non-linear path. Keep

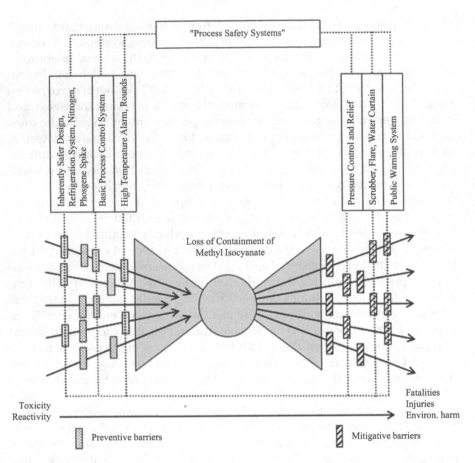

FIGURE 11.7
The barriers in the process design at Bhopal.

in mind that all of the safeguards noted in the Bhopal design were in a failed state at the time of the incident due to systemic weaknesses and failures. Just one of many potential non-linear paths resulting from systemic failures is represented on the bow tie diagram in Figure 11.8. Although the deterioration of each of the safeguards occurred during the months preceding the incident, the alignment of all the holes in these barriers at Bhopal was not a matter of *if*, it was simply a matter of *when*. The event was inevitable due to poor management and operational decisions months beforehand which degraded the original design intents and the equipment integrity [40,43–46].

Bhopal Discussion Part 5 – Understanding the foundational weaknesses

Before discussing how the foundational weaknesses led to the disaster at Bhopal, it is worth noting at this point that an audit was initiated by senior leadership in the United States to help better understand what issues were

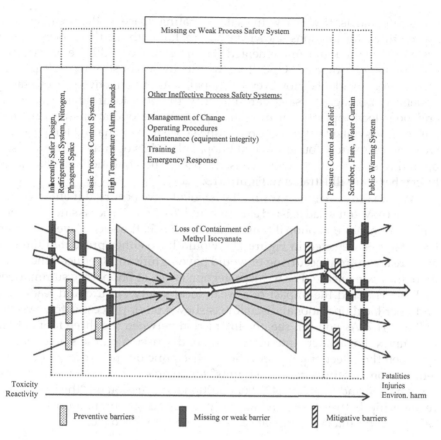

FIGURE 11.8
The non-linear path for the systemic failures at Bhopal.

keeping the facility in India from meeting its safety goals and its production schedule. In addition, unhappy company investors were applying considerable pressure on the senior leadership to improve Bhopal's productivity at the same time. Hence the audit was needed to help understand why Bhopal had failed to show a profit in India [46]. The report from the corporate safety audit, held well before the incident in December 1984, documented safety concerns which were not adequately addressed, if addressed at all, at the time of the disaster.

Collectively, the weaknesses in the process safety foundations and process safety systems deteriorated multiple barriers that were needed to manage the process safety risks at Bhopal. Based on the original design of the facility, there was a "safe operating space" for manufacturing this pesticide. This safe operating space relies on the equipment being designed per their specifications, the processes being operated per their safe operating limits, that changes which affect the safeguards are adequately reviewed and approved, and that the equipment is reliable, is tested and inspected, and is maintained per their specifications. It takes operational discipline to sustain these safeguards. If we view the safe operating space as the area bounded by the three foundations shown in Figure 1.3, with all the process safety systems designed and implemented (Figure 3.6), we will have an effective process safety program. Operating within these constraints will prevent catastrophic incidents. However, at Bhopal, without effective process safety systems to manage the safeguard integrity, the poor local safety leadership and poor local operational discipline pushed the operations and maintenance efforts outside of the safe operating space into uncharted territory. The weak process safety foundations at Bhopal placed its operations in an unsafe operating space with horrible consequences. The conditions at Bhopal on December 3 are illustrated in Figure 11.9.

We can only imagine – would the incident have occurred if effective process safety systems had existed at Bhopal in 1984? If the process safety systems had been in place at Bhopal (Barrier 2, Figure 3.6), the successive protection layer failures depicted in Figure 11.8 would have either prevented the runaway reaction from occurring or would have significantly reduced the severity of the toxic release. The weak protection layers at Bhopal are summarized in Table 11.6 and the impact of a weak local safety culture and leadership and poor local operational discipline is shown other process safety systems in Table 11.7. As you can see, the interaction between the three foundations is complex. For readers interested in more discussion on the safe operating space, on the effect of poor operational discipline on the process safety system, and on "safety system constraints" for complex processes, please refer to Leveson [28] and Vaughen [43]. For additional discussion on "the interwoven relationship between asset reliability, process design, human error, process safety and regulatory compliance," please refer to Bloch [46].

FIGURE 11.9
The foundational weaknesses for the Bhopal disaster.

TABLE 11.6

A Summary of the Bhopal Process Safety System Failures

	Protection Layers Designed for the Bhopal facility		Condition on December 3, 1984	Process Safety System Failures (Assuming that they had been in existence in 1984.) [Details provided in references 43 and 46]
	Preventive	Mitigative		
Inherently Safer Process (ISP) design	√		Carbon steel used instead of stainless steel for the vent gas scrubber header	Design Safe Processes – removed an inherently safer process design specification Change Processes Safely - did not fully understand and anticipate the impacts of design changes
Refrigeration system	√		No longer in service	Identify and Assess Process Hazards – did not fully understand impact of changes on the safeguards and the hazards being managed Evaluate and Manage Process Risks – did not fully understand impact of changes on how to manage the risks Change Processes Safely - did not fully understand and anticipate the impact of changes to how the runaway reaction was monitored
Variable rate nitrogen purge	√		Altered design and used Nitrogen pressure to transfer MIC due to chronic transfer pump failures	Evaluate and Manage Process Risks – did not fully understand impact of changes on how to manage the risks Operate Safe Processes – operators were conditioned to how things were getting done, not how they were supposed to be done Change Processes Safely - did not fully understand and anticipate the impacts of design changes
Phosgene spike	√		Not used	Evaluate and Manage Process Risks – did not fully understand impact of changes on how to manage the risks Operate Safe Processes - operators were conditioned to how things were getting done, not how they were supposed to be done

(Continued)

TABLE 11.6 (*Continued*)

A Summary of the Bhopal Process Safety System Failures

	Protection Layers Designed for the Bhopal facility			Process Safety System Failures
	Preventive	Mitigative	Condition on December 3, 1984	(Assuming that they had been in existence in 1984.) [Details provided in references 43 and 46]
Operator surveillance (local gauges)	√		Local gauges unreliable	Maintain Process Integrity and Reliability – maintenance personnel were conditioned to how things were getting done, not how they were supposed to be done Change Processes Safely – did not fully understand and anticipate the impacts of operating procedure changes
High temperature alarm	√		High temperature alarms no longer in service; None of the dependent administrative controls were used	Evaluate and Manage Process Risks – did not fully understand impact of changes on how to manage the risks (Note: The high temperature alarm triggered five different administrative controls; see discussion in text). Operate Safe Processes - operators were conditioned to how things were getting done, not how they were supposed to be done Maintain Process Integrity and Reliability – maintenance personnel were conditioned to how things were getting done, not how they were supposed to be done Change Processes Safely – did not fully understand and anticipate the impacts of operating procedure changes
Public warning system		√	Alarm silenced soon after release began	Manage Incident Response and Investigation – facility emergency response inadequate to contain release Manage Incident Response and Investigation – facility and community conditioned to small MIC releases which no longer elicited a safe response

(Continued)

TABLE 11.6 (*Continued*)

A Summary of the Bhopal Process Safety System Failures

	Protection Layers Designed for the Bhopal facility			Process Safety System Failures (Assuming that they had been in existence in 1984.) [Details provided in references 43 and 46]
	Preventive	Mitigative	Condition on December 3, 1984	
Tank vent valve (pressure control)		✓	Design modified for the nitrogen system being used to transfer MIC due to chronic pump failures	Operate Safe Processes - operators were conditioned to how things were getting done, not how they were supposed to be done Change Processes Safely - did not fully understand and anticipate the impacts of design and operating procedure changes
Pressure relief valve		✓	Worked as designed	None identified
Vent gas scrubber system		✓	Disconnected Placed online too late	Operate Safe Processes - operators were conditioned to how things were getting done, not how they were supposed to be done Maintain Process Integrity and Reliability – maintenance personnel were conditioned to how things were getting done, not how they were supposed to be done Change Processes Safely - did not fully understand and anticipate the impacts of operating and maintenance procedure changes
Flare tower		✓	Disconnected due to maintenance needs	Operate Safe Processes - operators were conditioned to how things were getting done, not how they were supposed to be done Maintain Process Integrity and Reliability – maintenance personnel were conditioned to how things were getting done, not how they were supposed to be done Change Processes Safely - did not fully understand and anticipate the impacts of operating and maintenance procedure changes
Fire monitors (water curtain)		✓	Inadequate and could not reach venting MIC	Design Safe Processes – did not design for maximum relief system venting rate and/or pressure/flow demands required at the fire monitors Change Processes Safely

TABLE 11.7

A Summary of the Bhopal Process Safety Foundation Weaknesses

Process Safety System (Assuming that they had been in existence in 1984.)	Effect of a Weak Safety Culture and Leadership and of Poor Operational Discipline at the Facility [Details provided in references 43 and 46]
Operate Safe Processes	Local leadership pressured to make a last "production" run to "simply" deinventory MIC supplies in storage tanks before December 31, when asset decommissioning was to be completed. Untrained inexperienced crew was unfamiliar with hazards and changes to the equipment design, to safeguard conditions and availability, and the originally safely designed operating procedures.
Maintain Process Integrity and Reliability	The transfer pump failed yet again on December 3, resulting in implementation of the maintenance workaround procedure which included a water wash to remove trimer in the vent gas scrubber. Maintenance crew remaining after severe personnel cutbacks did not recognize that trimer deposited in valves prevented them from closing tightly, resulting in potential continuous feed of water into MIC storage tank during workaround on vent gas header.
Change Processes Safely	Innovative corporate culture responsible for its business successes when solving problems began with MIC storage tank pump reliability issues, solved by identifying *and approving* the use of existing nitrogen pressure system to transfer MIC through vent gas scrubber piping which, based on original design, was never to see MIC (justified change from stainless steel to carbon steel during fabrication and installation to save on construction costs).
Manage Incident Response and Investigation	Local leadership biased investigations conclusions using a "blame" culture, focusing on human error as the root cause. Operations, maintenance and local management accepted small MIC/phosgene leaks as "normal operations" without investigation near misses in spite of some serious incidents at the facility (leaks causing a fatality – a hit – and multiple hospitalizations – a near miss). Local community desensitized to actual hazards of MIC due to frequent small MIC leaks.
Monitor Process Safety Program Effectiveness	Local leadership did not adequately address recommendations from the corporate safety audit.

The operations suffered from process safety systemic failures *before* the Bhopal facility was even fully commissioned, such that all of the safeguards in the original design were no longer fit for service when they were needed. Since the incident surprised and saddened the company leadership [43], it is apparent that there was a low perceived risk at the time of the incident (see Figure 4.3). The unfortunate result, common to all complex, systemic failures, is that once the problem is recognized, the people responding do not have sufficient time to fix the problem before it is too late [40].

We hope this analysis and discussion provides you with an additional perspective on why tragic incidents like the ones listed in Tables 1.1 and 1.6 continue to occur when the process hazards and risks are not properly recognized, evaluated and managed. The results far too often are deadly exposures to hazardous materials, fires, explosions, runaway reactions and other serious events that lead to fatalities, significant environmental harm, and property damage. The loss of the safety culture and leadership, allowing the operational discipline to lapse, resulted in the degradation of the process safety systems that had been established to maintain the equipment integrity and reliability. In summary, process safety was compromised at the beginning and throughout the Bhopal facility's brief life.

11.10 Measures of Success

For responding to emergencies

- ☑ Design and implement an effective process safety system managing incident responses, including planning for and responding to emergencies.

- ☑ Provide training to all personnel to ensure that they understand their role and how they respond safely during emergencies.

- ☑ Provide training to emergency response teams and ensure that they are staffed with experienced and capable members and provided appropriate emergency response equipment.

- ☑ Ensure that drills are regularly scheduled and are debriefed, including:
  - o Developing well-written and supported recommendations based on improving the response,
  - o Communicating the drill results to affected personnel, and
  - o Tracking the recommendations until closure.

For investigation incidents

☑ Design and implement an effective process safety system for managing incident investigations.

☑ Define the effort required to perform investigations of unwanted events in hazardous processes, ranging from near misses or small releases to loss of containment events requiring activation of emergency response teams.

☑ Ensure that experienced, trained, and multi-disciplinary investigation teams are properly sized to analyze unwanted events, including:

   o Qualifying leaders on investigation methodologies,

   o Developing well-written and supported recommendations based on reducing process risk by addressing protection layer or systemic gaps,

   o Authorizing a comprehensive incident investigation report,

   o Communicating the investigation results to affected personnel, and

   o Tracking the recommendations until closure.

References

1. Mannan, Sam, editor. 2005. *Lees' Loss Prevention in the Process Industries, Third Edition. Hazard Identification, Assessment and Control.* Volume 1 and Volume 2. Elsevier, New York.
2. U.S. Chemical Safety Board. 2016. *West Fertilizer Company Fire and Explosion.* Report No. 2013-02-I-TX. www.csb.gov.
3. Vaughen, Bruce K., and Kletz, Trevor A. 2012. Continuing our process safety management (PSM) journey. *Process Safety Progress* 31:337–342.
4. Center for Chemical Process Safety (CCPS). 2016. *Process Safety Glossary.*
5. Czerniak, John and Don Ostrander. 2005. *Nine Elements of a Successful S&H System.* www.nsc.org.
6. Center for Chemical Process Safety (CCPS). 2007. *Guidelines for Risk Based Process Safety.* John Wiley & Sons, Inc. Hoboken, NJ.
7. U.S. Environmental Protection Agency (U.S. EPA). 1999. Chemical Emergency Preparedness and Prevention Office (CEPPO). Use Multiple Data Sources for Safer Emergency Response, EPA-F-99-006. www.epa.gov/ceppo/.
8. U.S. OSHA. 2016. 29 CFR 1910.120, Hazardous waste operations and emergency response standard (HAZWOPER). www.osha.gov.
9. Kletz, Trevor A. 2009. Case histories of process plant disasters and how they could have been avoided. *What Went Wrong?* 5th ed. Butterworth-Heinemann/IChemE.

10. Washington State Department Labor & Industries. 2009. Accident Investigation Basics. http://www.lni.wa.gov/SAFETY/TRAININGPREVENTION/ONLINE/courseinfo.asp?P_ID=145.
11. Center for Chemical Process Safety (CCPS). 2003. *Guidelines for Investigating Chemical Process Incidents,* 2nd ed. John Wiley & Sons, Inc. Hoboken, NJ.
12. UK Health and Safety Executive (UK HSE). 2004. Investigation accidents and incidents; HSE245.
13. Hyatt, Nigel. 2006. *Incident Investigation and Accident Prevention in the Process and Allied Industries.* Taylor and Francis Group (CRCPress), Boca Raton, FL.
14. McKinnon, Ron C. 2012. *Safety Management: Near Miss Identification, Recognition, and Investigation.* Taylor and Francis Group (CRCPress), Boca Raton, FL.
15. Vaughen, Bruce K. 2012. A tribute to Trevor Kletz: What we are doing and why we are doing it. *Journal of Loss Prevention in the Process Industries* 25:770–774.
16. The Center for Chemical Process Safety (CCPS). 2008. *Incidents That Define Process Safety.* John Wiley & Sons, Inc., Hoboken, NJ.
17. Chemical Safety Board (CSB). 2016. *Completed Investigations.* www.csb.gov.
18. UK Health and Safety Executive (HSE). 2016. *COMAH Case Studies.* www.hse.gov.uk.
19. Safety and Chemical Engineering Education (SAChE) program. 2016. Center for Chemical Process Safety (CCPS)/American Institute of Chemical Engineers (AIChE). www.sache.org.
20. Sanders, Roy. 2015. *Chemical Process Safety, Learning from Case Histories,* 4th ed. Butterworth-Heineman (Elsevier), New York.
21. Center for Chemical Process Safety (CCPS). 2016. *Introduction to Process Safety for Undergraduates and Engineers.* John Wiley & Sons, Inc. Hoboken, NJ.
22. Crowl, D. A., and J. F. Louvar. 2011. *Chemical Process Safety, Fundamentals with Applications,* 3rd ed. Prentice Hall, Upper Saddle River, NJ.
23. UK Health and Safety Executive (UK HSE). 2001. *Root Causes Analysis: Literature Review.* Contract Research Report 325/2001. www.hse.gov.uk.
24. Dekker, Sidney. 2006. *The Field Guide to Understanding Human Error.* Ashgate Publishing Company, Burlington, VT.
25. Vaughen, B. K., and T. Muschara. 2011. A case study: combining incident investigation approaches to identify system-related root causes. *Process Safety Progress,* 30:372–376.
26. Kletz, Trevor A. 2001. *An Engineer's View of Human Error,* 3rd ed. Taylor and Francis Group, New York.
27. Center for Chemical Process Safety (CCPS). 2004. Guidelines for preventing human error in process safety. John Wiley & Sons, Inc., Hoboken, NJ.
28. Leveson, N. G., and G. Stephanopoulos. 2014. A system-theoretic, control-inspired view and approach to process safety. *AIChE Journal.* 60:2–14.
29. IEC 61025:2006. 2006. *Fault Tree Analysis (FTA).* webstore.iec.ch.
30. IEC 61511-1:2016 RLV, Redline version. 2016. *Functional safety—Safety instrumented systems for the process industry sector—Part 1: Framework, definitions, system, hardware and application programming requirements.* webstore.iec.com.
31. BP. 2010. *Deepwater Horizon Accident Investigation Report.* www.bp.com.
32. Rausand, Marvin. 2011. *Risk Assessment: Theory, Methods, and Applications.* John Wiley & Sons, Hoboken, NJ.

33. Center for Chemical Process Safety (CCPS). 2014. *Guidelines for initiating events and independent protection layers in layer of protection analysis.* John Wiley & Sons, Inc. Hoboken, NJ.

34. Saud, Yaneira E., Kumar (Chris) Israni, and Jeremy Goddard. 2014. Bow-tie diagrams in downstream hazard identification and risk assessment. *Process Safety Progress* 33:26–35.

35. Blanco, R. F. Understanding hazards, consequences, LOPA, SILs, PFD, and RRFs as related to risk and hazard assessment," *Process Safety Progress* 33:208–216 (2014).

36. CGE Risk Solutions. 2016. *The History of the Bow Tie.* http://www.cgerisk.com/knowledge-base/risk-assessment/the-bowtie-methodology. (Accessed August 24, 2016)

37. Perrow, Charles. 1999. *Normal Accidents.* Princeton University Press, Princeton, NJ.

38. Broadribb, Michal P., and Mervyn R. Currie. 2010. HAZOP/LOPA/SIL: Be Careful What You Ask For! Presentation for 2010 Spring Meeting, 6th Global Congress on Process Safety, San Antonio, Texas. AIChE/CCPS.

39. Center for Chemical Process Safety (CCPS). 2016. *Guidelines for Integrating Management Systems and Metrics to Improve Process Safety.* John Wiley & Sons, Inc. Hoboken, NJ.

40. Leveson, Nancy G. 2011. *Engineering a Safer World: Systems Thinking Applied to Safety.* MIT Press, Cambridge, MA.

41. UK Health and Safety Executive (HSE). Optimising hazard management by workforce engagement and supervision, RR637. 2008. HSEBooks. Norwich, UK.

42. Pitblado Robin., Potts Tony., Fisher Mark., and Greenfield Stuart. A method for barrier-based incident investigations, *Process Safety Progress* 34:328–334 (2015).

43. Vaughen, Bruce K. 2015. Three decades after Bhopal: what we have learned about effectively managing process safety risks. *Process Safety Progress* 34:345–354.

44. Willey, R. J. 2014. What are your safety layers and how do they compare to the safety layers at Bhopal before the accident? *Chemical Engineering Progress*, 111:22–27.

45. Institution of Chemical Engineers (IChemE). 2014. Remembering Bhopal—30 years on. *IChemE Loss Prevention Bulletin*, Issue 240.

46. Bloch, Kenneth. 2016. *Rethinking Bhopal: A Definitive Guide to Investigating, Preventing, and Learning from Industrial Disasters*, 1st ed. Elsevier Press/IChemE.

47. The Center for Chemical Process Safety (CCPS). 2011. *Guidelines for Auditing Process Safety Management Systems*, 2nd ed. John Wiley & Sons, Inc., Hoboken, NJ.

48. U.S. OSHA. 2016. 29 CFR 1910.119, Process safety management of highly hazardous chemicals. www.osha.gov.

12

Monitor Process Safety Program Effectiveness

I think it's very important to have a feedback loop, where you're constantly thinking about what you've done and how you could be doing it better.

Elon Musk

Why We Need to Monitor Process Safety Program Effectiveness

When working with a patient, medical professionals measure vital signs, such as temperature and blood pressure, to help evaluate the health of the patient and to provide potential early warning of health problems. Similarly, vital signs—key performance indicators (KPIs) or, simply, metrics—are used to measure the health or effectiveness of process safety programs and to provide early warning of potential performance problems. Are operating procedures being updated? Is training being completed on schedule? Are equipment inspections and tests being done at the right frequency? Are incident recommendations being tracked and completed? The answers to these and other questions reflect on the performance of the process safety program and the individual management systems described in this book. If performance is slipping, how would leadership know that the risk of process incidents might be increasing? What specific systems need to be improved? Is one plant or process area performing significantly better—or worse—than other plants or process areas? Obviously, an appropriate set of metrics is essential to understand if the process safety program is working as desired or if changes need to be made to make the program more effective. Trending of metrics over time to see if the results are getting better or worse, combined with periodic management review to identify strengths, problem areas, and improvement opportunities, is essential feedback to help ensure that process safety programs achieve performance goals and reduce the risks of potentially catastrophic incidents.

**Monitor process safety program effectiveness by ensuring
that appropriate metrics are selected, evaluated, and
used to measure and improve performance.**

12.1 Introduction

A method to monitor and provide feedback on process safety system
activities and performance is essential for an effective process safety program. Otherwise, how would anyone know what effective means? In a
general sense, it means both that process safety system implementation
is meeting system requirements (leading metrics), such as equipment
inspections being completed on time, and that process safety program
performance is or is not being achieved (lagging metrics), such as the
prevention of significant process incidents and injuries. Are systems functioning as designed in order to reduce the risk of incidents? Are the actual
number and severity of process incidents low or at least trending better
than past performance? Measurement is the essential aspect of feedback
systems. Without appropriate program feedback, many opportunities for
learning can be lost. The result of that can be even worse performance, as
warning signs are missed, and improvements that could have prevented
serious incidents are not made. More incidents. More injuries. More cost.
Well-designed feedback systems and learning plans ultimately pay for
themselves. The success then, over time, of an effective process safety
program is ensuring that the right things are measured, evaluated, and
acted on to promote learning and avoid complacency resulting from long
periods of good performance:

> In the absence of bad outcomes, the best way—perhaps the only way—to
> sustain a state of intelligent and respectful wariness is to gather the right
> kinds of data. This means creating a safety information system that col-
> lects, analyzes, and disseminates information from incidents and near-
> misses as well as from regular checks on the system's vital signs. All
> of these activities can be said to make up an *informed culture*—one in
> which those who manage and operate the system have current knowl-
> edge about the human, technical, organizational, and environmental
> factors that determine the safety of the system as a whole [1].

The importance of feedback to process safety program performance was
introduced in Chapter 2. The primary methods of feedback for process
safety are periodic audits of process safety system implementation and performance and the development of appropriate KPIs or metrics, which will
be discussed in this chapter. Metrics by themselves are of no real use. Rather,

the periodic collection, trending, and review of metrics by process safety resources and management provide the opportunity for evaluating system and program performance, identifying strengths and weaknesses, and initiating improvement efforts. Metrics must also be targeted correctly with a focus on process safety and not personal safety. The U.S. Chemical Safety Board determined as part of their investigation of the BP Texas City incident [2], for example, that the "safety campaigns, goals, and rewards focused on improving personal safety metrics and worker behaviors rather than on process safety and management safety systems." A focus on personal safety is naturally important but will not provide sufficient information on the effectiveness of process safety programs, as discussed in Chapter 1, or on the risk of major incidents.

Underlying successful process safety feedback systems are a sensitivity to operations—understanding *what* is really happening—and a propensity for learning—understanding *why* something is happening. With this knowledge, appropriate action can be taken both in response to information collected and in anticipation of larger potential performance issues to help prevent serious incidents and injuries.

12.2 Key Concepts

12.2.1 Sensitivity to Operations

Sensitivity to operations is a tenant of high-reliability organizations[*] (HROs), where a preoccupation with the possible failure of hazardous processes requires constant vigilance: "Knowing that the world they face is complex, unstable, unknowable, and unpredictable, HROs position themselves to see as much as possible [3]." Key activities include anticipating potential problems, monitoring operations, and mindfully responding to data as it is obtained. Data must be based on what is really occurring, rather than what people think may be occurring, so appropriate design of monitoring systems, such as metrics, audits, and field observations, and of review processes to ascertain the real meaning of the resulting data requires careful thought and development. Doing these tasks effectively is difficult when people are distracted, do not really know what they are looking for, or misunderstand the significance of what they are seeing. Other problems can be complacency, based on past successes, poor assumptions, and overreliance on safety system design, and simply the potentially boring day-to-day routine of operations. Sensitivity to operations therefore requires development

---

[*] Also more recently called high reliability organizing, emphasizing the active nature of the organization to monitor the current state/conditions and then update responses based on the new information/trends (i.e., to continuously improve).

of effective ongoing observation, measurement, communication, review, and response processes by trained and motivated observers.

Sensitivity to operations also requires recognition of "weak signals" that may not seem important, but when reviewed singularly or with other data can be warning signs of potential changes and problems [3]. Sometimes weak signals may be readily apparent, such as near miss incidents that indicate bigger problems may result if not appropriately investigated or responded to, as discussed in Chapter 11. Sometimes weak signals may be more subtle, such as a slight process temperature trend over several months, possibly warning of condenser fouling or other problems, or small changes in cycle times, material loading, or even process noises. Sometimes, an observant person may simply feel that something is not quite right without really knowing what, feelings that have been characterized as "leemers" [3]. The natural tendency is to minimize or ignore these feelings, but it will often make sense to trust these feelings and investigate further. Anticipation of possible problems, awareness of actual operations, and communication and review of both weak and obvious signals of potential issues are continually necessary to avoid complacency that may result from hopefully long term successful performance. Known operating problems and warning signs should be incorporated into training programs to raise and maintain awareness in personnel to provide for ongoing sensitivity to operations.

Without appropriate foresight and training, warning signs of larger problems can be missed, leading to major incidents and injuries [4]. Warning signs may not be fully recognized, and they may even become accepted as normal. Both the Challenger and Columbia space shuttle explosions, for example, resulted from continuing operational problems that were well known from previous missions [5]. Sensitivity to operations therefore requires recognition, or at least a strong reporting environment, of differences from expected operations allowing for additional review and interpretation. Learning from experience is an axiom of process safety and fundamental to development of an effective process safety program. Development of a learning organization, as discussed in the next section, overlaps and builds on sensitivity of operations efforts to maximize immediate learning from and response to potential warning signs and provides for continuing organizational learning and memory.

12.2.2 Learning Organizations

Development of learning organizations has been a subject of interest for organizations focused on learning from experience and learning from others in order to improve or grow new capabilities that help maintain or develop competitive advantages for continuing success [6]. A learning organization has been defined as "an organization skilled at creating, acquiring, interpreting, transferring, and retaining knowledge, and at purposefully modifying its behavior to reflect new knowledge and insights [7]." Some key aspects of learning organizations include:

- New ideas, information, and data must be acquired (collect)
- Knowledge must be interpreted and given relevant meaning (evaluate)
- Knowledge must be shared and available to the organization for use (apply)
- Learning must be retained in organizational memory as part of policies, procedures, and training to sustain the learning (remember).

Sensitivity to operations and effective organizational learning are both related and overlapping parts of the feedback systems involving purposeful monitoring and improvement of process safety program performance that must occur if serious process incidents are to be prevented. Reflecting this focus, a questioning/learning environment was discussed as an essential characteristic of a strong safety culture in Chapter 2. Weaknesses in this area, whether from failures to collect, evaluate, apply, or remember information important to safe and reliable operations, can lead to catastrophic incidents:

> ... virtually all these "accidents" were not what we normally mean when we use that term—that is, unpreventable random occurrences... these disasters had long buildups and numerous warning signs. What's more, they display a startling number of common causes... [5]

A learning plan is a good place to start when working on building or improving a learning organization. A learning plan consists of the thoughtful preparation of learning goals, analysis of competency needs, learning methods, and assessment needs [8]. For process safety, a learning plan (many may be needed for different parts of the overall process safety program) should include what information is desired/needed, how it will be obtained, how it will be used, who will need training, and how often and by what means will its effectiveness be evaluated. More directly, how will a facility maintain sensitivity to operations, what KPIs are needed, and how will they be collected, evaluated, shared, and used? Too often, incident investigations look backward for warning signs, when earlier recognition of the warning signs could have prevented the incident. An effective learning plan and feedback system can minimize this type of disconnect. Development and use of specific KPIs will be discussed in Section 12.3.

Steps for getting started in developing a learning organization are shown in Table 12.1. In assessing the current learning plans and environment, many things will already be working well, some things may need improvement, and some new things will need to be implemented. A facility may have already developed several KPIs to measure process safety

TABLE 12.1

Steps for Promoting Learning Organizations [Modified from 9]

- Assess your current learning culture
- Map out the learning vision and plan
- Promote a positive learning environment
- Train and help people begin learning and working together
- Implement and connect feedback and learning systems
- Apply learning to benefit the organization
- Assess results periodically and sustain

program performance. This system may be working well. Or, it may be found that KPIs are not really evaluated or used to define and implement improvements. Current KPIs may focus more on personal safety than process safety. New KPIs may be needed. Better evaluation or more frequent evaluation of KPIs may be desired. Openness for reporting deviations from expectations may need improvement, related to the need for open and effective communication and other essential features of safety culture, as discussed in Chapter 2. Some questions for evaluating the existence or quality of a learning organization include [7]:

- Does the organization have a defined learning agenda?
- Is the organization open to discordant information?
- Does the organization avoid repeated mistakes?
- Does the organization lose critical information when key people leave?
- Does the organization act on what it knows?

A more extensive assessment protocol for evaluating learning organizations is available [8].

TABLE 12.2

Potential Learning Disabilities [Modified from 7 and 8]

- Biased information
  - Blind spots, resulting from narrow or misdirected perspectives
  - Filtering, based on downplaying or ignoring data not considered acceptable
  - Information hoarding, based on selective or ineffective sharing of information
- Flawed interpretation
  - Illusory causation or correlation, based on false connections
  - Varying framing of information, based on different perspectives
  - Data bias, based on stereotypes, credibility, and ease of recall
  - Hindsight bias, based on using past experience to assume relevance
- Inaction
  - Failure to act, based on inability or unwillingness to accept new information
  - Inertia, based on fear of change or risk avoidance

Some of the problems that can hinder development of effective learning organizations have already been discussed, but some additional learning disabilities are provided in Table 12.2. Many of these problems ultimately relate to the safety culture and leadership in a facility, as discussed earlier and in Chapter 2, and are appropriately remedied by focusing improvement in these areas. In addition, steps can be taken specifically to develop a supportive learning environment [7]:

- Recognize and accept differences in people and organizations, accepting that one method of learning may not always work effectively for everyone. Learning mechanisms, though, must ultimately ensure that differing methods of data collection and analysis can be combined and resolved in ways that do not lead to ineffective or divergent learning.
- Provide timely feedback, based on assessment of learning organization implementation and practices. Is the learning plan being followed? Are learning goals being met? Are there any examples of where important information was missed or not acted on?
- Stimulate new ideas, ensuring that differing viewpoints from both inside and outside the organization are obtained and shared to challenge and improve current practices.
- Tolerate errors and mistakes, allowing for reasonable testing of learning norms and boundaries to develop new approaches in a low risk environment. To be clear, this refers to overall learning mechanisms, **not** to process safety programs, where management of change systems are required to evaluate and authorize all changes, as discussed in Chapter 10.

Retention of learning and knowledge is a continuing challenge due to organization change. How much information is lost when a highly experienced worker transfers, resigns, or retires? Is important information lost when a knowledge management system fails or is replaced? How organizational memory is maintained is therefore an important part of learning organizations as well as effective process safety programs [10–11]. As new knowledge is obtained, from targeted learning, from incidents, or from other sources, is it incorporated into improving process safety program policies, guidelines, and procedures? Are key learnings from incidents maintained in training materials, so workers know what has happened in the past and how to respond if it ever occurs again? Are process safety information and other technical documents updated to include new design, formulating, or operating learnings? Once information is updated, are there mechanisms for ensuring that it remains known or does it reside in a manual that is never looked at again? Some key practices for maintaining organization memory are shown in Table 12.3.

TABLE 12.3

Maintaining Organizational Memory [Modified from 10 and 11]

- Ensure that appropriate program activities are monitored, reviewed periodically, and documented in readily accessible locations to develop increasingly improved knowledge and understanding of process operations and safety.
- Maintain an open environment for reporting and sharing of information and communication with all employees.
- Use knowledge management systems to maintain well organized and readily available information required for safe and reliable operations. This includes information, for example, that previously could only be found in the missing file cabinet of the engineer who left two years ago.
- Investigate incidents and other operating problems to determine root causes and modify procedures and training as appropriate.
- Include key incident findings in training, using storytelling effectively to help workers understand what has happened before, that it could but should not happen again, and what to do if it does happen again.
- Develop effective systems for managing organizational change to ensure effective personnel turnovers, including documentation of potential program gaps and identification of training and other activities needed to manage the gaps. Many companies also include a proficiency demonstration for newly assigned supervisors and other personnel to ensure basic understanding of process safety program requirements.
- Incorporate external information into internal organization memory, such as learning from external incident investigations, training, publications, and conference presentations.

12.3 Process Safety Metrics (Key Performance Indicators)

It is an axiom of management that it is necessary to measure what you want to improve, both to know the current status and whether it is getting better or worse. Key performance indicators, which will be simply called metrics in this section, are used to measure the activities and performance of process safety programs. Metrics have been mandated and used for many years (e.g., lost work time), but following the BP Texas City incident in 2005 [2,12], the identification of appropriate metrics for process safety received increased attention. Metrics must be specific to process safety; personal safety metrics are not sufficient, as discussed earlier in this chapter. Also, metrics must be both leading and lagging [13–17]:

> The collection and assessment of a company's leading and lagging process safety indicators can measure operational performance and promote ongoing safety improvement, leading to the potential for enhanced accident prevention efforts [18].

Leading metrics proactively measure the performance of process safety activities, such as whether requirements such as equipment tests and inspections are being completed as scheduled, helping to determine if the program is functioning as intended to help reduce the risk of serious incidents.

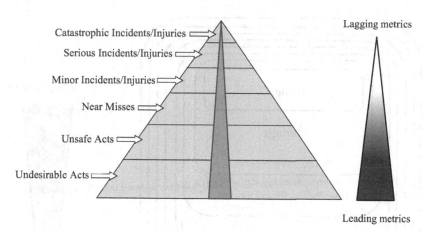

FIGURE 12.1
Safety triangle and metrics.

Lagging metrics retroactively measure the results of program outcomes, such as process incidents, helping to determine if program performance goals are being met. If leading indicators start to indicate that process safety systems are not working as desired, then the risk of injuries and incidents increases. Leading metrics help identify potential problems *before* severe incidents, directly reflecting sensitivity to operations, helping an organization identify undesired deviations and make appropriate adjustments or improvements. Lagging metrics indicate that serious malfunctions have occurred resulting in major performance deficiencies. Both types of metrics are useful for organizational learning and response.

Figure 12.1 shows the safety triangle that was discussed in Chapter 4. The bottom of the triangle represents the occurrence of undesirable and unsafe acts that indicate operational discipline and other problems, including failure to follow systems as well as potentially the design of poor or ineffective systems. At the top of the safety triangle are increasingly severe injuries and incidents that occur at a lower frequency and that usually result from the unsafe behaviors at the bottom of the triangle. For some high risk activities, such as electrical and other types of work where one mistake can potentially lead to a fatality, the base of the triangle can be very narrow. Leading metrics measure activities and behaviors at the bottom of the safety triangle, providing advance notice of problems that can lead to bigger problems at the top. Lagging metrics measure the results observed towards the top of the triangle, which have unfortunately already occurred. Use of appropriate leading metrics, therefore, allow for appropriate analysis and response to reduce the risk of the higher severity events at the top of the triangle. If only lagging metrics are used, then the ability to prevent injuries and catastrophic incidents is obviously greatly reduced.

Similarly, Figure 12.2 shows the protection layer hierarchy discussed in Chapter 3. Leading metrics targeted to the "higher" level (earlier)

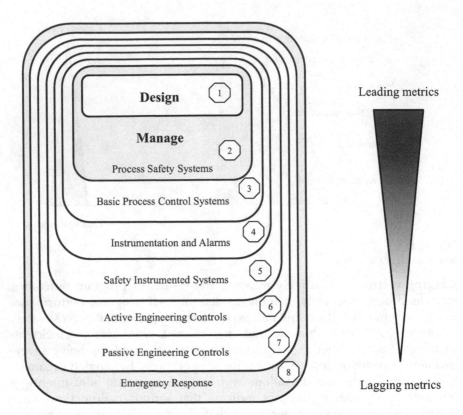

FIGURE 12.2
Protection layer hierarchy and metrics.

protection layers, such as process safety systems or response to instrumentation and alarms, provide advance warning of problems that allow for response and correction. "Lower" level (later) protection layers measure actions that indicate automatic or emergency response to near miss or real incidents, such as the action of safety instrumented systems or other engineering controls. The ability to prevent incidents again is greatly reduced if the higher level protection layers are functioning well, as measured by well-targeted leading metrics.

Classification of performance measures, such as incidents, can be established based on actual consequences [17], scoring systems [19], or other approaches to provide severity thresholds for analysis, trending, and reporting. Development and use of appropriate process safety metrics are discussed in the following sections.

TABLE 12.4

Steps for Establishing Process Safety Metrics [Modified from 14–16]

- Provide resources and establish methods for implementing and maintaining metrics systems
- Identify the activities and performance outcomes requiring development of leading and/or lagging metrics
- Establish and implement appropriate leading and lagging metrics
- Collect and report metrics
- Evaluate and act on metrics; revise metrics as needed
- Maintain metrics data for organizational learning

12.3.1 Development of Metrics

The metrics needed to evaluate the effectiveness of process safety programs vary somewhat with the types and levels of process risks being managed [13], but since the systems used to manage risks are similar, many typical metrics can be developed. The steps for developing metrics are shown in Table 12.4 and include [14–16]:

- Provide resources, including personnel, time, and funding, and establish methods for implementing and maintaining metrics. In some cases, methods for collecting metrics can be automated or developed easily within document or knowledge management systems (KMS) for reporting. In other cases, manual collection of metrics, either from KMS that currently lack the required reporting structure or from review of physical documents may be necessary. Spreadsheets have often been used to collect and report metrics data, but automation of data collection methods when possible is desirable.

- Identify the activities and performance outcomes where development of leading and/or lagging metrics is desirable. The process safety systems discussed in this book should typically have at least one leading metric. Process safety program goals, such as prevention of incidents, should have related leading and lagging performance metrics, as discussed in Chapter 2. Annual improvement goals, such as a reduction in overdue action items, should be tracked using available metrics, or additional metrics should be provided as necessary.

- Establish and implement leading and lagging metrics based on the program activities and outcomes that have been identified. The methods for collecting metrics should be developed, whether automated or manual. Training of affected personnel may be necessary to ensure that data will be available.

- Periodically collect and report metrics. Some metrics may be collected daily and some may be collected monthly or quarterly, depending on the specific metric. Reporting of metrics for use by system leaders and management is typically monthly or quarterly.

- Evaluate and act on metrics, based on appropriate frequencies and levels of review. Some sensitive data, related to process operations, may need to be looked at and acted on daily. Other data, such as the status of PHRA recommendations, may be looked at only monthly or quarterly. Collection of metrics without appropriate review and response obviously does not lead to timely recognition of potential problems. Trending of metrics to determine if the results are getting better or worse, or simply changing, can be the key to the recognition of patterns that impact performance.

- Maintain metrics data for organizational learning. Historical review of metrics trends over several years can be essential for raising awareness and understanding of potential high level changes reflecting major organizational changes, such as leadership, culture, and staffing. In addition, specific learnings, actions, and results can be the key for avoidance of similar problems in the future. Specific examples, or stories, of incidents or other actions can be incorporated into training to reinforce organizational memory as discussed earlier.

An example of one corporation's metrics is shown in Table 12.5. In this example, several leading metrics are required, implemented, and used at the site level, and then also collected and evaluated at the corporate level quarterly. Once collected in this way, management has the opportunity to see how well the systems are functioning and to respond to improve performance and/ or improve the metrics systems. If a large number of overdue activities are observed, for example, management can set goals, change priorities, and provide resources as needed to reduce the number to a lower, desired level. Some possible metrics related to the process safety systems, culture and leadership, and operational discipline foundations discussed in this book are

TABLE 12.5

An Example of Corporate Process Safety Metrics [Modified from 19]

Leading Metrics	Lagging Metrics
• Open/overdue audit recommendations • Open/overdue scheduled PHAs (PHRAs) • Open/overdue PHA (PHRA) recommendations • Open/overdue incident recommendations • Overdue scheduled procedure revisions • Overdue equipment tests and inspections • Number of action items with extended due dates	• Number and severity of process incidents

Note: Additional metrics for site use only were required.

TABLE 12.6

Example Process Safety System Metrics

Process Safety System	Leading Metrics	Lagging Metrics
Design safe processes	• Missing process safety information (from audits)	• Incident findings related to equipment design causes
Identify and assess process hazards	• Out-of-date or missing safety data sheets • Failure to complete hazard analysis for new chemicals	• Incident findings related to identification or evaluation of process hazards
Evaluate and manage process risks	• Overdue PHRAs • Open/overdue PHRA recommendations	• Incident findings related to PHRA quality issues
Operate safe processes	• Overdue procedure revisions • Overdue personnel training • Contractor performance issues	• Incident findings related to procedure or training causes
Maintain process integrity and reliability	• Overdue tests and inspections • Average time to complete equipment repair work orders	• Equipment downtime (hours) • Number of minor spills or loss of containment events • Incident findings related to equipment failure and loss of containment causes
Change processes safely	• Overdue change requests/action items • Overdue prestart-up reviews/action items	• Incident findings related to process change causes
Manage incident response and investigation	• Number of open/overdue incident recommendations • Number of open/overdue emergency drill action items	• Incident findings related to investigation quality issues • Incident findings related to failure to complete previous incident recommendations • Number and severity of process incidents
Monitor process safety program effectiveness	• Number of extended action items • Overdue audit recommendations	• Incident findings related to metrics system issues or failure to detect warning signs

provided in Tables 12.6 through 12.8. Many additional metrics are possible, and a wide variety of sources are available [13–17].

12.3.2 Use of Metrics

The primary uses of metrics have already been described. Metrics must be collected periodically, reviewed by management and other affected personnel, and based on absolute values or trends of the metrics data, either

TABLE 12.7

Example Safety Culture/Leadership Metrics

Leading Metrics	Lagging Metrics
• Number of management audits or management visits to operating areas • Number of improvement suggestions • Low equipment utilization • Low product quality • Number of overdue activities and action items • Number of communications related to process safety goals and metrics • Equipment repair backlogs • Project capital to support process safety program improvements • Number of audit findings indicating resource or system issues	• Number and severity of process incidents • Incident findings related to safety culture and leadership causes • Productivity and other costs associated with incident investigations and other responses to ineffective systems and programs at the facility • High turnover of personnel

TABLE 12.8

Example Operational Discipline Metrics

Leading Metrics	Lagging Metrics
• Overall poor metrics results indicating a large number of open and overdue action items • Audit findings related to operational discipline issues • Operating issues related to quality, low yield, waste, etc. resulting from human factors issues • Environmental reporting violations • Staff turnover rates • Customer complaint rates	• Incident findings related to operational discipline issues (e.g., failure to follow procedures)

supporting or corrective actions must be taken to maintain positive results or improve negative results. Involvement of management in reviewing metrics data is essential for awareness and to obtain resources for response as needed. Sharing of results with site and corporate personnel should be a priority. Many sites will post metrics at various site locations, send out summaries by e-mail, and discuss summaries and response in shift or safety meetings. Effective communication can enhance sensitivity to operations, raise awareness of potential problems and responses, help avoid complacency, solicit additional information from experienced personnel, and contribute to enhanced process safety program effectiveness.

Trending of data to understand how it is changing is essential to add perspective. The primary types of trend graphs include:

- Trend change with time (Figure 12.3) – useful for seeing how a specific metric is improving or worsening over time. If the number of overdue procedure revisions is increasing over several quarters or the equipment repair backlog is becoming older, appropriate corrective actions can be implemented.
- Comparison between different metrics (Figure 12.4) – useful for seeing how a collection of metrics compares. A large number of different

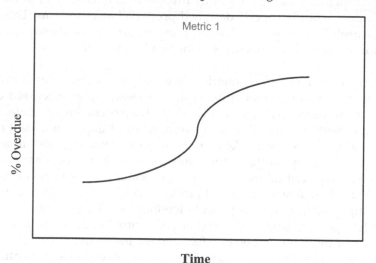

FIGURE 12.3
Typical trend chart for metrics.

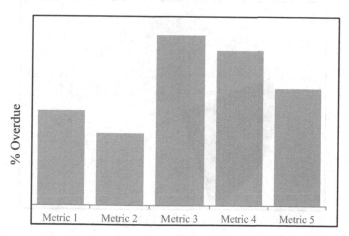

FIGURE 12.4
Typical comparison chart for metrics.

metrics all having many overdue action items is indicative of a larger problem than the overdue action items on one metric by itself. A sudden increase in the number of overdue items for many metrics may indicate resourcing, management, personnel, or distraction issues that are having a larger impact on a site than anticipated, requiring a higher level response.

- Percent share (Figure 12.5) – useful for determining the distribution of metric results between the different process safety systems. Typical examples could be used to determine that mechanical integrity issues are the leading causes of incidents or that most audit findings relate to poorly designed process change systems. This can be useful for assigning additional resources based on these priorities for making improvements, similar to a Pareto process.

The inclusion of results from metrics collection and review in organizational learning processes is essential for sustaining learning as discussed earlier. Simply the recognition that similar incidents keep occurring or similar causes are being identified can allow for development of improvements. Progress on annual goals can be tracked to see if improvement goals are actually being met. Observation that improvements made in the past are not being sustained can be critical for prevention of incidents or even create momentum for substantial modification of process safety systems, safety culture, or leadership practices. Metrics provide feedback on the performance of process safety programs over time and as such, provide the early warnings of problems or the confirmation of effectiveness. Ineffective use of metrics can predictably lead to missed opportunities for improvement and eventual slippage in performance. The result can be catastrophic.

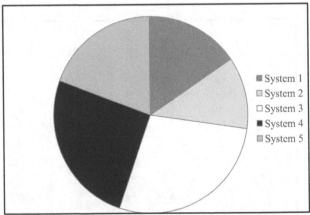

% Audit Findings

FIGURE 12.5
Typical percent distribution chart for metrics.

12.4 Process Safety Audits

Process safety audits are conducted to:

- evaluate process safety program implementation for compliance with regulatory and organizational requirements
- identify potential improvement opportunities based on industry or corporate's best practices.

Audits present a formal evaluation of and provide feedback on the basic design, implementation, and performance of the process safety program. Problems identified in audits are potentially early warning signs and leading indicators that process risk is increasing because the process safety program is not functioning as desired. As such, audit findings and metrics must be reviewed and corrective actions must be implemented, as discussed earlier in this chapter.

Audits are normally conducted to meet regulatory requirements, such as the U.S. OSHA PSM requirement to conduct and certify process safety audits every three years. In some cases, audits may be conducted more frequently to meet additional regulations or corporate requirements, to respond to serious incidents, or to evaluate process safety programs that have undergone significant changes. Companies may choose, for example, to conduct "first party" audits annually with site personnel, "second party" audits every three years with independent auditors, and "third party" audits of the overall audit program itself every one to three years [19]. The scope of the audit and the planned audit protocol must be established prior to the audit to ensure that the audit objectives are met. Consulting organizations are often used to conduct audits to help ensure that experienced auditors are available.

12.4.1 Preparing for an Audit

The primary steps for preparing for a process safety audit are shown in Table 12.9, which include:

- **Define the objectives, scope, and level of the audit** – The objectives establish the goals of the audit, based on the defined physical, regulatory, and historical scope. The physical scope provides the specific

TABLE 12.9

Preparing for a Process Safety Audit

- Define the objectives, scope, and level of the audit
- Confirm audit protocol
- Staff the audit team with qualified auditors
- Schedule audit activities and personnel
- Share and review program guidelines
- Discuss audit logistics

processes or physical boundaries of the audit at a site. The regulatory scope establishes the specific regulations and corporate requirements that the audit must review for compliance. The historical scope identifies the time period for the audit, such as the review of program documentation and activities since the last audit, which was conducted three years ago. The level of the audit relates to the desired depth of review, which in most cases, will be the detailed review of compliance requirements. In some cases, the audit level may be higher or less detailed, though, based on an interim review of program status or improvements separate from the primary regulatory compliance audits. In some facilities, all of the relevant processes may be included in the audit scope, and in some very large facilities, representative processes may be selected, depending on regulatory requirements, or multiple audits may be scheduled.

- **Confirm the audit protocol** – Different audit protocols are possible, varying with how an organization manages the process safety program. In some cases, a specific regulatory protocol may be used [20], and in other cases, the organization may have incorporated the regulatory requirements into corporate guidelines that include additional requirements [19]. The specific protocol to be used in the audit must be established in advance to ensure that an adequate number of auditors and sufficient time are provided. Some audit protocols may involve scoring of audit questions and systems [19]. A possible simple audit protocol based on the systems presented in this book could involve use of the "Measures of Success" sections at the end of the chapter, though in most cases, audits based on regulatory requirements will be needed.

- **Staff the audit team** – Depending on the facility and physical scope of an audit, the time required can range from one auditor over several days to many auditors over a week or longer. Careful consideration should be given to the backgrounds and experience of the auditors to ensure that the audit team and leader have the capabilities to review all process safety systems and requirements. The identity and availability of site personnel who will support and participate in the audit should also be reviewed to ensure that they can participate in the audit. The audit team must include someone knowledgeable in the process being audited. The personal characteristics of successful auditors are listed in Table 12.10. Training should be provided to auditors on best practices for planning, conducting, and documenting audits. Outside consultants are often used to help ensure experience and independence in conducting the audit.

- **Schedule the audit** – The audit schedule should be established providing appropriate blocks of time for evaluation of all program requirements. Some systems, such as process integrity and

TABLE 12.10

Characteristics of Successful Auditors

* Knowledgeable and experienced in systems being audited
* Good written, listening, and spoken communication skills
* Professional, consistent, and approachable
* Confident and persistent
* Thorough, logical, and curious
* Open-minded and flexible

reliability, may require several days with participation of specialized auditors. Other systems, such as manage incident response and investigation, may be completed in a shorter time. The backgrounds of auditors should be matched with the systems they will be auditing. Site personnel who manage the systems should also be scheduled for the required time blocks.

* **Share and review program guidelines** – The site being audited should provide information, such as system guidelines, process overviews, and site organizational charts, to the auditors in advance so that they can review and prepare for the audit.

* **Discuss audit logistics** – The site should discuss specific safety practices, including site entry and escort, safety orientation, and personal protective equipment requirements. Additional information related to travel arrangements, major site projects, labor agreements, etc., should also be discussed as appropriate.

With adequate preparation, the audit team can arrive for the audit with sufficient background information allowing them to maximize the time provided for the audit.

12.4.2 Conducting an Audit

The primary steps for conducting a process safety audit are shown in Table 12.11, which include:

TABLE 12.11

Conducting a Process Safety Audit

* Review site safety orientation and training
* Conduct an opening meeting
* Tour the facility
* Audit the program based on the audit schedule and protocol
* Conduct daily summary meetings
* Document audit findings
* Conduct a closing meeting

- **Review site safety orientation and training** – On arrival at the site, the auditors must review the site safety orientation training, usually a video, and obtain any personal protective equipment needed for accessing process areas. As a guest at the audited facility, it is important for the auditors to understand relevant safety rules and practices to ensure their safety during the audit.
- **Conduct an opening meeting** – The audit team should present a short overview of the audit objectives, protocols, schedule, and anticipated results for site personnel. Many people will be participating in the audit, but others will be managers interested in helping ensure that a high-quality audit is conducted and who may be responsible for following up on audit findings.
- **Tour the facility** – An escorted tour of the facility should be taken to understand the processes and boundaries of areas that are being audited, to review hazards and safety requirements, and to begin the audit by observing area activities and operations.
- **Audit the program** – The audit protocol and schedule should be followed to evaluate different parts of the process safety program. For each requirement, it will be typical to review the associated system guidelines to understand how the requirement will be met, including corporate-or site-specific requirements, to interview site personnel, to sample and review documents, and to field audit as needed, to determine if there are any gaps in the design, implementation, or performance of the requirement. Potential gaps may lead to findings in the audit report, although additional discussion may be needed to ensure that the site practices are clearly understood. Additional investigation may also be needed to determine the root causes of gaps to see if potentially deeper systemic, cultural, or leadership issues may be contributing to the existence of the gaps.
- **Conduct daily summary meetings** – A brief meeting should be held daily to review audit observations and possible findings to discuss their validity and to request more information if needed. Any auditing issues, such as schedule changes or the unavailability of information, should be discussed so adjustments can be made and communicated.
- **Document audit findings** – Preliminary audit findings and recommendations should be documented as discussed in Section 12.4.3. If a scoring protocol is used, audit questions and systems should be scored, and ranking of high to low scoring questions and systems can be developed.
- **Conduct a closing meeting** – The audit team should review all or at least significant findings, depending on the number, in a closing meeting with site management and interested personnel. Additional discussion of the findings at the closing meeting may provide

clarification on the identified gaps and possible responses, but all findings discussed should have already been reviewed during daily summary meetings so that there are no surprises. The draft audit report or findings may be presented after the closing meeting or may be sent at a later date.

Documentation of the audit report and resolution of the audit findings are discussed in the next section.

12.4.3 Documenting an Audit

Audit findings involve gaps observed in the design, implementation, or performance of process safety programs relative to both regulatory and corporate/business/facility requirements. Findings are written to document the gap, and depending on the audit protocol, specific recommendations may also be written to suggest possible approaches for improving the gap. Both findings and recommendations should be written carefully so that the gaps are clearly identified, specific improvements are suggested (but normally not required, since other approaches may also be feasible), and personnel ultimately assigned to correct the gaps will know what to do. Findings may be classified by type such as regulatory, policy, or observation and by priority such as 1, 2, or 3, if some findings have greater urgency for follow-up. A report of the findings must be documented, providing a summary of the audit protocol and results. In some cases, the report will primarily include the findings and recommendations, and in other cases, detailed descriptions of how the requirements are being met may also be provided, as shown in Table 12.12. In the United States, the U.S. OSHA requires that the facility certify that the audit verified that the process safety program procedures and practices are adequate and being followed in compliance with U.S. OSHA PSM requirements, as provided by the audit report. The two most recent audit reports must also be retained to demonstrate that the audits are being conducted on the required three-year frequency.

TABLE 12.12

Example of an Audit Protocol Worksheet

Requirement (Regulatory or Policy Requirements)	Discussion (How requirements are being met)	Finding/Recommendation (Specific gaps in compliance and suggested improvements)
Requirement #1		
Requirement #2		
Etc.		

Facility management must document a response to the audit report, usually by accepting the findings and recommendations as well as assigning responsibilities, resources, and timing for tracking and completion of the recommendations. The time to complete recommendations may vary with the scope involved, such as the need for capital improvement allocations, but in some cases, regulatory requirements or organizational policies may require specific timing based on the priority and/or risk associated with the finding. Tracking systems should be used to monitor the closure status of audit recommendations and to provide appropriate metrics for management review. Specific information should be provided to document the closure of recommendations to help ensure that the intents of the recommendations were met. Additional metrics related to the findings, such as the number, type, priority, or score of findings, if provided, can be developed. If most of the findings relate to process integrity and reliability, for example, additional review of this system may be desirable. The presence of repeat findings from one audit to the next audit should also be reviewed to determine if the recommendations were completed correctly or if the improvements were not sustained.

12.5 Measures of Success

- ☑ Design and implement an effective process safety system for monitoring process safety program effectiveness, based on concepts for:
  - o Sensitivity to operations
  - o Learning organizations
  - o Metrics selection and evaluation.
- ☑ Provide resources for collecting, evaluating, and responding to appropriate process safety metrics.
- ☑ Communicate metrics to affected personnel and use them to develop and track the progress of improvement goals.
- ☑ Schedule regular audits to monitor process safety program effectiveness.
- ☑ Ensure that experienced, trained, and multi-disciplinary auditing teams are assigned to monitor process safety program effectiveness, including:
  - o Developing well-written and supported recommendations based on addressing systemic or performance gaps
  - o Documenting a comprehensive audit report
  - o Communicating the audit results to affected personnel
  - o Tracking the recommendations until closure.

References

1. Reason, James. 1997. *Managing the Risks of Organizational Accidents*. Ashgate, Farnham, VA.
2. U.S. Chemical Safety Board. 2007. *Refinery Explosion and Fire*. Report No. 2005-04-I-TX. www.csb.gov.
3. Weick, Karl E. and Kathleen M. Sutcliffe. 2007. *Managing the Unexpected*, 2nd ed. Jossey-Bass, San Francisco, CA.
4. Center for Chemical Process Safety. 2012. *Recognizing Catastrophic Incident Warning Signs in the Process Industries*. Wiley, San Francisco, CA.
5. Gerstein, Marc. 2008. *Flirting with Disaster: Why Accidents are Rarely Accidental*. Union Square Press, New York.
6. Senge, Peter M. 1990. *The Fifth Discipline: The Art & Practice of the Learning Organization*. Doubleday, New York.
7. Garvin, David A. 2000. *Learning in Action: A Guide to Putting the Learning Organization to Work*. Harvard Business School Press, Brighton, MA.
8. Sarder, Russell. 2016. *Building an Innovative Learning Organization*. Wiley, San Francisco CA.
9. Kline, Peter and Bernard Saunders. 1998. *Ten Steps to a Learning Organization*. 2nd ed. Great Ocean Publishers.
10. Murphy, John F. and James Conner. 2014. Black swans, white swans, and 50 shades of grey: Remembering the lessons learned from catastrophic process safety incidents. *Process Safety Progress* 33:110–114.
11. Throness, Barry. 2014. Keeping the memory alive, preventing memory loss that contributes to process safety events. *Process Safety Progress* 33:115–123.
12. Baker, James A., Frank L. Bowman, Glenn Erwin, Slade Gorton, Dennis Hendershot, Nancy Leveson, Sharon Priest, Isadore Rosenthal, Paul V. Tebo, Douglas A. Wiegmann, and L. Duane Wilson. 2007. *The Report of BP US Refineries Independent Safety Review Panel*. www.bp.com/bakerpanelreport.
13. Center for Chemical Process Safety. 2007. *Guidelines for Risk Based Process Safety*. Wiley-AIChE, New York.
14. OECD. 2008. *Guidelines on Developing Safety Performance Indicators Related to Chemical Accident Prevention, Preparedness and Response*, 2nd ed. www.oecd.org.
15. HSE. 2006. *Developing Process Safety Indicators: A Step-By-Step Guide for Chemical and Major Hazard Industries*. www.hse.gov.uk.
16. Center for Chemical Process Safety. 2017. *Guidelines for Integrating Management Systems and Metrics to Improve Process Safety Performance*. Wiley-AIChE, New York.
17. API. 2010. *Process Safety Performance Indicators for the Refining and Petrochemical Industries*. RP-754.
18. U.S. Chemical Safety Board. 2016. *Tesoro Martinez Refinery: Process Safety Culture Case Study*. Report No. 2014-02-I-CA. www.csb.gov.
19. Cummings, David E. 2009. The evolution and current status of process safety metrics. *Process Safety Progress* 28:147–155.
20. U.S. OSHA. 1994. 29 CFR 1910.119, *Process Safety Management of Highly Hazardous Chemicals. Compliance Guidelines and Enforcement Procedures*. CPL 2-2.45A CH-1.

Section IV

Practical Approaches for Achieving Process Safety Excellence

13

Develop Personal Capability in Process Safety

For too many people, ten years of work experience is merely the first year's experience repeated ten times.

Robert E. Kelley

Why Develop Capability In Process Safety

To understand and become effective at process safety, there can be a lot to learn. At one company, there are over 2500 requirements in almost 500 pages of process safety standards. There are also thousands of pages of additional guidance documents, relevant regulatory requirements from around the globe, and industry consensus codes and standards. Not all companies will obviously have this many requirements, but it can be hard to be an expert in all aspects of process safety. This is not to discourage anyone from being interested in learning about process safety systems, developing capability and expertise over time. Like professionals in other industries, some level of specialization is possible depending on career interests.

For most people, though, learning enough about process safety system requirements in order to successfully understand and apply the systems daily as part of their overall work will be sufficient. Not everyone needs to be an expert in process safety, for example, to know how to correctly fill out a required change form to install new or modified equipment. If process hazards are present, they must be evaluated and managed appropriately, and the organization must ensure that qualified process safety resources are available to support this primary goal. Your personal goal may simply be to develop sufficient personal capability to understand and work with process safety systems, or it may be to become a full-time process safety professional with either broad, generalized knowledge or narrower, specialized knowledge. Either way, documenting a learning plan and exploring training opportunities are first steps for developing more extensive capability in process safety.

Develop a learning plan to develop process safety capability
and to become effective in understanding, working
with, and supporting process safety programs.

13.1 Introduction

The contributions of working as safety professionals can be highly rewarding. Serious injuries, catastrophic incidents, environmental harm, major property damage, and significant business loss can be prevented, contributing to overall business excellence and sustainability. There can be many challenges as well. Safety professionals often have many technical competencies to master, have heavy workloads, and may find that the demands of their work can be relatively stressful. Strategies that can help safety professionals be more productive can help promote career satisfaction and security, provide work/life balance, and help improve organizational safety.

Developing knowledge of process safety may be a necessary career goal, even if a career in process safety is not your primary goal. To work at many facilities today, use of process safety systems designed to manage hazardous materials and processes is fundamental to being able to get any work done at all. Personnel in management, design, engineering, maintenance, operations, contractor services, and various support functions must know at least some process safety system requirements to be able to complete their work correctly and safely. Lack of knowledge can impact process safety in many ways, potentially leading to serious injuries or incidents. What can happen if a purchasing clerk, for example, substitutes a cheaper product for use in a highly hazardous process? Probably nothing good, even if the intent is quite reasonable, the result may be a release of a hazardous chemical leading to a serious injury. Developing some level of capability in process safety, as appropriate to the work role and the type of process hazards at a facility, will be necessary to be successful in various careers, and who knows, an actual career in process safety may be rewarding in many ways. Some reasons for considering a career and developing capability in process safety are shown in Table 13.1.

13.2 Process Safety Roles and Responsibilities

The primary roles involving process safety are shown in Table 13.2. As described by DeBlois in 1918, the primary goal and function of someone in a process safety role are the "reduction or elimination of accidents [1]," and this

TABLE 13.1

Why Develop Process Safety Capability?

- Contribute to the process safety program
  - Help prevent serious injuries and incidents that can impact you, your coworkers, your facility, your organization, and your community
  - Help improve the program, so process safety systems are more effective and efficient
- Achieve professional growth
  - Develop understanding, knowledge, and capability in process safety
  - Improve personal technical and/or leadership skills
  - Be recognized as an expert
  - Potentially work with many different manufacturing operations
  - Network with other facilities and companies

TABLE 13.2

Process Safety Roles

- **Supervisor/Manager**—Develops a high level of understanding and capability in process safety, typically becoming a site or organizational leader who manages the process safety program. The leader typically develops broad knowledge of process safety systems and requirements, but does not necessarily need capability in all process safety systems and does not necessarily work full time in process safety. This role typically includes process safety managers, leaders, or facilitators who lead the process safety program at the site or organizational levels.
- **Consultant**—Develops specialized technical knowledge and expertise in one or more areas of process safety, such as process hazard and risk analysis or reliability engineering. Consultants also include process safety resources with comprehensive knowledge of process safety systems and requirements who typically work full time in process safety roles and may often be in corporate, site leadership, or consulting positions, resourcing or leading activities at a site or broader organization.
- **System Leader**—Develops more in-depth knowledge of one or more process safety systems, such as managing change or incident investigation, and is responsible for monitoring and improving system performance. System leaders are typically in a technical or operations support role who develop expertise and leadership on a part-time basis.
- **Pocess Safety/System Specialist**—Develops a good understanding of process safety in one or more areas and may assist the supervisor/manager in managing the process safety program. The specialist may be a in a full-time or part-time process safety role from different positions in the organization.
- **User**—Develops a practical understanding of process safety systems to meet requirements as needed to complete work tasks, but process safety is not their primary role. The user is simply someone at a facility who needs to work with process safety systems in support of their full-time role in engineering, operations, maintenance, etc. They do not need to develop specific capability in pocess safety, but they do need to follow process safety system requirements to help maintain safe and reliable operations at their facility.

certainly remains true today. Process safety roles provide a significant opportunity to help prevent serious injuries and potentially catastrophic incidents, ultimately contributing to overall business excellence and sustainability. Many people find the process safety role provides great personal and professional satisfaction as well as career success. Many technical, engineering, operations, maintenance, and support roles at a facility need to obtain both general and more specific process safety training depending on their specific roles and how frequently they need to work with different process safety systems.

The key responsibilities of the process safety roles obviously vary, but most will include the following activities at some level:

- Developing process safety knowledge and experience
- Understanding applicable regulations and industry codes and standards
- Implementing and revising corporate/facility process safety requirements
- Monitoring and ensuring process safety regulations and requirements are met
- Sharing, leading, and championing process safety requirements and systems
- Providing training and coaching on process safety requirements and systems
- Collecting and analyzing performance data
- Developing strategies for continuous improvement.

Given the large scope of most process safety programs, it is increasingly difficult for one person to be an expert in all aspects of process safety, so many large organizations share these responsibilities among several people who are part of a part-time process safety committee or team, as shown in Figure 13.1. Typically, someone in a process safety role will have at least a

FIGURE 13.1
Typical Site Process Safety Organization.

few years of experience working with site processes, will have developed practical experience with most site process safety systems, and often will have developed detailed knowledge around one or more specific process safety systems. In most cases, process safety roles are not line supervisory roles, so professionals assist line leadership by supporting process safety program compliance, by evaluating process safety system performance via appropriate metrics, and by identifying and helping to achieve continuous improvement goals.

13.3 How to Be Effective in Process Safety Roles

What does it mean to be effective [2,3]? The dictionary defines effective as "producing a decided, decisive, or desired effect [4]." Peter Drucker said "efficiency is doing things right; effectiveness is doing the right things [5]." In order to be effective, therefore, someone in a process safety role must correctly identify the right things to work on to produce a desired result, which in this case typically involves maintaining or improving process safety systems and performance. The ability of people in process safety roles to select the right work, focus on it, and proceed in ways that positively influence their organization to impact process safety performance is critical in preventing injuries and, especially in a process safety role, in avoiding potentially catastrophic incidents. Although most process safety training is targeted to improving technical knowledge and capability, an additional focus on how to be effective in a process safety role by working successfully within an organization is also desirable. What should be emphasized may vary depending on organizational priorities, safety culture, and process risks.

A focus on safety leadership and effectiveness is nothing new. DeBlois in the early 1900s, for example, discussed the need for "technically trained and experienced [resources] who are capable of studying an industry or process from all its angles and of making constructive recommendations for changes in design and operation which strike to the root of the unnecessary accident evil [1]," and "the safety engineer is, or should be, the technical inspirational leader. It is largely within [their] power to make or mar the local movement; consequently, [their] selection, [their] training, and the opportunities afforded [them] of becoming more proficient in [their] work are of no slight importance [6]." Many books and articles have since discussed personal effectiveness [7,8] and leadership [9,10]. Articles in Process Safety Progress have discussed various aspects of process safety roles and effectiveness [11–14]. The Center for Chemical Process Safety (CCPS) has also included personal competency as an element of risk-based process safety [14], and the European Process Safety Centre has provided guidance on how to set up a process safety competence management system [15]. A focus on developing strong knowledge and competence in process safety both through learning and influence is essential for achieving superior, effective performance in a process safety role.

13.3.1 Developing Technical Process Safety Knowledge

To develop technical competence and capability, the process safety professional must learn, understand, and apply basic process safety information and obtain a variety of technical and operational experiences. Several approaches for accomplishing this are shown in Table 13.3, which include:

- **Reading** – When new in a position, it is necessary to read and understand the basic process safety regulations (e.g., U.S. OSHA, EPA), corporate policies and guidance, consensus industry standards and codes (e.g., NFPA, ASTM, ISA), books, journals (e.g., Process Safety Progress, Journal of Loss Prevention, etc.), and news, such as monthly safety beacons from the CCPS and incident reports (and videos) from the U.S. Chemical Safety Board and other organizations. Experienced resources may continually review these as references as they are revised or as their job needs change to stay current and to expand their knowledge and capabilities. Review of intrinsic hazard assessments and other process safety information as well as internal incident investigation and process hazards and risk analysis reports can also help provide deeper knowledge for equipment design, safety systems, and operations.

- **Training** – Some of the knowledge listed above can be obtained by attending process training courses offered internally in your organization or externally by consulting firms and other organizations. Basic training should include relevant regulations, process safety systems, and practices implemented by the organization,

TABLE 13.3

Methods for Developing Technical Process Safety Knowledge

- **Reading**–including regulations, standards, codes, guidelines, incidents, books, and journals
- **Training**–attending both internal and external process safety and safety training courses
- **Certifications**–studying and attaining specialized certifications
- **Networks**–participating in both internal and external process safety teams, networks, and conferences
- **Experiences**–including working in plants, visiting "sister" sites, and auditing other facilities
- **Leading**–developing training for and providing assistance to facility employees
- **Monitoring**–reviewing and improving process safety system requirements, documentation, and metrics

---

and methods for identifying and evaluating process hazards. More specialized training on process hazards and risk analysis, incident investigation, auditing, reliability engineering, and other subjects should be taken as appropriate.

- **Certifications** – Both as knowledge increases and as a way of increasing knowledge by preparing for the exams, certification in process safety (CPSP, PPSE) [16,17], safety (CSP) [18], and specialty certifications in hazard analysis, quality assurance, auditing, fire systems, etc., as appropriate, can enhance technical capability and support career success.

- **Networks** – Both internal and external networks and teams of process safety professionals and leaders provide the opportunity for learning from other experienced professionals, for sharing of new information and incidents, and for identifying issues that may also be relevant at your facility. Networking is a key skill for helping to quickly provide answers to sometimes complex questions, given the broad scope of process safety programs and is discussed further in the following section.

- **Experiences** – Sometimes the best way to learn things is do them, even if they are normally outside your main job description. This may include spending a lot of time in process areas to better understand day-to-day operations. Participating in shift meetings, table-top discussions of possible incidents and response, safety meetings, etc., can provide practical knowledge to supplement review of process documentation and procedures. It can also be helpful to visit "sister" sites with similar processes to learn from their process designs and operations. Participating in audits, incident investigations, process hazards and risk analysis reviews conducted internally and at other sites can provide in-depth knowledge of process safety systems and operations not available from other sources. Finally, attending corporate process safety team meetings or external conferences, such as the Global Congress on Process Safety (GCPS), promote networking as well as exposure to new ideas and approaches.

- **Leading** – As knowledge on process safety is gained, leading related training sessions or developing new training can be a good way to strengthen your knowledge of process safety systems. The day-to-day interaction with other personnel who have questions about process safety can also lead you to research new information to help solve problems or simply to share your expertise with others.

- **Monitoring** – The daily requirements of working with process safety systems include collecting, reviewing, and evaluating performance metrics; developing or revising facility or corporate standards or guidelines; and working on improvement activities that can help deepen knowledge about process safety. Seeing what problems occur, seeing what questions other personnel have, and seeing what other sites and companies are doing can all help develop practical personal knowledge and capability.

Process safety resources for universities are available from the Safety and Chemical Engineering Education (SAChE) website [19] and from a description of "conservation of life" programs that have been implemented at one university [20]. Development of a learning plan, as discussed in Section 13.4, to plan and obtain support for learning opportunities is key.

13.3.2 Influencing the Organization

Working to champion and improve process safety systems and performance by influencing an organization are fundamental activities of process safety roles. This requires "soft" skills that are rarely taught as part of engineering and process safety training, but that are keys to being effective. DeBlois commented in 1926 on these aspects of safety roles:

> The safety engineer must possess initiative and be to a reasonable degree aggressive, though unintelligent aggressiveness is certainly no asset. Too often one encounters safety enthusiasts who are so vigorously and tactlessly aggressive, so prolific in stunts and schemes, so vociferous in their demands for "safety first" that they do as much harm as good [6].

Many general books on being successful and working within organizations are available [21,22]. Robert E. Kelley [3,23] investigated what factors made some workers highly productive or star performers. Kelley studied various cognitive (e.g., reasoning, creativity) personalities (e.g., self-confidence, risk-taking) and social (e.g., interpersonal skills, leadership ability) factors through testing and interviews with workers and managers to understand what differentiates star performers and other workers. No significant differences were observed. The study showed, though, that: "It wasn't what these stars had in their heads that made them standouts from the pack, it was how they used what they had... Star performers do their work very differently than the solid, average performing pack [23]." In particular, Kelley was able to identify nine work strategies or behaviors that star performers use effectively to increase their individual productivity, as shown in Table 13.4. The reader is referred to Kelley's book [23] for additional details on his study as well as to other useful materials on related topics outside the scope of this chapter [3,24–30].

Given the large scope of process safety programs and the sometimes overwhelming amount of information and potential problems that arise today, the importance of networking should be emphasized. Kelley [23]

TABLE 13.4

Star Organizational Behaviors [Modified from 23]

1. **Initiative**–Going above and beyond the job description for the benefit of the organization	6. **Teamwork**–Actively participating in group goal setting, commitments, activities, and accomplishments
2. **Networking**–Finding routes to knowledge experts who can help solve problems quickly	7. **Leadership**–Using expertise and influence to convince a group of people to accomplish a substantial task
3. **Self-management**–Directing work activities and career choices to help ensure high job performance	8. **Organizational savvy**–Navigating competing interests in the organization to address conflict and promote cooperation
4. **Perspective**–Seeing projects and problems through the eyes of customers to develop better solutions	9. **Show-and-tell**–Selecting the right information to communicate and developing persuasive presentations
5. **Followership**–Focusing on helping the organization succeed by working cooperatively with the organization's leaders	

and others [31,32] discuss the need for effective networking, since it is generally difficult to know everything you might need to solve every problem you encounter. Having access to a network of professionals therefore becomes critical. As Kelley notes:

> Networking, more than any other skill in the star performer model, can have dramatic impact on the speed, quality, and quantity of your output. Without a high-quality network, you are unlikely to become a star performer. With one securely in place, you can leverage your knowledge base and give it a tremendous boost [23].

Star performers have access to a broad group of experts with knowledge in a variety of areas, who are often connected to other networks of their own. Although most employees have networks available to them, star networks stand out due to the overall quality of network experts and the speed in which responses are obtained. Participation in star networks is usually earned by (1) having expertise yourself that the network values and (2) achieving credibility through past participation. Star performers proactively build and participate in networks before they need them. Organizations should focus on helping develop effective internal networks and teams to support process safety professionals as well as encouraging external networking, such as attending process safety conferences or participating on industry teams, where appropriate.

13.3.3 Thinking and Communicating Independently

In addition to developing basic process safety knowledge and capability and understanding how to influence an organization, being effective in a process safety role requires the following characteristics:

- **Sensitivity to operations** – Process safety professionals must be observant and responsive to process operations and systems looking for, investigating, and understanding any variances from normal operations that could be warning signs of potential problems [33], as discussed in Chapter 12. This will allow communication and earlier troubleshooting of problems to help prevent more significant problems and potential incidents.

- **Independent thinking** – Independent thinking is required to avoid groupthink [34,35] and similar influences, where organizational dynamics can lead group members who work together to try to minimize conflict and reach consensus without necessarily considering and critically evaluating different perspectives and approaches. A variety of reasons can lead to this behavior, including a desire not to look foolish or anger other group members, as well as basic cultural issues related to leadership, openness, and trust. The result, though, is that important concerns about process safety may not be brought up or might be overlooked rather than discussed in detail to understand and respond to potential process safety issues that can lead to injuries or incidents if nothing is done.

- **Direct communication** – In order to help avoid groupthink and to clearly present possible process safety issues and concerns to leadership, avoidance of mitigated speech [36] through direct, honest communication is critical. Mitigated speech is a practice of downplaying the importance of something by being overly polite or deferring to someone with leadership authority or a strong personality. The result can be poor communication that can mask important issues, limit awareness, and lead to ineffective follow-up.

Process safety professionals and others may be the difference between a safe and reliable operation and a potentially catastrophic incident. By being sensitive to operations, focusing on independent thinking, and directly communicating any operational and safety concerns to leadership, process safety professionals can avoid becoming organizational bystanders [37]. Bystanders sometimes withhold or minimize their concerns and input due to organizational or cultural pressures or a belief that someone else will deal with a problem. Ultimately, they may have important information that could make a real difference in helping prevent serious injuries and incidents if it is effectively communicated. If the process safety professional is not effective in expressing and acting on their concerns on process safety issues, then who will be? Being effective ultimately means developing appropriate capability, paying close attention, and speaking up to help leadership make sure that process safety concerns are known and addressed.

13.4 Developing a Learning Plan

Without a learning plan, development of expert knowledge and capability in process safety may be constrained by lack of management agreement or funding. A learning plan should be documented as part of annual performance reviews, agreed to by management, and actively pursued to support training, conference attendance, and other activities discussed earlier. Kelley [23] notes that star performers seek out perspective on their work priorities and career direction through multiple sources by asking:

- What is going on inside and outside the organization?
- Where do your activities fit in?
- How can you add value through your work?
- How should you plan and prioritize projects?

By adding to this list consideration of what learning, training, experiences, etc., as discussed in this chapter, are needed to support the resulting activities, a learning plan can help lead to improved capability, enhanced contribution, and career progression. Some key elements of a learning plan, as part of an annual performance review or equivalent, include:

- Career direction and goals
- Learning goals (e.g., training, conferences, certifications, audits)
- Approval of schedule and funding for planned learning activities
- Identification of management or technical mentors/champions.

The result should be annual goals and plans for personal development to ensure that process safety capability, whether as a career path or to enhance related career paths (e.g., engineering or operations roles), is increased over time. Most companies already have career-planning processes, but process safety professionals, like others, should actively think about what capabilities are needed for their role and how they can be best developed.

> Productivity is never an accident. It is always the result of a commitment to excellence, intelligent planning, and focused effort.

> Paul J. Meyer

References

1. DeBlois, L. A. 1918. *The Safety Engineer*, American Society of Mechanical Engineers, Hagley Museum and Library, Wilmington, DE.
2. Klein, James A. 2012. How to be effective in a process safety role. *Process Safety Progress* 31:271–274.
3. Cole, Bruce C. and James A. Klein. 2003. *How to be a Safety Star*, ASSE Professional Development Conference. Denver, CO.
4. Webster's New Collegiate Dictionary. 1979. G. & C. Merriam Company.
5. Drucker, P. F., www.brainyquote.com.
6. DeBlois, Lewis A. 1926. *Industrial Safety Organization for Executive and Engineer.* McGraw-Hill, New York.
7. Covey, Stephen R. 1989. *The 7 Habits of Highly Effective People: Restoring the Character Ethic.* Simon & Schuster, New York.
8. Canfield, Jack. 2005. *The Success Principles: How to Get from Where You Are to Where You Want to Be.* HarperCollins, New York.
9. Kouzes, James. M. and Barry Z. Posner. 2008. *The Leadership Challenge*, 4th ed. Jossey-Bass.
10. Maxwell, John C. 2011. *The 5 Levels of Leadership to Maximize Your Potential.* Center Street, New York.
11. Hendershot, Dennis C. and John E. Murphy. 2007. Expanding role of the loss prevention professional: Past, present, and future. *Process Safety Progress* 26:18–26.
12. Dowell, Arthur M. 2016. A career in process safety: 50 years of LPS. *Process Safety Progress* 35:8–11.
13. Baybutt, Paul. 2016. The meaning and importance of process safety competency. *Process Safety Progress* 35:171–174.
14. Center for Chemical Process Safety. 2007. *Guidelines for Risk Based Process Safety.* Wiley-AIChE, New Yorks.
15. European Process Safety Center. 2013. *Process Safety Competence: How to Set up a Process Safety Competence Management System.* Report No. 35.
16. Center for Chemical Process Safety. No date. *Certified Process Safety Professional.* www.aiche.org/ccps/resources/ccpsp-certified-process-safety-professional.
17. IChemE. No date. www.icheme.org/communities/countries/professional-process-safety-engineer-register.aspx.
18. Board for Certified Safety Professionals. No date. *Certified Safety Professional.* www.bcsp.org/CSP.
19. SAChE. No date. www.aiche.org/ccps/community/technological-communities/safety-and-chemical-engineering-education-sache.
20. Davis, Richard and James A. Klein. 2012. Implementing conservation of life across the curriculum, *Chemical Engineering Education* 46(3):157–164.
21. D'Alessandro, David. 2008. *Career Warfare: 10 Rules for Building a Successful Personal Brand on the Business Battlefield.* 2nd ed. McGraw-Hill, New York.
22. Asher, Donald. 2007. *Who Gets Promoted, Who Doesn't, and Why: 10 Things You'd Better Do If You Want to Get Ahead.* Ten Speed Press, California, CA.
23. Kelley, Robert. E. 1999. *How to Be a Star at Work: 9 Breakthrough Strategies You Need to Succeed.* Crown Business.

24. Scholtes, Peter. 1988. *The Leader's Handbook: Making Things Happen, Getting Things Done.* McGraw-Hill, New York.
25. Scholtes, Peter. 2000. *The Team Handbook,* 2nd ed. Oriel.
26. Fisher, Roger, William L. Ury, and Bruce Patton. 1997. *Getting to Yes: Negotiating Agreement Without Giving In.* Penguin.
27. Camp, Jim. 2002. *Start with NO ... The Negotiating Tools that the Pros Don't Want You to Know.* Crown Business.
28. Dawson, Roger. 1999. *Secrets of Power Negotiating.* Career Press, Oakland NJ.
29. Kotter, John P. 2012. *Leading Change.* Harvard Business Review Press, Watertown MA.
30. Reynolds, Garr. 2011. *Presentation Zen: Simple Ideas on Presentation Design and Delivery,* 2nd ed. New Riders Press, San Francisco CA.
31. Mackay, Harvey. 1997. *Dig Your Well Before You're Thirsty: The Only Networking Book You'll Ever Need.* Doubleday, San Francisco CA.
32. Gitomer, Jeffrey. 2006. *Little Black Book of Connections: 6.5 Assets for Networking Your Way to Rich Relationships.* Bard Press.
33. Weick, Karl E. and Kathleen M. Sutcliffe. 2007. *Managing the Unexpected.* 2nd ed. Jossey-Bass.
34. Janis, Irving L. 1982. *Groupthink: Psychological Studies of Policy Decisions and Fiascoes.* 2nd ed. Cengage Learning, Boston MA.
35. Harvey, Jerry B. 1988. *The Abilene Paradox and Other Meditations on Management.* Jossey-Bass.
36. Gladwell, Malcolm. 2008. *Outliers: The Story of Success.* Back Bay Books, New York.
37. Gerstein, Marc with Michael Ellsberg. 2008. *Flirting With Disaster: Why Accidents are Rarely Accidental.* Union Square Press.

14

Commit to a Safe Future

Concern for man himself and his safety must always form the chief interest of all technical endeavors. Never forget this in the midst of your diagrams and equations.

Albert Einstein

14.1 Maintaining an Effective Process Safety Program

Everyone goes to work each day expecting to come home safely. In far too many cases, process safety incidents have occurred resulting in fatalities, life-changing injuries, destruction of property, and lost business opportunities. Process safety incidents can truly be catastrophic in terms of potential impacts on people, facilities, companies, and communities. Process safety program effectiveness requires daily focus and discipline by everyone in an organization to understand, implement, follow, and improve systems for successfully managing process hazards. As shown in Table 14.1, process safety programs must necessarily focus on:

- **Safety culture/leadership** – Safety culture influences the daily behaviors of leadership, who either reinforce and improve culture over time or allow it to degrade. A process safety program cannot be effective without a strong safety culture and continuous daily attention by leadership at all levels to achieve appropriate focus and priority for safety as fundamental for business success and sustainability.
- **Process safety systems** – Comprehensive and well-designed process safety systems ensure daily activities include the necessary actions to maintain safe and reliable operations.
- **Operational discipline** – Process safety systems only work reliably if everyone is aware of and follows the requirements, every time. Poor operational discipline at a facility can greatly increase the risk of major injuries and incidents. Just one shortcut, one time, can potentially lead to catastrophic consequences, especially if appropriate risk management practices have not been implemented.

TABLE 14.1

The Effective Process Safety Program

---
1. **Safety culture/leadership**
2. **Process safety systems**
   2.1 Design safe processes
   2.2 Identify and assess process hazards
   2.3 Evaluate and manage process risks
   2.4 Operate safe processes
   2.5 Maintain process integrity and reliability
   2.6 Change processes safely
   2.7 Manage incident response and investigation
   2.8 Monitor process safety program effectiveness
3. **Operational discipline**

Ultimately, process safety excellence, conservation of life, and sustained operating success require a continuous process safety focus through the entire process life cycle, from initial design to eventual removal, to manage process risks and to help ensure that everyone goes home safely at the end of the day.

The primary goals of process safety systems are shown in Table 14.1, which are based on the following simple and practical principles:

- Identify and understand the process hazards in your facility, regardless of the type or size of facility; if process hazards are present, you must ensure that they are managed appropriately, including going beyond minimum essential practice defined by regulations, if needed.

- Manage the risks associated with process hazards to develop and maintain appropriate process design, safeguards, and process safety systems, based on multiple protection layers, to achieve safe and reliable operations.

- Ensure that everyone understands process safety system concepts and requirements, as appropriate, so they know their role in working daily with these systems and meeting system requirements.

- Monitor process safety system performance to identify and evaluate improvement opportunities and to continually improve performance.

For many people, process safety programs have been implemented for at least 20 years or more [1–5]. For some, process safety programs may be a new effort. In all of these cases, a focus on process safety system goals, as shown in Table 14.1, and the requirements, as discussed in this book, will support effective process safety programs and improved process safety performance.

Your individual role in process safety may vary due to many factors, but ultimately, you do have an important role in achieving sustained, excellent process safety performance. It may be designing safe processes or implementing new or revised process safety system requirements. It may be making sure you understand and follow requirements as part of your daily work. A personal commitment to zero injuries and zero severe incidents, a continuing sense of vulnerability and respect for process hazards, and a focus on learning and understanding process safety systems all underlie your ability to contribute to safe and reliable operations at your facility. There really is no acceptable alternative.

14.2 Looking to the Future of Process Safety

In thinking about the future, the one certainty is that change will be continual. Businesses will grow, acquisitions will be made, and new technologies will be developed [5]. Process safety will need to change as well [6–10] and will continue to be a focus for engineering and industrial education [11]. New acquisitions may present challenges due to different corporate or regional safety cultures. New technologies, perhaps in industrial biotechnology, nanotechnology, or some other new innovation, may benefit from the application of process safety but at the same time may require that new approaches be developed. Continuous improvement will always be necessary:

> Systems and organizations continually experience change as adaptations are made in response to local pressures and short-term productivity and cost goals... A corollary of this propensity for systems and people to adapt over time is that safety defenses are likely to degenerate systematically through time, particularly when pressure toward cost-effectiveness and increased productivity is the dominant element in decision making... The critical factor here is that such adaptation is not a random process—it is an optimization process... [12].

In effect, if we are not moving forward in process safety, we are falling behind. The challenge is to recognize this fundamental principle and to ensure the safety culture and leadership commitment function to maintain robust process safety programs.

This concept of probable drift, or normalization of deviation, in managing risk and the resulting erosion of process safety systems, has also been described as "robbing the pillar [13]." This term describes an underground mining practice of removing material, usually coal, from pillars that supported mine shafts. Removing the coal added to profit, but eventually enough could be taken away that the ceiling collapsed. As various changes are made as the

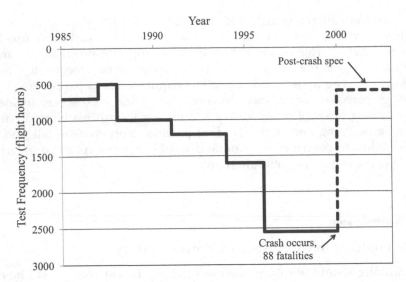

FIGURE 14.1
The extended lubrication interval for Alaska Airlines, Flight 261 [Modified from 14].

result of cost, productivity, and other factors, possibly robbing the pillars of our process safety programs, we must ensure that process safety programs are maintained strong and effective. Increasingly, this will mean that we must learn to improve process safety, not from experiencing serious incidents as we often have in the past, but from an improved ability to predict possible consequences of changes. Improved use of quantitative methodologies may be needed to ensure that adequate safety margins are maintained as changes are made. Just as importantly, a renewed emphasis on safety culture and human factors will be necessary to ensure that possibly smaller safety margins are not negatively impacted by human error.

An example of drift is shown in Figure 14.1. Alaska Airlines flight 261 crashed in 2000, resulting in the deaths of the 88 crew and passengers aboard [14]. The National Transportation Safety Board determined that the probable cause of the incident was a loss of control resulting from failure of the horizontal stabilizer due to excessive wear resulting from insufficient lubrication. The investigation showed that several extensions of the stabilizer lubrication schedule, all of which were approved through normal procedures, increased the likelihood that a missed or inadequate lubrication would result in excessive wear and the potential to progress to failure without the opportunity for detection. Figure 14.1 shows that, for a variety of reasons, the lubrication schedule over 10 years was changed from every 700 hours to 2550 hours. After the incident, the lubrication schedule was changed to 650 hours.

Unfortunately, with most of the catastrophic incidents, it is not unusual to see drift and the normalization of deviance:

> The normalization of deviance is present in all industrial disasters. It allows undetected hazards to persist until an accident exposes it [15].

As drift occurs often through normal processes, a greater dependence on technology may not provide the complete solution [13,16–18]. While incidents in the past have been generally related to equipment or human failures, greater reliance on technology for process design and control introduces new complexity:

> We are increasingly experiencing a new type of accident that arises in the interactions among components (electromechanical, digital, and human) rather than the failure of individual components [12].

The result is a need for increased reliance on systems engineering approaches that evaluate individual component failures (e.g., equipment breakdown, software faults, and human error) and also failures resulting from more complex component interactions [18]. As operators rely on software for process control, for example, they can move further away from the process and the understanding of what can go wrong and what should be done to fix problems in the process or in the software. How information is communicated and how instructions are entered becomes more critical, as does the training challenge to develop and maintain experienced, knowledgeable personnel [19]. Preventative maintenance of critical equipment, good design, and testing of process automation systems, and high levels of operational discipline will always be important for safe operations. In the future, systems engineering capabilities to manage the growing complexity of our processes will be increasingly important.

The primary goal of process safety is the prevention of injuries and severe incidents—the goal and commitment must be zero—and this will not change. Knowledge of process safety system requirements and awareness of how drift in managing risk can occur over time require a continued focus on achieving process safety goals. What a DuPont safety manager wrote during the Depression in 1933 remains true today and in the future:

> However, past successes will not prevent present or future accidents. If we are to maintain our position in this field of work, we must not only continue our efforts, but we must increase them [20].

Continually seeking to understand and manage the process hazards in our facilities to help ensure safe and reliable operations today and in the future will always be our challenge and must always be our commitment.

References

1. Center for Chemical Process Safety. 2007. *Guidelines for Risk Based Process Safety.* Wiley-AIChE, New York.
2. Klein, James A. and S. Dharmavaram. 2012. Improving the Performance of Established PSM Programs. *Process Safety Progress.* 31:261–265.
3. Crowl, Daniel A. 2009. A Century of Process Safety—and More. *Process Safety Progress.* 28:103–104.
4. Hendershot, Dennis C. 2009. A History of Process Safety and Loss Prevention in the American Institute of Chemical Engineers. *Process Safety Progress.* 28:105–113.
5. Klein, James A. 2009. Two Centuries of Process Safety at DuPont. *Process Safety Progress.* 28:114–122.
6. Mary Kay O'Connor Process Safety Center, Texas A&M University. 2011. *Process Safety Research Agenda for the 21st Century.*
7. Reniers, Genserik and Paul Amyotte. 2012. Prevention in the Chemical and Process Industries: Future Directions. *Journal of Loss Prevention* 25:227–231.
8. Knegtering, B. and H. J. Pasman. 2009. Safety of the Process Industries in the 21st Century: A Changing Need of Process Safety Management for a Changing Industry. *Journal of Loss Prevention* 22:162–168.
9. Vaughen, Bruce K. and Trevor A. Kletz. 2012. Continuing Our Process Safety Management Journey. *Process Safety Progress.* 31:337–342.
10. Center for Chemical Process Safety. 2014. *Vision 20/20.* www.aiche.org/ccps/about/vision-2020.
11. Klein, James A. and Richard A. Davis. 2011. Conservation of Life as a Unifying Theme for Process Safety in Chemical Engineering Education. *Chemical Engineering Education* 45:126–130.
12. Leveson, Nancy. 2004. A New Accident Model for Engineering Safer Systems. *Safety Science* 42:237–270.
13. Chiles, James R. 2002. *Inviting Disaster: Lessons From the Edge of Technology.* Harper Business, New York.
14. National Transportation Safety Board. 2002. *Loss of Control and Impact with Pacific Ocean Alaska Airlines Flight 261 McDonnell Douglas MD-83, N963AS About 2.7 Miles North of Anacapa Island, California, January 31, 2000.* Aircraft Accident Report NTSB/AAR-02/01. Washington, DC.
15. Bloch, Kenneth A. 2016. *Rethinking Bhopal: A Definitive Guide to Investigating, Preventing, and Learning from Industrial Disasters.* Elsevier.
16. Perrow, Charles. 1999. *Normal Accidents: Living With High-Risk Technologies.* Princeton University Press, Princeton NJ.
17. Foord, A. G. and W. G. Gulland. 2006. Can Technology Eliminate Human Error? *Process Safety and Environmental Protection* 84(B3):171–173.
18. Levenson, Nancy G. 2011. *Engineering a Safer World: Systems Thinking Applied to Safety.* MIT Press, Cambridge, MA.
19. U.S. Chemical Safety Board. 2011. *Bayer CropScience.* Report No. 2008-08-I-WV. www.csb.gov.
20. Miner, H. L. 1933. *Accidents Are Increasing,* letter, Hagley Museum and Library. Wilmington, DE.

Epilogue

Why Process Safety?

By Bruce K. Vaughen

Why process safety?
    To prevent hazardous materials and energies from
    causing harm to people, the environment and the business.

    Why do we design safe processes?
        To control the hazardous materials and energies.

        Why do we identify and assess process hazards?
            To understand and evaluate their risks.

            Why do we evaluate and manage process risks?
                To identify safeguards for the equipment design.

                Why do we operate safe processes?
                    To control the process hazards and risks.

            Why do we maintain process integrity and reliability?
                To control the process hazards and risks.

        Why do we change processes safely?
            To control the process hazards and risks.

      Why do we manage incident response and investigation?
         To safely respond to and learn from them.

    Why do we monitor process safety program effectiveness?
      To control the hazardous materials and energies.

Why process safety?

Because

Index

Note: Page numbers followed by f and t refer to figures and tables, respectively.

A

Amuay case study, 260
Amuay Incident, 260–266
As low as reasonably practicable
  (ALARP), 58, 58f, 110,
  114–115, 251
Assess process hazards, 66, 131–167
  chemical compatibility matrix
    development, 160–164
  combustible dust, 151–155
  explosion, 148–150
  flammability, 144–151
  IHA, 134–137
  toxicity, 137–143
Audit(s), 353–358
  characteristics of successful,
    354, 355t
  team, 354, 356

B

Barrier protection layer models,
  71–76, 72f
  bow tie diagrams, 74–76, 75f, 77f
  Swiss Cheese Model, 72–74
Bhopal case study, 127, 321
Bhopal incident, 10, 21, 127–128,
  321–333
BLSR case study, 150
Bow tie diagram(s), 74–76, 75f, 77f, 186,
  189, 315–320
  visualization of
    barrier weaknesses, 316–317, 317f
    systemic weaknesses,
      319–320, 319f
BP Texas City case study, 49, 76
BP Texas City incident, 30, 36, 49
  process safety system issues for, 78t
Buncefield case study, 275

C

Case studies
  Amuay, 260
  Bhopal, 127, 321
  BLSR, 150
  BP Texas City, 49, 76
  Buncefield, 275
  Deepwater Horizon, 127
  DPC Enterprises, 142
  Formosa Plastics Corp., 100
  Large companies, 20
  Napp Technologies, 164
  Small companies, 21
  T2 Industries, 124
  Universities, 22
  West Pharmaceutical Services, 155
Cause-and-effect approach,
  312–313, 313f
Change processes safely, 271–288
  defining, 273
  management, 273–274, 281–287, 282f
    MOC checklists, 285t–286t
  organizational, 276–279, 278f, 279f
  technology, 279–281, 280t
Chemical
  compatibility
    interaction, 157
    matrix, 160–164
  reactivity hazards, 155–165
Chemical Compatibility Database
  (CCD), 164
Code of ethics, 19
Combustible dust hazards, 151–155
  definition, 153–154
Competency training, 221
Consequence analysis, 173, 182–186
Consequence modeling
  explosion overpressure contours, 188f
  toxic dispersion contours, 187f

Conservation of life (COL), 20
Critical equipment, 116, 117t, 248,
    258–259
    identification of, 243–244
Culture. *See* Safety Culture

D

Deepwater Horizon case study, 127
Deepwater Horizon incident, 4, 5f, 17, 73,
    73f, 77f, 127
Design safe processes, 66, 107–128
    process and equipment design,
        112–128
    technology documentation, 111
Deviations evaluation, HAZOP, 193
Domino effect, 156, 233, 300
DPC Enterprises case study, 142
Dust Hazards Assessment
    (DHA), 154

E

Emergency Action Plan (EAP), 297
Emergency response plan (ERP),
    296–297
Employee participation, 37
Equipment
    critical, 116, 117t
    integrity, 121–122
    types of, 117, 118f
Explosion hazards, 148–150

F

Facility life cycle, PHRA of, 175–176
Facility siting, 173–174, 200–203
    checklist topics, 201t
    steps involved, 201f
Failure Mode and Effect Analysis
    (FMEA), 197–198
Fault Tree Analysis (FTA), 198, 315
Fire triangle, 144f
Fishbone diagrams, 312–313, 313f
Flammability hazards, 144–151
Formosa Plastics Corp. case study, 100
Foundations. *See* Process Safety
    Foundations

G

Globally Harmonized System of
    Classification and Labelling of
    Chemicals (GHS), 132
    acute toxicity effects on humans, 139t
    chemical hazards, 133f
    self-reactive substance types, 159f
Guide words, HAZOP, 193, 193t

H

Hazard identification and risk analysis
    (HIRA), 172
Hazardous Event Evaluation (HEE), 173,
    189–198, 208
    FMEA methodology, 197–198
    FTA methodology, 198
    HAZOP, 190–195
    methodologies, 190t
    section, PHRA, 186
    What-if/Checklist methodology,
        195–197, 196f
    worksheet, 191t
HAZards and OPerability study
    (HAZOP), 190–195, 192f, 193t
Human factors (HF), 173, 198–199

I

IHA. *See* Intrinsic Hazard Assessment
Incident response and investigation,
    289–334
    bow tie diagrams, 315–320
    changing and sustaining after,
        310–311
    emergencies, planning and
        responding, 296–301
    incidents, 297–301, 298t
    investigation methods, 311–315
        fishbone diagrams, 312–313, 313f
        FTA, 315
        RCA, 314–315
        why-tree model, 312
    key concepts, 291–293
    managing incidents, six phases,
        292, 293f
    types of emergencies, 293–295
    types of incidents, 295–296, 296t

Incidents, global nature, 18f
Independent protection layers (IPLs),
    209, 258
Industry injury rates, 15t
Inherently safer design (ISD), 60
Inherently Safer Process (ISP), 108–109,
    174–176, 203
  approaches, 60, 60t
  design, 112
Inherently safer process technology
    (IST), 60
Inspection, testing, and preventive
    maintenance (ITPM) programs,
    240–241, 317–318
  critical equipment, identification,
    243–244
  goals, 245
Integrated management systems,
    13, 66
Integrity and reliability, 239–267
  effective maintenance system
    development, 247–258
  ITPM programs, 240–241
Intrinsic Hazard Assessment (IHA),
    112–113, 132, 134–136, 135f, 173,
    182, 368
IPLs (independent protection layers),
    209, 258

K

Key performance indicators (KPIs), 35,
    337, 342, 344

L

Layer of protection analysis (LOPA), 189,
    198, 205, 208–210
Leadership
  commitment and focus, 47
  process safety, 12, 43–45
Leading metrics, 344–346
Learning organizations, 340–344,
    343t
Life cycle phases, process/equipment,
    108, 108f
Loss of containment, 228, 239, 245,
    254, 299

M

Maintenance capability and
    competency, 242–243
Maintenance system development,
    247–258
  combining programs, 256–258
  Integrity programs, 255–256
  QA program, 254–255
  Reliability programs, 251–254
  questions, 247–250
  RBI program, 250–251
Management of Change (MOC), 176,
    272, 274, 279
Management systems, 64, 66
Management systems, scope, 56, 56t

N

Napp Technologies case study, 164
Near-miss incidents, 297
Networking, 369, 371
Nodes identification, HAZOP,
    192–193
Normalization of deviance, 29, 381

O

Operate safe processes, 219–237
Operating limits, different types
    of, 119f
Operational discipline (OD), 14, 49, 310
  benefits of strong, 83–86, 84t
  characteristics, 87f
  data sources for evaluation, 94t
  definition, 82
  evaluation, 94–98
  improvement, 92–95, 93f
  leadership focus in, 88
  organizational OD, 86–89
  overview, 82–83
  personal OD, 89–92
  program characteristics, 86–92
    organizational, 86–89
    personal, 86, 89–92
  risk equation, 84
  safety triangle, 83–84, 84f
  self-assessment, 95
Operational risk, 278, 279f

Organizational change, 276–279
    operational risk, 279f
    process safety risk, 278f
Organizational memory, maintaining,
        343, 344t
Organizational OD, 86–89
    characteristics, 88, 88f, 89t
OSHA PSM, 68

P

Personal OD, 86, 89–92
    characteristics, 90, 90f, 91t
    evaluation, 96–98, 97t, 98t
    potential issues, 98t
Personal safety, 14–15
PHRA. *See* Process hazard and risk
        analysis
Physical property data, 134
Plan, Do, Check, Act (PDCA) lifecycle,
        70, 71f
Process
    and equipment design, 112–128
    definition, 7
    hazards, 3, 5, 7, 11, 11t, 182
        identification, 182
    incidents, potential consequences, 17t
    risk, 12
        management, 59–64, 61f, 61t, 62f, 63f
        variables, HAZOP, 193, 193t
Process hazard and risk analysis
        (PHRA), 113, 132, 171–214
    definition, 173–175
Process hazards analysis (PHA),
        172, 279
Process hazards definition, 7
Process safety, 3–22, 363
    application, 6–9, 23t
    audits, 353–358
    business impact, 18f
    capability, 363–373, 365t
    definition, 9, 9t, 10f, 11, 14
    effective programs, 12
    foundations, 13f
    future, 379–381
    global, 17, 18
        operational discipline, 14
        process safety systems, 13
        safety culture and leadership, 12

    identifying and foundational
            weaknesses, 320
    importance of, 16
    integrated framework, 6
    key questions, 23
    leadership, 43–45, 43f, 45t
    metrics, 344–352, 345f, 346f
    performance, 46–49, 46t, 47t, 48f
    personal safety, 14–16
    risk, 278, 278f
    roles and responsibilities, 364–367,
            365t, 366f
Process safety information (PSI), 132
Process safety metrics, 344–352
Process safety system, 13, 48–49, 55–78, 67t
    barrier protection layer
            models, 71–76
        bow tie diagrams, 74–76, 75f
        Swiss Cheese Model, 72–74
    compared to element-focused
            frameworks
        CCPS RBPS, 70t
        EU Seveso Directive, 69t
        US OSHA PSM, 68t
    identifying systematic weaknesses,
            319
    life cycle, 71f
    protection layer model, 64–65, 65f
Project PHRA, 176
Protection layer(s), 64, 204–205
    hierarchy, 345, 346f
    model, 64–65, 65f

Q

Quality Assurance (QA) program, 244,
        247, 254–255
    equipment, 254–255
Quantitative risk analysis (QRA), 202,
        210–211

R

Recognized and Generally Accepted
        Good Engineering Practices
        (RAGAGEP), 46, 57, 110,
        114–115, 251
Reliability Centered Maintenance
        (RCM) programs, 251

Revalidation PHRA, 176
Risk
    definition, 12
    equation, 84–85
    management systems, 35, 59–64
    matrix, 58, 59, 205–207, 206f, 206t
    mitigation strategy, 110
Risk analysis, 205–211
    LOPA, 208–210
    PHRA, 175
    QRA, 210–211
    qualitative, 205–207
    quantitative, 210–211
Risk Based Inspection (RBI) approach/
        program, 250–251
Risk-based process safety approach, 57
Risk matrix, 58, 59f, 62
Root cause analysis (RCA), 38, 245,
        314–316, 318
Runaway reaction(s), 20–21, 61–62,
        114, 124
    conditions for, 156

S

Safe operating space, 327–328
Safe process operations
    capability development, 231–236
    procedures development, 223–230
    procedures development
        contractor, 230
Safety
    cultural change, 42, 42t
    culture, 12, 28, 33, 377
        behaviors, 33, 33t

    characteristics, 34–37
    definition, 31–34, 32t
    evaluation and improvement,
        37–42, 39t, 41t
    improvement grid, 40f
    themes, 30, 30t
    triangle, 345, 345f
    OD, 83–84, 84f
Safety hazard index (SHI), 142
Safety instrumented system (SIS),
        62, 65, 114, 244, 252,
        315, 318
Safe work practices (SWP), 228, 229t
Sense of vulnerability, 34–37
Sensitivity to operations, 339–340
Seveso directive, 69
Startup PHRA, 176
Swiss Cheese Model, 72–74
System-Theoretic method, 318

T

T2 Industries case study, 124
Technology change, 279–281, 280t
Toxicity hazards, 137–143
    assessment, 138–142
    IHA of, 137

W

West Pharmaceutical Services case
        study, 155
What-if/Checklist methodology,
        195–197, 196f, 197f
Why-tree model, 312

Printed in the United States
by Baker & Taylor Publisher Services